Advances in Electrochemical Science and Engineering

Volume 6

WILEY-VCH

Advances in Electrochemical Science and Engineering

Advances in Electrochemical Science and Engineering

Volume 6

Edited by Richard C. Alkire
and Dieter M. Kolb

Contributions from

E. Spohr, Ulm
L. M. Peter, D. Vanmaekelbergh, Bath, Utrecht
D. Lincot, M. Froment, H. Cachet, Paris
D. J. Economou, Houston, Texas

WILEY-VCH

Weinheim · New York · Chichester · Brisbane · Singapore · Toronto

Editors:
Prof. Richard C. Alkire
University of Illinois
Vice Chancellor for Research
601 East John Street
Champaign, IL 61820-5711
USA

Prof. Dieter M. Kolb
University of Ulm
Department of Electrochemistry
Albert-Einstein-Allee 47
D-89081 Ulm
Germany

e-mail (for orders and customer service enquiries): sales-books@wiley-vch.de

Visit our Home Page on http:/www.wiley-vch.de

Other Editorial Offices

John Wiley & Sons Ltd
Baffins Lane, Chichester
West Sussex PO19 1UD, England

Jacaranda Wiley Ltd, 33 Park Road, Milton,
Queensland 4064, Australia

John Wiley & Sons (Asia) Pte Ltd, 2 Clementi Loop #02-01,
Jin Xing Distripark, Singapore 129809

John Wiley & Sons (Canada) Ltd, 22 Worcester Road,
Rexdale, Ontario M9W 1L1, Canada

A catalogue record for this book is available from the British Library

Die Deutsche Bibliothek – CIP-Einheitsaufnahme
Advances in electrochemical science and engineering:
Weinheim ; New York ; Chichester ; Brisbane ; Singapore ; Toronto : WILEY-VCH
ISSN 0938-5193
Erscheint unregelmässig. – Aufnahme nach Vol. 1 (1990)
Vol. 1. (1990) –
ISBN 3-527-29515-1

Typesetting: Asco Typesetters, Hong Kong
Printing: betz-druck gmbh, D-64291 Darmstadt
Bounding: Wilhelm Osswald & Co, D-67433 Neustadt
This book is printed on acid-free paper responsibly manufactured from sustainable forestation, for which at least two trees are planted for each one used for paper production.
Printed in the Federal Republic of Germany

Introduction

This sixth volume continues, with new editorship, the series established by the late Heinz Gerischer and Charles W. Tobias. The favorable reception of the first five volumes and the steady increase of interest in electrochemical science and technology provide good reasons for the continuation of this series with the same high standards. The purpose of the series is to provide high quality advanced reviews of topics of both fundamental and practical importance for the experienced reader.

Richard C. Alkire
Dieter M. Kolb

Preface

Spohr describes in detail the use of computer simulations in modeling the metal/electrolyte interface, which is currently one of the main routes towards a microscopic understanding of the properties of aqueous solutions near a charged surface. After an extensive discussion of the relevant interaction potentials, results for the metal/water interface and for electrolytes containing non-specifically and specifically adsorbing ions, are presented. Ion density profiles and hydration numbers as a function of distance from the electrode surface reveal amazing details about the double layer structure. In turn, the influence of these phenomena on electrode kinetics is briefly addressed for simple interfacial reactions.

In their chapter on time- and frequency-resolved studies of photoelectrochemical kinetics, Peter and Vanmaekelbergh give an extensive survey of how modulation techniques such as photoelectrochemical impedance spectroscopy or intensity-modulated photocurrent spectroscopy can yield valuable information on the time dependence of reactions at semiconducting surfaces over a broad range of time scales. Kinetic studies with single crystals as well as porous or nanocrystalline material reveal the important role that is played by the bulk structure of semiconductor electrodes.

Lincot, Froment, and Cachet review the chemical and mechanistic aspects of chemical bath deposition of chalcogenide compounds with special emphasis on structural properties associated with epitaxial growth. Present applications of chemically deposited films are reviewed and several characteristic advantages are identified that may be exploited in the future for applications such as small band gap semiconductors, large area electrochromic devices, electroluminescence, quantum sized films, and films with spatially modulated composition and structure.

Economou reviews the engineering fundamentals of plasma etching, and illustrates many similarities with the field of electrochemical reactor engineering. These include the consideration of convective mass transport, the potential distribution, complex homogeneous and heterogeneous chemistry, complex boundary conditions involving surface films, a space charge region near electrodes and surfaces, and surface chemistries that are not temperature-driven. Because these phenomena exert themselves over a wide range of length scales, including the noncontinuum region, sophisticated engineering modeling procedures have been developed. Such procedures may in the future be adapted for use in "wet" systems, particularly where quality control is a consequence of molecular scale features such as in electrodeposition and corrosion applications.

Contents

List of Contributors

Hubert Cachet
Laboratoire de Physique des
Liquides et Electrochimie
Unité Propre du CNRS n° 15
Université Pierre et Marie Curie
4 place Jussieu
75252 Paris cedex 05
France

Demetre J. Economou
Plasma Processing Laboratory
Department of Chemical Engineering
University of Houston
Houston, TX 77204-4792
USA

Michel Froment
Laboratoire de Physique des
Liquides et Electrochimie
Unité Propre du CNRS n° 15
Université Pierre et Marie Curie
4 place Jussieu
75252 Paris cedex 05
France

Daniel Lincot
Laboratoire d'Electrochimie et
de Chimie Analytique
Unité Associée au CNRS
Ecole Nationale Supérieure
de Chimie de Paris
11 rue Pierre et Marie Curie
75231 Paris cedex 05
France

L. M. Peter
Department of Chemistry
University of Bath
Bath BA2 7AY
UK

E. Spohr
Department of Theoretical Chemistry
University of Ulm
D-89069 Ulm
Germany

D. Vanmaekelbergh
Debye Institute
University of Utrecht
PO Box 80000
3508 TA Utrecht
NL

Computer Simulations of Electrochemical Interfaces

E. Spohr

Department of Theoretical Chemistry, University of Ulm, D-89069 Ulm, Germany

Contents

Introduction

Computer simulations have contributed substantially to our understanding of the liquid state during the past 40 years. In many cases they can provide essentially exact results for problems in statistical mechanics that would otherwise only be solvable by approximate methods, or might be intractable. Therefore, one of their most prominent applications is the investigation of "realistic" models of physical and chemical systems, which cannot at all, or not easily, be treated by analytical theory. The interface between two bulk phases is one subject that is notoriously hard to approach theoretically. The extension of computer simulation methods from the bulk phase to interfaces was straightforward and occured as soon as computers were powerful enough to handle the complexity of inhomogeneous two-phase systems (for an early review see, e.g., Ref. 1).

On the other hand, electrochemistry has for a long time been a part of physical chemistry with a strong footing in thermodynamics and in kinetics. During the last decades the advent of surface sensitive techniques shifted the interest more and more towards the interface. This trend led, e.g., to the title "Interfacial Electrochemistry" for a recent textbook [2], where the interface is defined as "those regions of the two adjoining phases whose properties differ significantly from those of the bulk", a definition that is also adopted here.

One of the goals of computer simulation studies of electrochemical interfaces which are reviewed here is to find out how large these regions are and to what extent they are different from the bulk phases on the molecular level. In what follows, the term "electrochemical interface" shall denote more specifically the interface between pure water or an ionic conductor (the liquid electrolyte solution) and an electronic conductor (the metal). The interface is usually charged: the metal phase carries an excess charge which is balanced by an equal amount of charge of opposite sign in the electrolyte solution; together, these charges are known as the "electric double layer".

Over the years, double layer models have evolved towards increasing complexity. The first model by Helmholtz [3, 4] treated the double layer as a rigid arrangement of positive and negative charges on both sides of the interface, comparable to a parallel-plate capacitor. Gouy [5] and Chapman [6] considered the fact that the applied potential and electrolyte concentration influence the double layer capacity. With the ions free to move, the thickness of the *diffuse double layer* varies with concentration, and the electrostatic potential varies exponentially in the solution. Stern [7] combined the Helmholtz and Gouy-Chapman models. In this model, the double layer consists of a compact and a diffuse layer and the boundary between them is the so-called

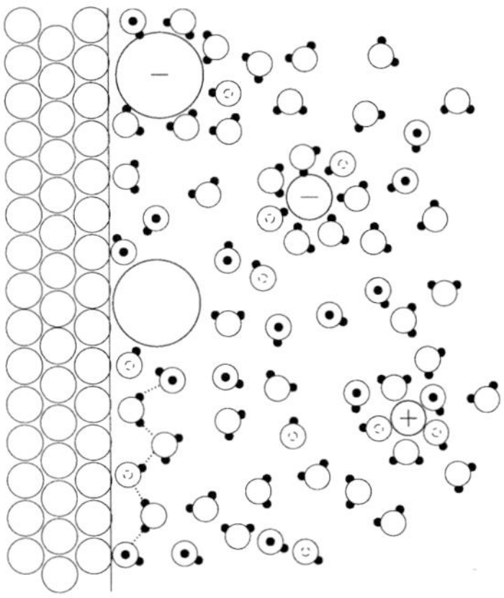

Fig. 1. The metal/solution, or electrochemical, interface.

Outer Helmholtz Plane (OHP). Later Grahame [8] distinguished between specifically adsorbed ions (those who loose their solvation shell) and non-specifically adsorbed ions. The double layer consists of three different regions, separated by the *Inner Helmholtz Plane* (IHP), which passes through the location of specifically adsorbed ions, and the OHP.

The importance of the dipolar nature of the solvent and of the interactions between solvent and electrode were recognized in the double layer model by Bockris, Devanathan and Müller [9]. Water 'hydrates' the electrode, which is regarded as a giant ion, and so contributes to the electric fields near the interface. Starting with the work of Damaskin and Frumkin [10] the differences between *sp* and transition metals were described by a series of "chemical" models. More details on double layer models can be found, e.g., in Refs. 2, 11, 12.

The properties of the metal phase have been successfully described by rather simple models, most notably the jellium model. In many theoretical treatments of the liquid/metal interface, the liquid electrolyte in contact with the metal has been described, to first order, as an external field, acting on the jellium model (see Ref. 13 and references therein). In many simulation studies, the reverse approach is taken. The focus is on the description of the liquid phase and the effect of the metal on the aqueous phase is approximated, to first order, by an external potential acting on the ions and molecules in the liquid phase. This is done within the framework of classical mechanics and classical statistical mechanics. The models for the interparticle interactions will consist of distributed point charges in combination with soft interatomic repulsions and dispersive attractions. Some of the models can also be considered chemical models; they can be regarded as a first step towards electrochemical modeling, very much in the spirit of "molecular modeling".

Each generation of textbooks of electrochemistry shows a sketch of the interface like Fig. 1 in its introductory chapters. The complexity of the sketch increases with our increasing knowledge of the molecular detail of the electrochemical interface. The characteristic difference between Fig. 1 and similar ones in, e.g., the books by Bockris and Reddy [14, 15], Brett and Brett [12], or by Schmickler [2] is the fact that water molecules are not represented as dipolar spheres. Instead, they are depicted as three-atomic molecules that can form intermolecular hydrogen bonds, some of which are indicated by dotted lines. Hydrogen bonding determines many of the properties of bulk liquid water and aqueous solutions. The same is true for water and aqueous solutions near interfaces, and the hydrogen bond theme will shine through the discussion of results in many places. A realistic description of hydrogen bonding is the essential aspect introduced by the computer modeling of electrochemical interfaces. It is also present in some recent analytical theoretical treatments. Details in Fig. 1 hint to the issues that will be discussed in this work: adsorbed and non-adsorbed ions and atoms incorporated into the hydrogen-bonded network of water at the interface.

In the next section a brief layout of simulation methods will be given. Then, some basic properties of the models used in computer simulations of electrochemical interfaces on the molecular level will be discussed. In the following three large sections, the vast body of simulation results will be reviewed: structure and dynamics of the water/metal interface, structure and dynamics of the electrolyte solution/metal interface, and microscopic models for electrode reactions will be analyzed on the basis of examples taken mostly from my own work. A brief account of work on the adsorption of organic molecules at interfaces and of liquid/liquid interfaces complements the material. In the final section, a brief summary together with perspectives on future work will be given.

2 Computer Simulations

Computer Simulations are a statistical mechanical method to sample configurations from the *phase space* of a physical or chemical *system* of interest. Formally, they generate a *sequence of states* of a system in some statistical mechanical *ensemble*. The system is uniquely described by specifying the interactions between the particles (which are usually described as interacting points), the masses of all these particles, and the boundary conditions. The set of potential energy hypersurfaces describing the interactions between *all* particles for all relative arrangements together with molecular geometries and their masses is called the *model* of the system. The models relevant to the present work will be discussed in section 3.

Following the construction of the model is the calculation of a sequence of states (or a *trajectory* of the system). This step is usually referred to as the actual simulation. Simulations can be stochastic (Monte Carlo) or deterministic (Molecular Dynamics); or they can combine elements of both, like force-biased Monte Carlo, Brownian dynamics or general Langevin dynamics (see Ref. 16 for a discussion). It is usually assumed that the physical system can be adequately described by the laws of classical mechanics. This assumption will also be made throughout the present work.

The Monte Carlo (MC) scheme was first used by Metropolis, Rosenbluth, Rosenbluth, Teller and Teller [17] in 1953 to calculate the equation of state for a simple hard sphere model of a liquid. Each configuration in an MC simulation is generated stochastically in such a way that the molecular configuration (specified by the set $\{\vec{r}_i\}$ of all particle positions $\vec{r}_i$) depends only on the previous configuration. The purely stochastic MC method is often performed on a fixed number of molecules N placed in a fixed volume V and maintained at a constant temperature T. During the MC simulation configurations are generated in such a way that, after many ones have been obtained, each configuration occurs approximately with the appropriate probability in the *canonical ensemble*, given by the *Boltzmann factor* $\exp[-U(\{\vec{r}_i\})/kT]$ where U is the interaction potential and k is Boltzmann's constant.

The Molecular Dynamics (MD) method was first used by Alder and Wainwright [18]. In the standard MD scheme for equilibrium systems, the atom positions $\vec{r}_i$ are obtained by solving Newton's equations of motion

$$m\ddot{\vec{r}}_i(t) = \vec{F}_i(t) = -\vec{\nabla}_i U(\{\vec{r}_i\}). \tag{1}$$

$\vec{F}_i$ is the force on particle i caused by the other particles, the dots indicate the second time derivative and m is the molecular mass. The forces on particle i in a conservative system can be written as the gradient of the potential energy, $\vec{\nabla}_i U$, with respect to the coordinates of particle i. In most simulation studies, U is written as a sum of pairwise additive interactions, occasionally also three-particle and four-particle interactions are employed. The integration of Eq. (1) has to be done numerically. The simulation proceeds by repeated numerical integration for tens or hundreds of thousands of small *time steps*. The sequence of these time steps is a set of configurations, all of which have equal probability. The completely deterministic MD simulation scheme is usually performed for a fixed number of particles, N in a fixed volume V. As the total energy of a conservative system is a constant of motion, the set of configurations are representative points in the *microcanonical ensemble*. Many variants of these two basic schemes, particularly of the Monte Carlo approach exist (see, e.g., Ref. 19–23).

Typical particle numbers N are of the order of several hundred to a few thousand, which are located in a regular, usually tetragonal, cell. To avoid unwanted surface effects beyond those that one explicitly wishes to study, *periodic boundary conditions* are used. This means that the cell and the particles replicate themselves infinitely in two or three directions of space. The cells are open and particles can move freely from one cell to the next. Since for each particle leaving the cell one of its replicas enters the cell, the overall particle number in the cell, and thus the density, remain constant. Furthermore, the *minimum image convention* is usually applied. This prescription states that from all the interactions of a particle i with particle j and its images only the one is used that is associated with the smallest distance. In this way the number of interactions to be calculated is finite. This approximation renders the calculation possible on a computer. For the treatment of long-range forces, like Coulomb forces, special procedures are needed, the most prominent being the Ewald summation method [24]. Several groups [25–27] have recently brought forward strong arguments against the use of truncation methods. Ewald summation schemes, or the fast multipole method [28, 29] for very large systems, are deemed to be an essential prerequisite for a realistic description of microscopic charge distributions, electric fields and potentials.

After the simulation, the trajectories or configurations are analyzed in order to obtain the *observables*. The theory of statistical mechanics provides the formalism to obtain observables as ensemble averages from microscopic configurations. From both the MC and the MD trajectories, ensemble averages can be formed as simple averages of the properties over the set of configurations stored on computer disk or tape during the simulation. Most properties displayed in the figures in the text, except those specifically denoted as 'snapshots', are ensemble averages over thousands of configurations from one simulation; some properties, like the free energy profiles, are functions of the configurations of many simulations. For a detailed discussion of computer simulations the reader is referred, for instance, to the books by Allen and Tildesley [19], Haile [16], Hoover [30, 31], Binder [20–22], Haberlandt *et al.* [32] or Smit and Frenkel [23], and the references therein.

3 Models

In this section some basic properties of molecular models of the electrochemical interfaces, as used in computer simulations, are summarized. Such models consist of a geometrical setup and a specification of the various interactions in the system.

3.1 Geometries

Like in electrochemical experiments, in computer simulations of two-phase systems there are at least two interfaces to be considered. The simulation of a single isolated interface is not possible. Various system setups are possible. For the liquid/solid two-phase systems consisting of a liquid electrolyte solution and a solid electrode, an alternating arrangement of liquid and solid phases is conceptually most simple. It has the minimum number of two equivalent interfaces (Fig. 2a). The properties of both interfaces can be averaged in the absence of an external electric field. Furthermore, from the differences between the two interfaces the magnitude of systematic errors, introduced into the system by either incomplete equilibration or inadequate length of the simulation, can be estimated. Periodic boundary conditions (see section 2) are applied in all directions of space including the direction normal to the interface (z-direction). Figure 3 shows as an example a snapshot of the geometrical arrangement used by various groups of authors in simulations of water and aqueous solutions in the vicinity of platinum surfaces with (100) surface structure.

The simulation of electrochemical systems with two equivalent interfaces also has its disadvantages. It is usually more difficult to equilibrate the system. While the bulk density of the liquid phase is usually known, no *a priori* knowledge about the density in the interfacial region exists. Rather, one of the objectives of the simulation is to

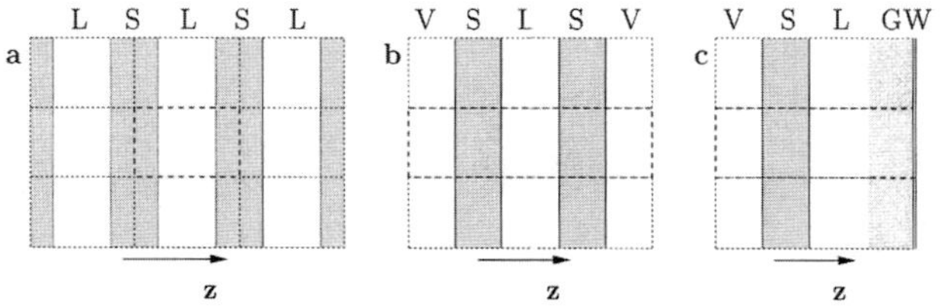

Fig. 2. Geometries for simulations of liquid/solid interfaces. (L: liquid, S: solid, V: vacuum, G: gas phase. W: wall; for further details see text.)

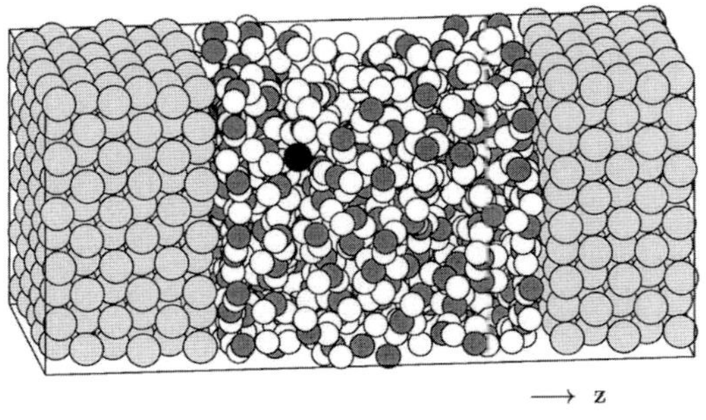

Fig. 3. Snapshot from a simulation of a lamina consisting of 304 water molecules and 1 ion (black) at 300 K, confined on two sides by a platinum crystal. Only the basic cell is shown; periodic boundary conditions are applied in all three directions of space.

establish the range and magnitude of the density oscillations. Consequently, the position of the phase boundaries has to be determined self-consistently in a simulation/adjustment cycle that is both costly in terms of computer time and cumbersome in terms of manpower. These problems can only partially be avoided by simulations at constant pressure or constant chemical potential [19]. Achieving hydrostatic equilibrium can be extremely difficult.

One of the objectives of a realistic model of an electrochemical system should be that a bulk phase with more or less isotropic properties is established or, equivalently, that the interfaces bounding the liquid phase are effectively decoupled. When the interfacial width is large, the minimum system size necessary to decouple the interfaces can also become quite large, leading to an excessive need for computational resources. The collective influence of the metal electrons on the liquid phase properties is often modeled by the classical image charge model (see below). In this case, the existence of two equivalent interfaces makes the calculation of long-range multiple image interactions necessary, which also leads to strong coupling between the interfaces. However, this is usually not desired, as a small slab of 20–50 Å thickness between two ideally polarizable metallic phases is not an adequate model of the electrochemical interface, except, e.g., to describe liquid junction experiments [33].

Therefore, alternative arrangements have also been used in the literature. If the solid phase is rigid and thick enough so that the water-metal and ion-metal interactions are fully converged such a system is equivalent to one with four interfaces,

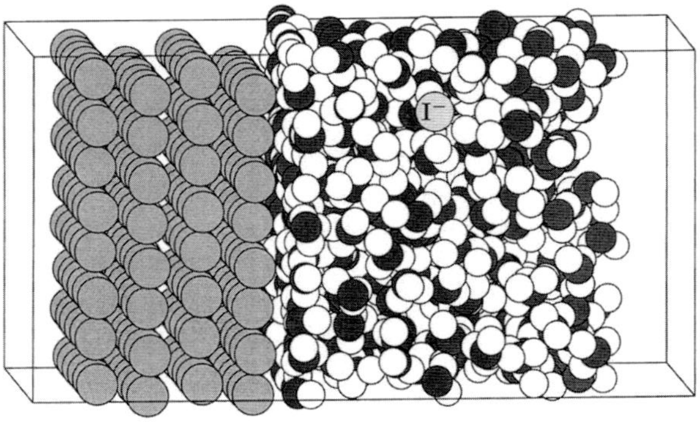

Fig. 4. Snapshot from a simulation at room temperature of 259 water molecules, one I^- ion and 4 non-mobile layers of a face-centered cubic mercury crystal with (111) surface structure. Only the basic cell is shown; periodic boundary conditions are applied in the x and y directions only.

two vacuum/solid and the two solid/liquid interfaces of interest (as shown in Fig. 2b). In many cases it is convenient to simulate the liquid/solid interface together with a liquid/gas interface. In order to avoid the 'loss' of molecules from the liquid state into an infinitely large gas phase, the latter is then confined on one side by a smooth wall (W in Fig. 2c). Such a system also consists of 4 interfaces, namely vacuum/solid, the electrochemical solid/liquid interface, a liquid/gas and a gas/wall interface (Fig. 2c). As an example for such a system, a snapshot from a simulation of an electrochemical interface between mercury and an aqueous solution of one ion in water is shown in Fig. 4.

Typical systems contain between several hundred and a few thousand water molecules. The number of ions is much smaller. In practice, it is determined by two contradicting requirements, namely (i) the need to achieve small concentrations, as they are typical for most experiments, and (ii) the statistical efficiency of the simulation, which requires the average over as many ions as possible. As a compromise, electrolyte concentrations are usually in the range between 0.5 and 3 mol/l. As more powerful computers become available and thus the simulation of larger systems becomes possible, systems at lower concentrations can be investigated. Typical time scales for Molecular Dynamics simulations reach up to about 2 nanoseconds at present.

3.2 Interaction Potentials

3.2.1 Water-Water Interactions

A large number of models for liquid water has been developed in the past 25 years (for a brief review, see, e.g., Ref. 34). Three classes of models can be distinguished: "rigid" models, "flexible" models, and "polarizable" models. The most simple ones, the rigid models, consider the molecule as an undeformable unit with various interaction sites specified within a molecule-fixed coordinate system. The majority of the rigid water molecules assumes only three partial charges, which can coincide with the atom sites (as, e.g., in the SPC/E model [35]) or are displaced from the atomic positions (as in the TIP4P model [36]). The partial charges are chosen in order to reproduce the molecular dipole and quadrupole moments (either the gas phase values or some 'effective' liquid state values) and other liquid state properties. For the SPC/E model, the partial charges on oxygen and hydrogen atoms are −0.8476 and 0.4238 elementary charge units, respectively. In addition to the electrostatic interactions, all water models incorporate a nonelectrostatic contribution that is usually modeled by a Lennard-Jones function, but in some cases, as in the MCY model [37], also by exponential functions.

In the presence of strong intermolecular forces, as they occur in hydrogen bonded liquids, the assumption of undeformable molecules is fairly restrictive. Flexible water models have been developed in order to describe the response of intramolecular geometries to intermolecular forces. Among others, the BJH model [38] has been used frequently in a variety of studies of electrochemical system by Heinzinger and coworkers and other groups.

Polarizable water models attempt to modify the internal charge distribution in response to the electric field produced by the environment. They are usually extensions of rigid or flexible models on the basis of various isotropic and anisotropic approximations to the molecular polarizability of the water molecule.

All successful water models make use of a distribution of point charges rather than of point multipoles. The main reason is that the directional properties of intermolecular hydrogen bonds can be obtained efficiently with only 3 or 4 point charges. Furthermore, Goldman and Backx [39] have shown that model molecules with such distributions of point charges are more effective as solvents (for instance in their ability to dissociate ion pairs) than molecules with equivalent point dipoles and quadrupoles.

The choice of a water model is usually not a completely rational decision. It rather reflects personal preferences or convenience (e.g., the existence of computer code for a particular model) in addition to objective criteria. Most of the models reproduce the properties of liquid water in an approximate fashion; all have their shortcomings. Rigid water models are often used for reasons of simulation economy, especially in inhomogeneous environments like the liquid/solid electrochemical interface. In computer simulation studies one objective is always to cover as much "relevant" phase space (equivalent to real simulation time in Molecular Dynamics) as possible at a reasonable cost of computer time. The highest frequency motions (usually the internal molecular vibrations) limit the length of the time step for the numerical integration of the equations of motion in MD simulations. Because of the absence of intramolecular degrees of freedom rigid models permit longer integration time steps and thus minimize the amount of computer time required to follow the intramolecular dynamics for a specified amount of real time. Consequently most of the simulations of interfacial properties so far have been carried out using rigid models. Since geometry changes of molecules in the vicinity of the interface are usually small, the use of rigid models is often adequate. A recent comparison showed no significant difference between the properties of (flexible) BJH and (rigid) TIP4P water in contact with a model mercury surface [40].

Polarizable water models are even more costly because the internal charge distribution of each molecule has to be determined self-consistently together with the external electric fields. However, with recent developments in computer technology, these models are being more frequently used nowadays, and expertise about their most efficient use is building up. A variety of polarizable models have been derived on the basis of the SPC/E and TIP4P models, some of which have been used in interfacial simulations [41–43]. With the strong electric fields and field gradients occuring in the vicinity of an electrode, the use of polarizable models in interfacial simulations may turn out to be essential in some systems. Therefore, comparative studies of polarizable and unpolarizable models in the vicinity of realistic metal surfaces have recently been started [44]. First results indicate, however, that simulations of liquid/gas interfaces will benefit more from the use of polarizable models than simulations of liquid/metal interfaces (see also section 4.1). Recent algorithmic advances (notably the RESPA scheme [45] and the fluctuating charge force fields [46]) are likely to overcome some of the problems associated with flexibility and polarizability and will further the use of flexible and polarizable models in the future.

3.2.2 Water-Surface Interactions

A variety of different models of the interface between water and a solid phase have been used in computer simulations. As far as the solid is concerned, a basic distinction can be made between smooth solid phases without atomic structure on the one hand and corrugated surfaces on the other. The latter surfaces have been modeled as rigid (frozen) or flexible atomic lattices representing the solid phase [47–51] or as a corrugated external potential that describes the effect of the solid phase by a more or less elaborate potential function $V(x, y, z)$ [52–56]. The generic metallic features are modeled by treating the metal phase as a medium of infinite dielectric constant or by using the jellium model (e.g., Ref. 57–59). In several cases, the results of semi-empirical and ab initio quantum chemical calculations have been parametrized [40, 48, 55].

In the so-called "image charge" model, surface-mediated forces are introduced by adding the interactions of a charge in the liquid phase with the images of all charges in the metal phase. The positions of the image charges are obtained by reflexion at the static image plane. When, like in electrochemical simulations, the electrolyte solution in vacuum is in contact with a conducting metal electrode of infinite dielectric constant, the values of the image charges are just the negative values of the real charges. Figure 5 illustrates the application of the image charge model. All pairs of real charges interact with each other, those image terms originating at the Cl^- ion are outlined by dashed lines. Calculations of the position of the static image plane (e.g., using a charged jellium surface [60]) show that it is shifted from the surface plane of atom centers towards the liquid phase by about 0.5–1.0 Å.

The one-dimensional jellium model of a metal [61–64] has been used quite fre-

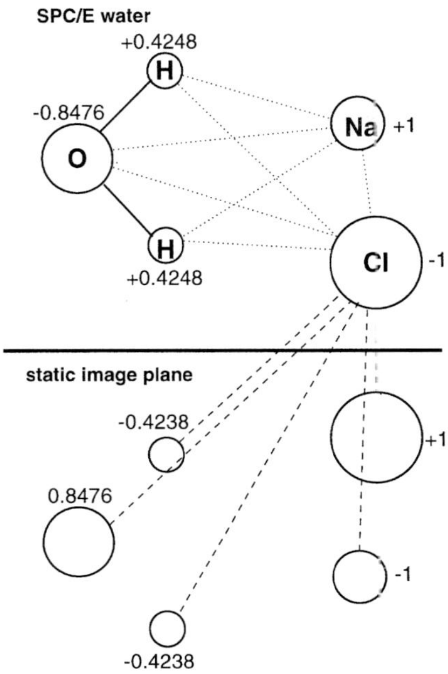

Fig. 5. Charges and image charges of an SPC/E water molecule, a Na^+ and a Cl^- ion in the vicinity of an interface. Dotted: direct interactions between charges; dashed: image terms affecting the Cl^- ion. Numbers denote (partial) charges in units of the proton charge.

quently to model the response of a metal to electric fields or the adsorption of ordered atomic layers on metal surfaces (see Ref. 65 for a recent review) and in analytical double layer studies of the metal/electrolyte solution interface (e.g., Ref. 58–60, 66–71). In the jellium model, the ionic charges of the metal ion cores are smeared out into a constant positive background charge, which drops abruptly to zero at the metal surface. The electrons are modeled as a quantum mechanical plasma within the framework of density functional theory. They interact with the background charge and with any external fields such as those caused by surface charges or ions and solvent molecules in the liquid phase. The electrons can penetrate into the region outside the positive background, i.e., into the solution.

Schmickler and Leiva [72] investigated the changes in the jellium charge distribution by explicitly incorporating the interaction of the jellium with liquid water (described by a central force model). They proceeded by specifying a surface charge density on the electrode, then performed a simulation at that surface charge, and used in turn the calculated average potential of the liquid phase (averaged over several configurations) as the external potential for the determination of the electronic density of the jellium. More recently, Shelley *et al.* [73] improved this approach by using an iterative scheme and applied it to water/mercury interfaces. They first solved for the jellium density near a vacuum; then they performed a grand-canonical Monte Carlo simulation of water under the influence of the electrostatic potential of the jellium. The average electrostatic potential produced by the aqueous phase was then used as the external potential to determine the jellium density in the next iteration cycle. The authors claim that two to three cycles are usually sufficient to obtain convergence. A logical extension of this scheme is to solve for the jellium density self-consistently and simultaneously with the equations of motion. With this approach the importance of fluctuations of the jellium density can be investigated; it is indeed found that the fluctuations play only a minor role [74].

A completely different description of the water/metal interface that is capable of taking into account some specific properties of various metals and surface structures is the approach taken first by Singer and coworkers [75] for salt surfaces and later by Spohr and Heinzinger [47] for metal surfaces. Here, the solid electrode is modeled as a regular lattice of atoms which interact with the molecules in the liquid phase through potential functions that are specific for the given metal. The collective effect of the delocalized metal electrons can be taken into account by means of the image charge model. These models of the interface permit a more realistic description of structure and processes at a real electrochemical interface. Taking into account the atomic structure of the solid does not increase the computer time significantly. These models belong to the class of chemical models.

The local interactions between the metal atoms and water molecules or ions have been obtained from semiempirical and *ab initio* quantum chemical cluster calculations (see below). This technique has been used by the groups of Spohr and Heinzinger, and later by Berkowitz and coworkers, for platinum surfaces with (100) and (111) surface geometry [48, 49, 52–54, 76] and also for mercury surfaces [40, 77–81].

Spohr [49] used the parametrized [48] extended Hückel cluster calculations for the interaction of water with a 5-atom platinum cluster [82] representing the quadratic Pt(100) surface. The key results of the semiempirical calculations, namely that adsorption occurs on top of a surface atom with the oxygen end of the molecule pointing towards the surface, were incorporated into the model. As normal, pairwise additive, distance-dependent potential energy functions tend to maximize the coordination number and, therefore, lead to preferential adsorption on the hollow site, the pairwise oxygen-platinum interaction was taken to depend not only on the interatomic distance but also on its projection onto the surface plane. The strength of the interaction in this model thus changes as a function of the lateral distance between oxygen and platinum sites. It can be adjusted to result in the desired form of the interaction potential surface. The correct orientation is achieved by adding a weakly repulsive hydrogen-platinum pair potential function.

The coefficients of a pairwise additive water-platinum potential

$$\phi_{\mathrm{H_2O-Pt}} = \phi_{\mathrm{O-Pt}}(\mathrm{r_{O-Pt}}, \rho_{\mathrm{O-Pt}}) + \phi_{\mathrm{H-Pt}}(\mathrm{r_{H_1-Pt}}) + \phi_{\mathrm{H-Pt}}(\mathrm{r_{H_2-Pt}}), \tag{2}$$

have been adjusted to reproduce the energies from extended Hückel calculations in several geometries [82]. The resulting functions are

$$\phi_{\mathrm{O-Pt}} = [1894.2\,\mathrm{e}^{-1.1004\mathrm{r}} - 1886.3\,\mathrm{e}^{-1.0966\mathrm{r}}] \cdot \mathrm{f}(\rho) + 10^6 \cdot \mathrm{e}^{-5.3568\mathrm{r}} \cdot [1 - \mathrm{f}(\rho)] \tag{3}$$

and

$$\phi_{\mathrm{H-Pt}} = 1.7142\,\mathrm{e}^{-1.2777\mathrm{r}} \tag{4}$$

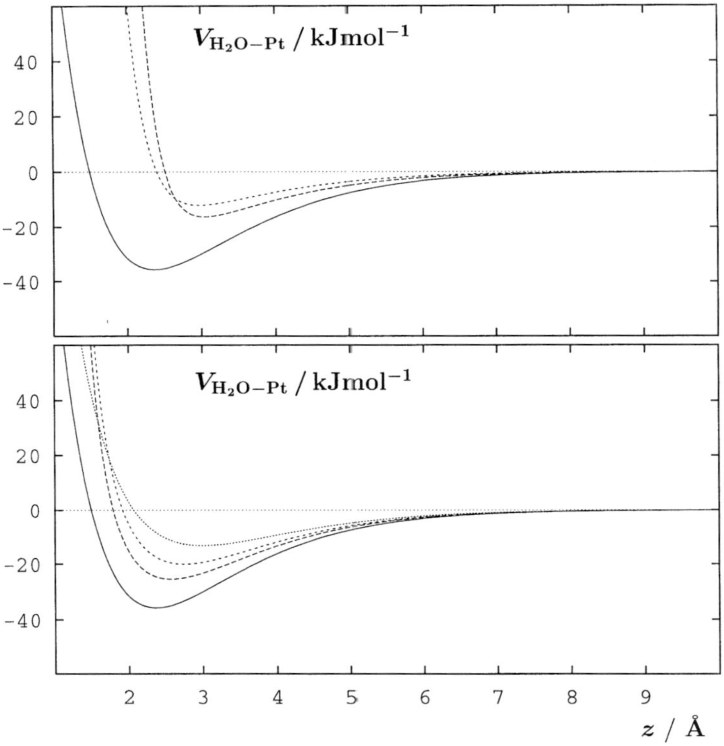

Fig. 6. Top: Total water-platinum(100) interaction energy, from eqns. 2–5, as a function of the oxygen-surface distance z for water on top of a platinum atom (full line), on a bridge site (long dashes), and on a 4-fold hollow site (short dashes). The dipole moment vector of the water molecule points away from the surface in all three cases. Bottom: Total water-platinum(100) interaction energy as a function of the oxygen-surface distance z for different orientations of water on top of a platinum atom: dipole vector pointing away from the surface (full line) and pointing towards the surface (dotted); dipole vector in the surface plane and proton-proton vector parallel to the surface (long dashes) and perpendicular to the surface (short dashes).

with

$$f(\rho) = e^{-0.5208\rho^2}. \tag{5}$$

The energy is given in units of 10^{-19} J, and r and ρ are given in Å. The site-site distances are denoted by r. ρ is the length of the projection of the distance vector onto the surface plane.

The (pair) potential ϕ_{O-Pt} is attractive for larger Pt-O distances only if ρ is small, leading to the preferential adsorption site on top of a platinum atom with an adsorption energy of 35.7 kJ mol^{-1} on the (100) surface (see below), in agreement with the extended Hückel calculations [82]. The distance dependence of the total interaction energy of one water molecule with an entire platinum crystal is shown in Fig. 6 for several adsorption sites and orientations. Quantum chemical calculations of water interactions with Al [83], Ni [84], Cu [85], Cu, Ag, and Au [86], and Hg [87, 88] clusters using different methods and cluster sizes yield binding energies in the same order of magnitude. These adsorption energies agree reasonably well with experimental estimates obtained from thermal desorption spectra, which range from 40 to approximately 70 kJ mol^{-1} for most transition metals (see, e.g., Ref. 89–92). The interaction energies on *sd* metals are significantly larger than the adsorption energies of water on simple *sp* metals. The ρ-dependence of the potential leads to repulsive interactions between water and platinum atoms with small overlap between oxygen and platinum orbitals (ρ large). This is consistent with the finding by Ribarski *et al.* [85] that the binding energy decreases approximately by a factor of 2 in going from the H_2O–Cu to the H_2O–Cu_5 cluster.

This approach formed the basis of various later modifications, extensions, and improvements by several groups of authors. In a similar manner, the results of *ab initio* calculations of mercury-water clusters at the SCF level [88] were parametrized [40]. These results, in turn, then formed the basis for the non-jellium part of the jellium model in [73]. Berkowitz [52–54] and later Zhu and Philpott [55] and Spohr [56] followed the approach by Steele [93] and Fourier-expanded the lattice sum of all (pairwise) interactions between the atoms in the solid and one molecule or ion in the liquid. Only the lowest order corrugation terms are kept in the expansion but in principle the summation can be extended to any desired accuracy. The procedure is adequate as long as there is no substantial coupling between liquid and metal motions that could influence the liquid structure and relaxation phenomena. Spohr [56] used a corrugated Morse potential for the oxygen-metal interactions and an exponentially repulsive potential for the hydrogen metal interactions in the form

$$V_{\text{water-surface}} = \phi_O(x_O, y_O, z_O) + \phi_H(z_{H1}) + \phi_H(z_{H2}) \tag{6}$$

with

$$\phi_O(x,y,z) = D_O[\mathrm{e}^{-2\beta_O(z-z_1)} - 2\cdot\mathrm{e}^{-\beta_O(z-z_1)}] + \alpha\cdot D_O\mathrm{e}^{-2\beta_O(z-z_1)}\cdot\left[\cos\left(\frac{2m\pi x}{L_x}\right) + \cos\left(\frac{2n\pi y}{L_y}\right)\right] \tag{7}$$

and

$$\phi_H(x,y,z) = \gamma\cdot D_O\mathrm{e}^{-2\beta_H(z-z_2)}. \tag{8}$$

D_O, α, and γ are parameters that characterize the adsorption energy, the energetic corrugation, and the anisotropy, respectively. z_1 and z_2 describe the equilibrium distance of the molecule from the surface, β_O and β_H the curvature of the potential. x, y and z are the atom coordinates and L_x and L_y are the box dimensions. m and n denote the number of elementary cells along the x and y directions of the periodic box; thus, L_x/m and L_y/n are the lattice constants of the surface elementary cell. For the simulation of bulk water in contact with a crystal, this assumption does not lead to serious defects, considering the very approximate nature of the currently used models. The mobility of the crystal atoms is, however, crucial for energy relaxation studies that can be performed with equilibrium and non-equilibrium molecular dynamics and trajectory techniques (e.g., Ref. 94, 95).

Recently, Siepmann and Sprik [50, 51] introduced a water-metal model potential consisting entirely of two-body and three-body terms involving oxygen and metal atoms. It does not require the construction of an artificial plane to describe image interactions or to modify interaction functions laterally; it can thus be extended to interfaces of arbitrary shapes. The nonelectrostatic part of the model describes the local weak chemisorption interaction in a manner similar to that used by Stillinger and Weber for liquid silicon [96]. The repulsive three-body term is constructed to reduce the coordination number and thus achieve preferential binding on top of a metal atom, in agreement with experiment [97]. An angular interaction term produces the preferential orientation where the oxygen atom points towards the surface. The charge distribution in the metal is described by a set of overlapping Gaussians with variable "induction" charge, centered at the sites of the metal atoms. The charges are calculated dynamically during the simulation by an extended Lagrangian technique. They are obtained from the Poisson equation by the requirement that the electrostatic potential is constant inside the metal and has the value of the applied potential.

All models of water-metal interactions used in computer simulations are limited by various restrictions. The generic models, while properly describing the asymptotic behavior of the long-distance interactions, fail at short distances and have to be augmented heuristically by short-range repulsive interactions of some sort. The models based on quantum chemistry do not exhibit the proper long-range asymptotics. They are useful to estimate short-range interactions like surface corrugation and orientational anisotropy of interactions but should not be considered as quantitative on an absolute scale, since the calculations usually show a substantial size dependence and the treatment of electronic correlation is often inadequate. Furthermore, the simulations are usually performed on adiabatic Born-Oppenheimer type ground state potential energy surfaces, which are certainly inadequate in many electrochemical situations. Much more work needs to be done and will certainly be done in the future on improving the interaction potential functions used in electrochemical modeling.

Presently, the simulation of "real" metal surfaces with defects of all sorts is hampered by the restriction of system size and also because only little quantitative information is available concerning the specific electronic structure near these defects and its consequences for the binding of water molecules and ions. However, first steps into this direction were made recently Siepmann and Sprik [51], who investigated the influence of surface topology on water/electrode systems, and by Nagy and Denuault [98, 99], who extended the platinum-water potential to defect surfaces. However, since they did not take into account the electronic structure of the metal, their approach is purely geometrical.

The modeling of electrochemical experiments requires, in general, the incorporation of the effect of surface charges on the metal. This is often done by applying a homogeneous electric field, $\vec{E}$, across the liquid phase. Ionic charges and solvent dipoles interact with this field. In real electrolyte solutions, the electric field from the surface charge is screened by the ions in the interfacial region. In pure water, this is not the case. Gardner and Valleau [100] modeled the system therefore in the following way: They applied the diffuse layer potential that had been derived from the Gouy-Chapman theory [5, 6] at a predetermined ion strength. This takes screening into account in an elegant way. Especially at high ionic strength this leads to a fast decay of the electrostatic potential and permits the simulation of systems with a field-free liquid region at moderate system sizes. However, the Gouy-Chapman theory is not valid at high ionic strengths. Furthermore, in reality the solvent behavior is not only affected by the mean field of the ions but also by localized hydration phenomena. Philpott and Glosli [101–103] and Spohr [104] recently approached the screening problem more adequately by calculating the ionic density distribution in response to an external electric field as produced by a homogeneous surface charge density on a metal electrode (see section 5.3 below). In these studies, positive and negative surface charges are produced by the images of an electrolyte solution with an excess of anions and cations, respectively. The image interactions are augmented in the usual way by Lennard-Jones and corrugated Morse potentials.

3.2.3 Ion-Water and Ion-Ion Interactions

Ion-ion interactions are often described by a simple combination of Coulomb and Lennard-Jones type potential functions [105, 106] which have also been employed in simulations of electrolyte solutions near interfaces [107–109]. It was noted that these Lennard-Jones functions are inadequate for LiCl solutions at high concentrations [32, 110], as they are too repulsive at intermediate distances. Born-Mayer-Huggins potential functions of the functional form

$$V_{ij}(r) + \frac{q_i q_j}{4\pi\varepsilon_0 r} + bA_{ij} \cdot \exp[(\sigma_i + \sigma_j - r)/p] - \frac{C_{ij}}{r^6} - \frac{D_{ij}}{r^8} \tag{9}$$

have been used alternatively; here, r is the ion-ion distance, q_i and q_j are the charges, σ_i and σ_j the Pauling radii of the ions, and b, p, A_{ij}, C_{ij}, D_{ij} further parameters. As the ion concentrations in the vicinity of the interface can be quite high, the form of the short-range ion-ion repulsion can have an influence on the ionic density distribution in the interphase. These potential functions have not yet been applied to electrolyte solutions near interfaces. Currently, however, not enough data from simulations of electrolyte solutions are available to assess the importance of the details of the ion-ion potential functions in simulations of the interface.

The ionic charges also interact with the distributed point charges of the water model (for more information see, e.g., the reviews by Heinzinger and coworkers [105, 106, 111]). The equilibrium structure of the hydration shell and the energetics of hydration result from these terms and from Lennard-Jones type or exponentially repulsive potential terms acting between the oxygen and hydrogen atoms of the molecule and the ion. The models are either empirical in nature, reproducing experimental data like hydration energies or X-ray and neutron scattering studies, or they are derived from *ab initio* quantum chemical calculations. In a few simulation studies of bulk electrolyte solutions (e.g., Ref. 112, 113) the importance of three-body terms in the ion-water interaction potentials has been demonstrated; no corresponding studies have been made near interfaces.

3.2.4 Ion-Metal Interactions

Currently, the realistic modeling of the ion-metal interactions is the most severe limitation of simulation studies of electrochemical systems. Unfortunately, it may be the most important aspect of the model in order to describe many experimental observations, like, e.g., specific adsorption and partial charge transfer. The simplest model for the interaction of an ion with an ideal metal surface is the image charge model (see above). At short range, the electronic structure of both the metal and the ion comes into play and leads to a modification of the potential function, thus compensating the divergent attraction term by the repulsion between the atomic and ionic cores. A shortcoming of the image model is that it is neither specific for the ion nor for the metal. However, since the model incorporates some generic properties of the metal surface, it has been widely used. Ion-jellium interactions, in analogy to the water-jellium interactions described above, have so far not been used; however, it is a natural extension of existing models and will incorporate at least some metal specificity.

An alternative to these simple, more or less unspecific, models is the use of *ab initio* electronic structure calculations, in a similar spirit as in the corresponding water-metal models [40, 49]. Most studies so far have been performed with the cluster model approach [114]. In this approach, a finite number of atoms is chosen to model a local site on the crystal surface. Usually, the geometry of the substrate is kept fixed. The size of the substrate cluster varies between 4 and 20 metal atoms. The geometrical arrangement of the surface atoms is usually taken from the unreconstructed surface.

In order to obtain a potential energy surface from *ab initio* calculations, the interaction energies for many different ion positions relative to the metal cluster have to be calculated. Therefore the use of small clusters and approximate methods is necessary. Seitz-Beywl *et al.* [115] investigated Li^+ and I^- on 5-atom and 9-atom platinum clusters, respectively, cut out from a Pt(100) surface on the SCF level. Bagus, Pacchioni and coworkers studied the adsorption of halogens on small silver clusters extensively [116–118] and analyzed the ionicity of the bond by means of a projection operator approach. With this method, they did not find significant charge transfer between metal cluster and halogen in the neutral clusters. Blanco *et al.* investigated the binding of halogen atoms and halide ions to small mercury clusters [119, 120]. Tóth *et al.* [121] compared the interactions of the electrochemically abundant alkali and halide ions with 9- and 10-atom mercury clusters cut out from a (111) surface by means of SCF calculations. Recently, Ignaczak and Gomes [122] studied halide adsorption on copper, silver, and gold surfaces. While Blanco *et al.* [119, 120] investigated only the binding in the on-top position of a seven atom mercury cluster, in Ref. 121 the distance variations above the on-top, hollow, and bridge positions have been obtained. Analytical potential functions have been fitted to the binding energies in order to obtain potential energy hypersurfaces suitable for MD simulations. Using these analytical potential functions, the interactions of the ions with the extended surface have been calculated. For most ions, the hollow position with the maximum coordination was found to be energetically slightly favored over the bridge position. Only in the case of Li^+ and Na^+ the bridge position was slightly more favorable than the hollow site.

It should be noted here that the results of cluster calculations can at present only be qualitative in character. Energies and partial charges are not converged with respect to metal cluster size and level of approximation. The value of the calculations lies more in the opportunity to compare different ions with each other (in a given group) and the relative stability of different adsorption sites. The prediction of absolute adsorption energies is hardly possible. More promising for the future are calculations of adsorbates on periodic surfaces within the framework of the local density approximation of density functional theory (e.g., Ref. 123).

3.3 Long-range Interactions

The potential functions discussed in the previous section include a Coulomb term. Since this term is of very long range, the electrostatic interactions do not decay to zero over the length scale of the system under study. Therefore, these interactions either have to be truncated, or appropriate long-range corrections have to be made. Truncation methods usually give reasonable (in the sense of being qualitatively correct) results in polar bulk liquids, owing to the fact that they are isotropic and that the intermolecular correlations decay rapidly.

Near the interface, however, the liquid is anisotropic and, due to the interactions with a structured substrate, the correlations can be of much longer range. Therefore, spherical truncation methods are often inappropriate. This is especially the case, when the focus of interest is on charge distributions, electrostatic fields, and potentials, as it usually is in simulations of the electrochemical interface. It was recently demonstrated convincingly by several authors [124, 26, 27], that the use of lattice summation methods based on the Ewald method [24] or the fast multipole method [29, 28], are essential for an adequate model of an electric double layer. Specifically, it was demonstrated that the interfacial potential drop obtained from simulations depends on system size and, furthermore, that no 'bulk' region of vanishing electrostatic potential is established in the simulation cell, when using spherical truncation schemes. The absence of a field-free bulk region will lead to spurious features in ion density profiles and other properties. Ewald summation methods on the other hand, especially those for systems periodic in two dimensions (with no coupling between equivalent interfaces, as in Fig. 2b), but also three-dimensional summation schemes (which couple equivalent interfaces in a three-dimensional crystal, see Fig. 2a) lead to the establishment of a bulk region with constant electrostatic potential in a simulation cell. The use of Ewald summation methods is, therefore, strongly recommended in order to avoid artefacts of the calculations.

4 Water/Metal Interfaces

The interface between a metal and pure water is, for reasons discussed previously, much better studied than the corresponding electrolyte solution/metal interface. A wealth of information, most notably on the density variation with distance, arrangements and correlations within the adsorbate layer, and orientational properties, but also on hydrogen bonding and the electrostatic potential drop have been generated. From it, some general conclusions can be derived, that do not depend on the fine details of the interaction potentials used. In this section, I will try to elucidate the emerging picture of the ion-free double layer. Various facets of the problem will be illustrated by taking examples from my own work.

4.1 Density Profiles

Density profiles are the central quantity of interest in computer simulation studies of interfacial systems. They describe the correlation between atom positions in the liquid and the "interface" or "surface". Density profiles play a similarly important role in the characterization of interfaces as the radial distribution functions do in bulk liquids. In integral equation theories this analogy becomes apparent when formalisms that have been established for liquid mixtures are employed. Results for interfacial properties are obtained in the simultaneous limit of infinitesimally small particle concentration and infinite radius for one species, the "wall particle" (e.g., Ref. 125–129). Of course, this limit can only be taken for a smooth surface that does not contain any lateral structure. Among others, this is one reason why, up to now, integral equation theories have not been able to move successfully towards realistic models of the double layer.

For real corrugated surfaces, in principle the full single particle density $\tilde{\rho}(x, y, z)$, where (x, y, z) denote the Cartesian coordinates, has to be considered. The one-dimensional normalized density function $\rho(z)$ can be regarded as the lateral average of the full density function $\tilde{\rho}(z)$ according to

$$\rho(z) = \frac{1}{A}\int_A \mathrm{d}A\,\tilde{\rho}(x, y, z), \tag{10}$$

where A is the area of the surface within the basic box employed in the simulation. Usually, only $\rho(z)$ is calculated from computer simulation studies, mainly for two reasons: most importantly, (i) because in liquids the lateral order vanishes usually beyond the first, in a few cases beyond the second layer of molecules, it is more instructive to view the lateral structure in the adsorbed layer separately (see below); and, (ii) because $\tilde{\rho}$ is a position-dependent single-particle property, statistically meaningful averaging is only possible with very long simulations when the particles have probed a large portion of space by diffusion.

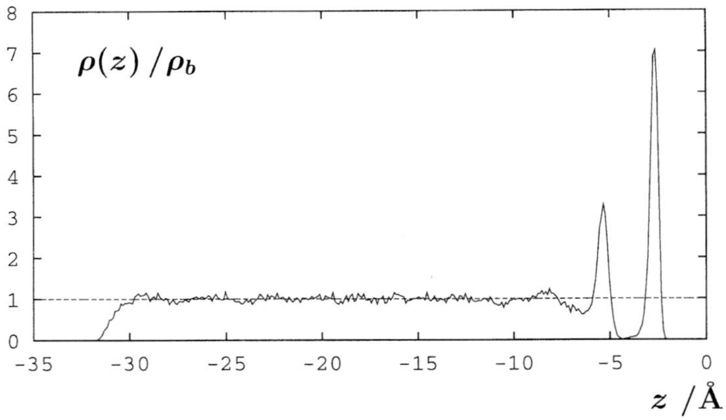

Fig. 7. Oxygen density profile from a 50 ps simulation of 385 water molecules that are confined on the left side by a simple (9-3) Lennard-Jones potential [137] and on the right side by 4 layers of mercury with a (111) surface structure.

Figure 7 shows the oxygen density profile for a lamina of water that is confined on the left side by a smooth (9-3) Lennard-Jones surface and on the right side by a rhombohedral mercury crystal with (111) surface structure. The mercury phase was modeled as a rigid crystal in order to avoid the complications in the simulation of the liquid state of mercury (see also Ref. 81, 130, 131). The atom-atom distances are chosen in accordance with the x-ray study by Bosio *et al.* [132]. These authors conclude that the structure of liquid mercury can be described in a good approximation as a solid α-mercury lattice with a larger lattice constant in z direction. The surface of such a crystal shows rather low corrugation. The changes in water structure in contact with a liquid mercury phase are discussed below.

The figure demonstrates clearly the major differences between the water structure next to a metal surface and near a free or unpolar surface. Due to the significant adsorption energy of water on metal surfaces (typically of the order of 20 to 40 kJ mol^{-1}; see, e.g., Ref. 97) strong density oscillations are observed next to the metal (right side of the lamina) which are absent near the smooth wall on the left side. Between three and four water layers have been identified in most simulations near uncharged metal surfaces, depending on the model and on statistical accuracy. Beyond about 10–12 Å from the surface the density is typically constant and equal to the bulk value. In strong unscreened electric fields several authors [133–136] have observed a phase transition towards a ferroelectric crystalline state in their simulations. However, it should be kept in mind that these systems, because of the absence of ionic screening, are rather unphysical in nature.

4.1.1 Dependence on Adsorption Energy

Spohr investigated the dependence of the shape of the oxygen density profiles on the adsorption energy [56]. The model contains a simple scaling parameter, D_O, that determines the adsorption energy (see Eq. (7) and Eq. (8)).

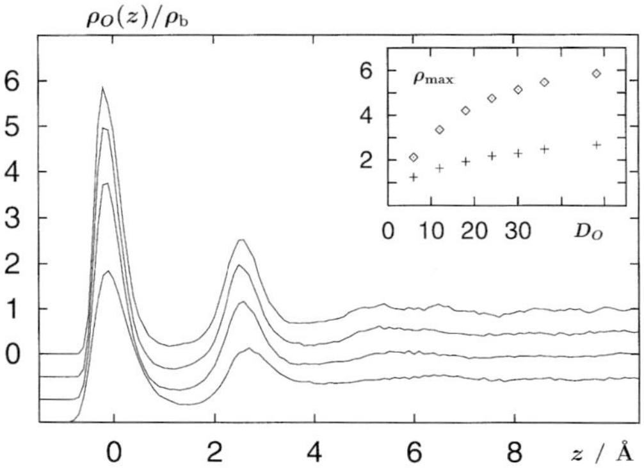

Fig. 8. Normalized oxygen density profile perpendicular to the surface from simulations of pure water with adsorption energies of 12, 24, 36, and 48 kJ/mol (from bottom to top). The lower curves are shifted downwards by 0.5, 1.0, and 1.5 units. The inset shows the height of the first (diamonds) and second peak (crosses) as a function of adsorption energy.

Figure 8 shows the oxygen density profiles for four values of D_O equal to 12, 24, 36, and 48 kJ/mol. In the inset, the height of the first (diamonds) and second peak (crosses) of the density profile are shown for the pure water simulations corresponding to these values of D_O and to several intermediate ones. A monotonic increase of peak heights is observed with increasing adsorption energy. Both curves level off at high adsorption energies. At low adsorption energies, peak heights approach 1 for $D_O \to 0$. This is in keeping with the observation that only small density variations are observed in simulations of water in contact with weakly attractive surfaces (see e.g. Ref. 137, 138). The experimentally accessible adsorption energy of water on a transition metal surface can thus be regarded as an indicator for the degree of local structuring in the vicinity of the interface.

The density profiles obtained from atomic models (platinum [49] and rigid mercury surfaces [40]), where water-metal interactions are described by pairwise additive atomatom interaction potentials, are similar in shape. Height and width are correlated with the depth and force constant of the interaction potential. A similar correlation between peak height and interaction energy holds for water near a variety of different smooth model surfaces (see Ref. 139 and references therein).

4.1.2 The Liquid Water/Liquid Mercury Interface

In experimental studies the liquid mercury electrode has been prototypical for a long time. Modeling the liquid mercury electrode is complicated by the fact that knowledge of mercury-mercury interactions is required. The liquid mercury/liquid water interface was investigated by Heinzinger and coworkers [78, 81] using a pseudopotential ansatz.

Figure 9 compares the oxygen and hydrogen density profiles for the interface between pure water and rigid mercury (solid lines; taken from Ref. 40) and water and liquid mercury (short dashes; taken from Ref. 78). The features of the water density profiles at the liquid/liquid interface are washed out considerably relative to those at the liquid/solid interface. However, in the first layer this effect is almost entirely due to the roughness of the mercury surface: not all mercury atoms at the interface are in the same plane (at $z = 0$) but cover a range of approximately 1.3 Å (see Figure 1 in

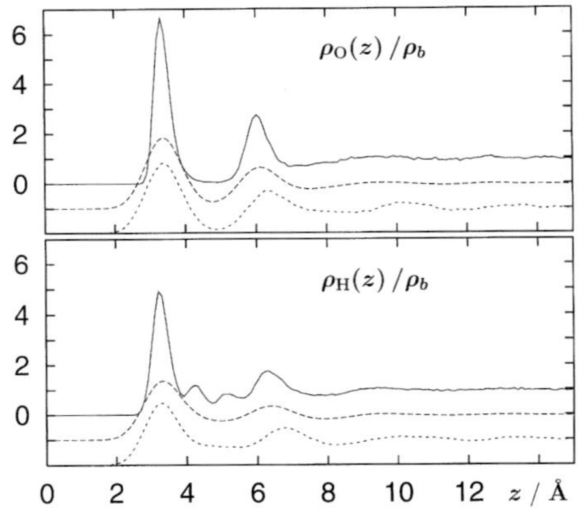

Fig. 9. Oxygen (top) and hydrogen (bottom) density profiles of water near a rigid mercury crystal (full line) and near liquid mercury (short dashes). The long-dashed curves are the convolution of the density profiles near the rigid crystal and the first maximum in the liquid mercury density profile according to Eq. (11) and 12. The dashed curves are shifted downward by one or two units for better legibility.

Ref. 78). The width is larger than the width of the oxygen and hydrogen peaks near the solid surface. Consequently, in order to compare the liquid/liquid with the liquid/solid interface, the density distribution near the rigid surface can be convoluted with a width function $w(z)$ due to the mercury motion according to

$$\rho'(z) = \int_{-\infty}^{\infty} \rho(z')w(z-z')\,\mathrm{d}z'. \tag{11}$$

The width function is the normalized shape of the first mercury peak in the density profile, approximated as a Gaussian distribution of width $\sigma = 1.3$ Å

$$w(z) = \frac{\sqrt{2}}{\sqrt{\pi}\cdot\sigma}\exp[-2(z/\sigma)^2]. \tag{12}$$

The convoluted density profiles are plotted in Figure 9 as the long-dashed curves. The convolution accounts for almost all the structural differences in the range $z < 5$ Å. The heights of the first peaks of the oxygen and hydrogen density profiles after convolution are identical to those of the respective functions at the liquid/liquid interface. The intermediate maxima in the hydrogen density profiles around 4.3 and 5.1 Å vanish. Only in the second layer there are slight structural differences; the second peak is further away in the simulation of the liquid/liquid interface than in the corrected function.

4.1.3 Dependence on External Electrical Field

Recently, Toney *et al.* [140] published the results of X-ray reflectivity measurements on dilute aqueous sodium fluoride solutions near positively and negatively charged silver electrodes with (111) surface geometry. The analysis shows an oscillatory density profile with three or four density maxima near the interface, qualitatively similar to the results in Fig. 7. They also found a very large shift of the distance of the first

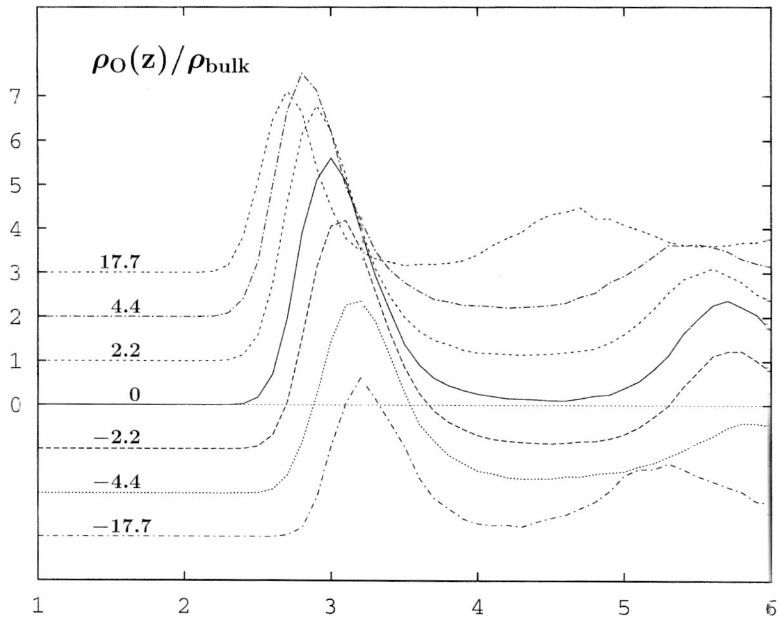

Fig. 10. Oxygen density profiles from 100 ps simulations of 700 water molecules between mercury surfaces with (111) surface structure in homogeneous external electric fields. The surface charge density (in units of $\mu C\ cm^{-2}$) is indicated.

maximum of the oxygen density profile away from the electrode when going from positive to negative field strengths. The data could only be explained by an enormously large water coverage [140] of 1.1 and 1.8 water molecules per silver atom at negative and positive surface charge densities, respectively.

Figure 10 shows the oxygen density profile from MD simulations of pure water at various surface charge densities. Since there is no ionic screening in the liquid, the surface charges translate into homogeneous electric fields across the liquid phase. Since no quantum chemical calculations for water adsorption on silver are currently available, the simulations were run for water near mercury with (111) surface structure. Indeed, a shift of the first maximum of the oxygen density is observed in analogy to the results in Ref. 140. However, no drastic increase in total particle density is found in the adsorbed layer. Contrary to experiment, at the high surface charges of $\pm 17.7\ \mu C\ cm^{-2}$, the density decreases. However, like in the simulations by Watanabe, Brodsky and Reinhardt [133, 134] and, more recently, Xia and Berkowitz [135, 136], the liquid phase is unstable at this high field strength: there is a spontaneous transition to a ferroelectric crystalline phase.

At all field strengths, the surface coverage is slightly less than 1 water molecule per mercury atom. The surface spacing of mercury atoms is 3.0 Å, which is larger than the one for Ag(111) of 2.88 Å (neglecting surface reconstruction and relaxation). Therefore, the surface area of one silver atom on the Ag(111) surface is smaller by about 10% than in the simulation. Consequently, a coverage of not larger than about

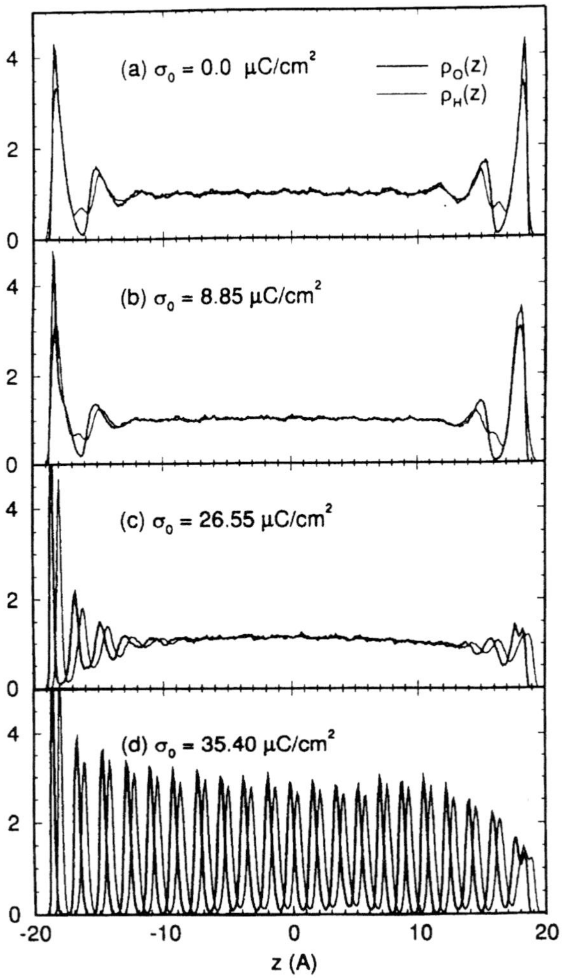

Fig. 11. Normalized density profiles of oxygen and hydrogen atoms for systems with (a) no surface charge density on the Pt surface, (b) surface charge density of 8.85 μC cm^{-2}, (c) 26.55 μC cm^{-2}, and (d) 35.40 μC cm^{-2}. The positively charged surface is on the left, the negatively charged surface is on the right. (Reproduced with permission from Ref. 135.)

0.9 molecules per surface atoms would be expected from MD simulations. In spite of the simple, or even simplistic, nature of the interaction models, it seems very unlikely from the simulations and geometric considerations that water coverages as high as the ones quoted in Ref. 140 can occur. A simple calculation based on the jellium model by Schmickler *et al.* [68] appears to be able to reproduce qualitatively the observed change in metal oxygen distance. This work, however, does not address the problem of the extremely high densities.

Xia and Berkowitz [135] analyzed the structural changes of a slab of water similar to the one in Fig. 3 between uncharged and charged Pt walls. Figure 11 from their work shows the oxygen and hydrogen density profiles for 4 different surface charge densities. The figure quite clearly shows that, starting at $\sigma_0 = 26.55$ μC cm^{-2}, the surface charge and the concomitant electric field lead to a more pronounced ordering near the positively charged surface on the left (c) than near the uncharged surfaces (a). At the highest surface charge density investigated by the authors (d) all oxygen and all hydrogen atoms are arranged in distinctive layers. The ordering is ferroelectric, as all the maxima of the hydrogen density are shifted by about 0.5 Å to the right of the corresponding oxygen maxima, indicating almost complete dipole orientation. The authors illustrate the ordering by snapshot configurations and present further evidence for a reduction of surface coverage with increasing surface charge density as a consequence of changes in lateral structure and dynamics.

The experiments by Toney and coworkers are very important since they provide at the present time the only direct experimental evidence for the layering of water molecules in the vicinity of metal surfaces. However, they also have been highly controversial because of the extreme increase in density. Simulations show an increase in correlation, not in total density. The discrepancies are probably an indication for specific adsorption or another yet unknown electrode process in the experiments. In the near future another series of experiments will be performed by Ocko and coworkers which will hopefully resolve the controversy.

4.1.4 Other Parameters

When the corrugation of the surface is not too large, i.e., the energy difference between the most favorable and the most unfavorable adsorption site is less than about 20% of the overall adsorption energy, it does not influence the density profile significantly in the direction perpendicular to the surface, in spite of the fact that substantial lateral correlations exist within the adsorbed layer of molecules. Weak correlations between favorable adsorption sites and preferred locations for molecules in the second layer have been observed (see Fig. 5 in Ref. 49). They are a consequence of the interactions between second-layer molecules and the molecules in the ordered adsorbate layer.

It is increasingly realized that many-body induction interactions should be included in computer models, especially in inhomogeneous environments. Kohlmeyer *et al.* [44] therefore investigated the role of molecular polarizability on the density profiles of a slab of water in contact with several different metal surfaces. They employed the polarizable TIP4P model by Rick and Berne [46]. It was found that the density profiles are almost identical near a metallic surface; the liquid/gas interface appears to become slightly wider. Earlier studies of polarizable water at a hydrophobic wall by Wallqvist [141] and near the liquid/gas interface by Motakabbir and Berkowitz [142] also concluded that polarization effects are of secondary importance.

Finally, a note on the dependence on the electrostatic boundary conditions of the simulation is in order here. The atom density profiles obtained from simulations with and without proper treatment of the long-range interactions look very similar. However, when the charge density, i.e., the difference between properly weighted oxygen and hydrogen densities, is calculated, small discrepancies become evident, once the statistical accuracy of the density profiles is sufficiently high. In the running integrals over the charge densities, which correspond to the average electric field and potential, these discrepancies lead to qualitatively different potential distributions, namely a field-free bulk-like region in the case of Ewald summations and an unphysical situation in the case of spherical truncation methods (see section 3.3 and Ref. 26, 27, 124).

4.2 Lateral Density Distribution

Lateral density fluctuations are mostly confined to the adsorbed water layer. The lateral density distributions are conveniently characterized by scatter plots of oxygen coordinates in the surface plane. A dot is made at the atom position at a specified interval of time (here 0.1 ps). Figure 12 shows the scatter plots of water molecules in the first (top) and second layer (bottom) near the Pt(100) (left) and the Hg(111) surface (right). The figure demonstrates that the lateral inhomogeneity is more pronounced for the Pt(100) surface than for the Hg(111) surface, in keeping with the

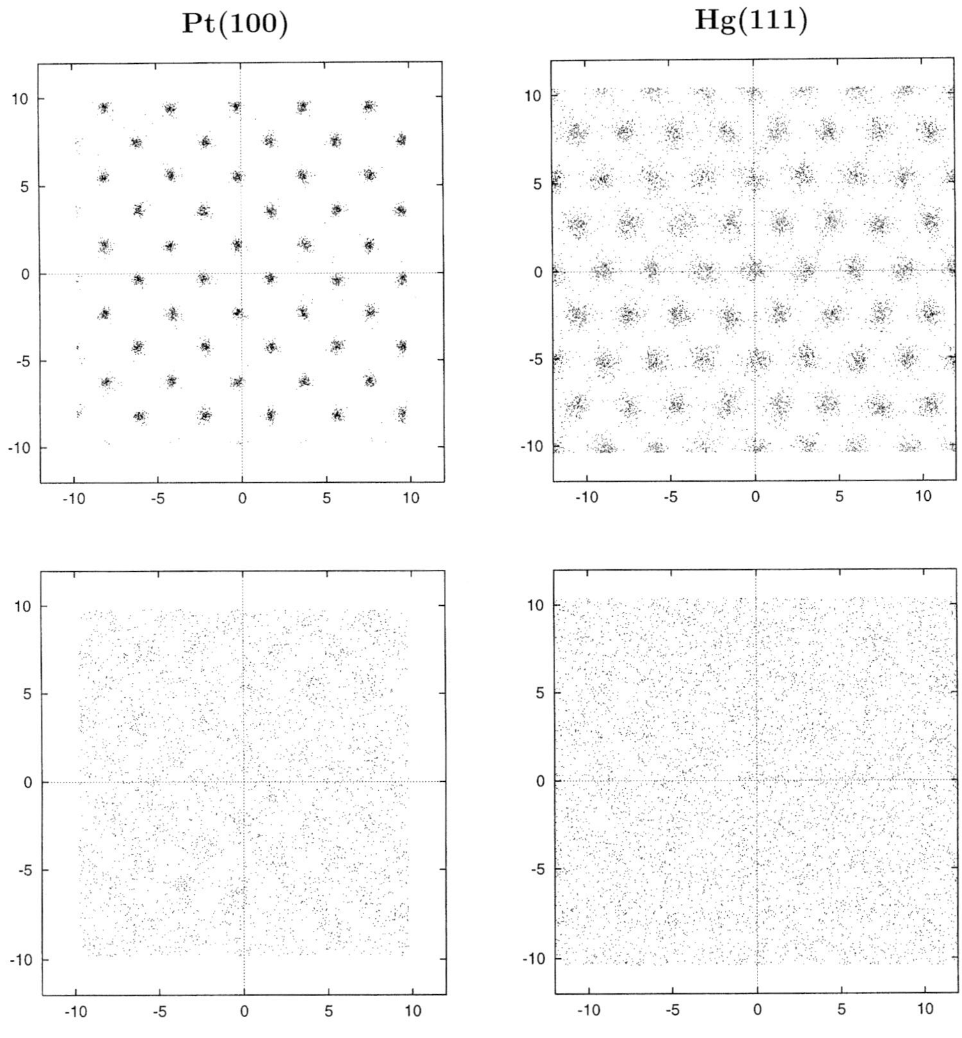

Fig. 12. Lateral density distribution of oxygen atoms in the first (top) and second layer (bottom) on Pt(100) (left) and Hg(111) (right). A symbol is plotted at time intervals of 0.1 ps.

larger corrugation on the Pt(100) surface. In the first layer, the oxygen distribution clearly shows the structure of the substrate lattice. The oxygen motion is predominantly oscillatory rather than diffusive. In the second layer, the distribution is almost random, except for a trace of order in the case of Pt(100) (see Ref. 49 for further discussion).

The data in Fig. 12 are qualitatively similar to those obtained in the group of Berkowitz [52, 54, 53, 143] with a modified version of the platinum-water interaction potential described in section 3. They compared the structure near Pt(100) and Pt(111) in detail and noted that the motion of water in the first layer is oscillatory about equilibrium positions and thus characteristic of a solid phase, while the motion in the second layer has more liquid-like character. These authors also point out that on Pt(100) all surface sites are occupied while there are vacant site on the Pt(111) surface; this is a consequence of the fact that adsorbed water has a tendency to form hexagonal rings. The coverage of adsorption sites is only about 0.8 on the (111) face; however, the number of water molecules per surface area is roughly equal on both surfaces. The vacant sites make hopping motions between neighbouring sites more likely on the (111) surface than on the (100) surface and thus lead to a larger lateral mobility on the (111) face [53].

4.3 Pair Correlation Functions

The strength of the water-metal interaction together with the surface corrugation gives rise to much more drastic changes in water structure than the ones observed in computer simulations of water near smooth nonmetallic surfaces. Structure in the liquid state is usually characterized by pair correlation functions (PCFs). Because of the homogeneity and isotropy of the bulk liquid phase, they become simple radial distribution functions (RDFs), which do only depend on the distance between two atoms. Near an interface, the PCF depends not only on the interatomic distance but also on the position of, say the first, atom relative to the interface and the direction of the interatomic distance vector. Hence, considerable changes in the atom-atom PCFs can be expected close to the surface.

Corrugation effects give rise to pronounced lateral correlations, which can also be characterized by the scatter plots shown in section 4.2. The corresponding PCFs have been investigated by Spohr [49] and Heinzinger [144, 145] for Pt(100) and by Böcker *et al.* [81] for mercury surfaces.

The influence of density inhomogeneities in the vicinity of the interface on the PCF is visualized in the most simple way by calculating the RDFs separately when one particle, call it i, is located in a certain distance interval from the surface. These functions then predominantly characterize the effect of inhomogeneity but also some aspects of the anisotropy of the correlations. Figures 13 and 14 show the atom-atom RDFs g_{ij} for the "i particle" located in the first (full) and second (dashed) water layer near the Pt(100) surface and the Hg(111) surface, respectively.

In both systems the RDFs in the second layer are very similar to the ones in the center of the lamina (not shown), which, in turn, are identical to the RDFs in bulk liquid water at the same temperature. The layer of adsorbed water molecules (full

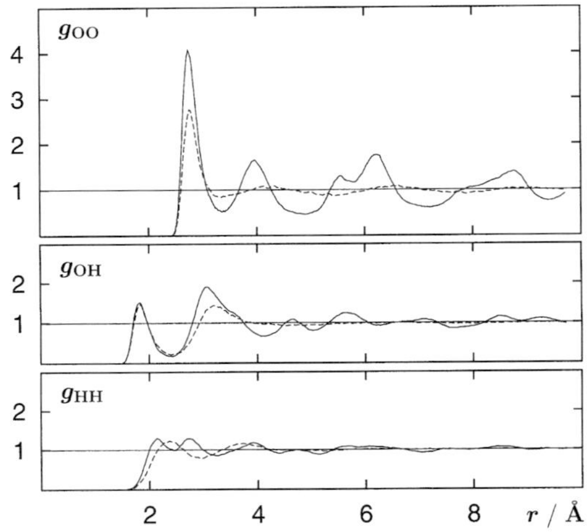

Fig. 13. Intermolecular oxygen-oxygen, oxygen-hydrogen, and hydrogen-hydrogen distance distribution functions with the reference atom in the first layer (full lines) and second layer (dashed lines) near the Pt(100) surface.

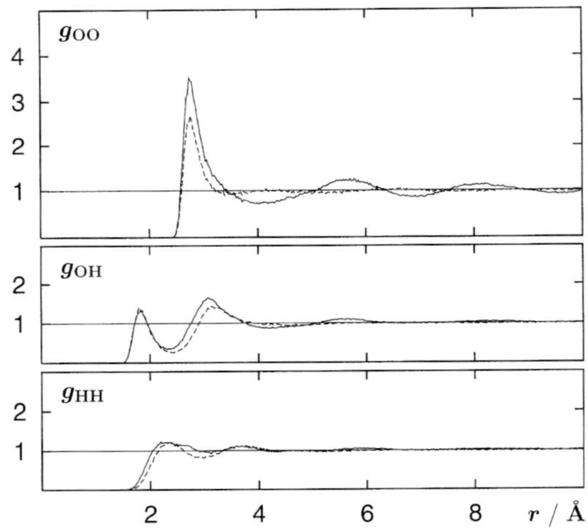

Fig. 14. Intermolecular oxygen-oxygen, oxygen-hydrogen, and hydrogen-hydrogen distance distribution functions with the reference atom in the first layer (full lines) and second layer (dashed lines) near the Hg(111) surface.

curves) behaves significantly different from the bulk. The oxygen-oxygen RDFs show a much more pronounced structure. In both systems the first neighbor peak is strongly enhanced but its position is not shifted from the bulk distance, 2.85 Å. The Pt(100) and Hg(111) crystal matrices obviously support the nearest-neighbor oxygen-oxygen arrangement in spite of the mismatch between water-water distance and nearest neighbor distances of the substrate.

Near Pt(100) several new maxima occur beyond the first neighbors at about 3.9, 5.6, 6.3, and 8.7 Å together with a shoulder around 8 Å while the maximum found in bulk water at 4.5 Å vanishes. The new maxima are induced by the periodicity of the surface adsorption sites on the quadratic lattice, which are positioned at $2^{1/2}$, 2, $5^{1/2}$, $8^{1/2}$, 3, and $10^{1/2}$ times the nearest neighbor distance d. Similarly, the pronounced maximum around 5.5 Å near Hg(111) is induced by the adsorption sites at $3^{1/2}d$ and $2d$, and the maximum near 8 Å originates from the adsorption sites at $7^{1/2}d$ and $3d$

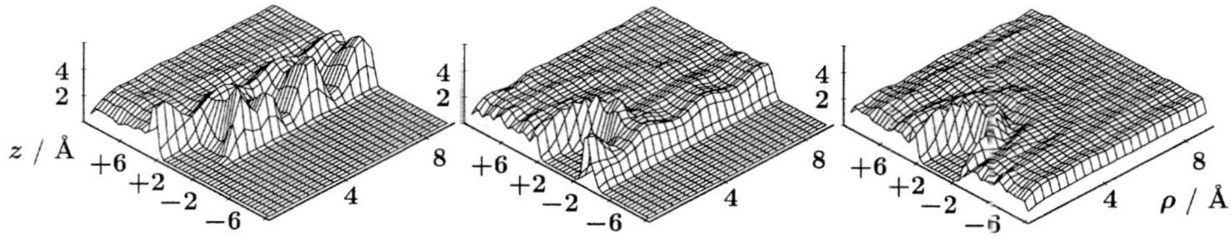

Fig. 15. Anisotropic oxygen-oxygen pair correlation functions $g_{OO}(\rho, z)$ for the adsorbate molecules (left), the molecules in the second layer (middle), and the molecules in the bulk-like center of the water lamina (right) between Pt(100) surfaces. ρ is the transversal and z the normal part of the interatomic distance.

on the hexagonal lattice. The oxygen-hydrogen and hydrogen-hydrogen RDFs in the vicinity of the Pt(100) surface also exhibit additional maxima. The distances at which they occur can be related to hydrogen bonds between adsorbed water molecules on neighboring adsorption sites [49].

Figures 13 and 14 show the effect of density inhomogeneities on the atom-atom PCFs, averaged over all possible orientations of the atom pairs relative to the interface. In Fig. 15 the degree of anisotropy of the oxygen-oxygen PCF is further analyzed for water near the Pt(100) surface in the surface layer (left), in the second layer (middle), and in the center of the lamina (right). The figure shows $g_{OO}(\rho, z)$ which is a representation of the PCF as a function of two components of the distance vector. $\rho = r_{OO} \sin \vartheta$ is the component of the interparticle distance parallel to the surface and $z = r_{OO} \cos \vartheta$ the component perpendicular to the surface. ϑ is the angle between the surface normal vector (z-direction) and the O–O vector r_{OO}. From the left part of Fig. 15 one concludes that the changes beyond the first neighbor shell of $g_{OO}(r)$ in the adsorbate layer are due mostly to correlations within the adsorbate layer. The PCFs show the characteristic long-range oscillations along the ρ axis that were seen in Fig. 13. Near the surface there is an increased probability of finding a neighbor in the direction away from the surface as indicated by the peak at $z \approx 3$ Å and $\rho \approx 0$ Å. This observation is reflected by the increased probability for second layer molecules to find a neighbor at $z \approx -3$ Å and $\rho \approx 0$ Å (Fig. 15, middle). The two peaks are consistent with the formation of interlayer hydrogen bonds between the directly adsorbed water layer and the second layer (see below). In the center of the lamina the oxygen-oxygen PCF is isotropic (left part of Fig. 15).

4.4 Orientational Structure

The orientational structure of water near a metal surface has obvious consequences for the electrostatic potential across an interface, since any orientational anisotropy creates an electric field that interacts with the metal electrons. The anisotropy of the orientational distribution of water has therefore been investigated in most studies of aqueous systems in inhomogeneous environments. The results can be summarized as follows. In almost all studied systems, a preference for orientations in which the water dipole moment is more or less parallel to the interface has been observed. The driving force for the avoidance of orientations that can lead to surface electrostatic

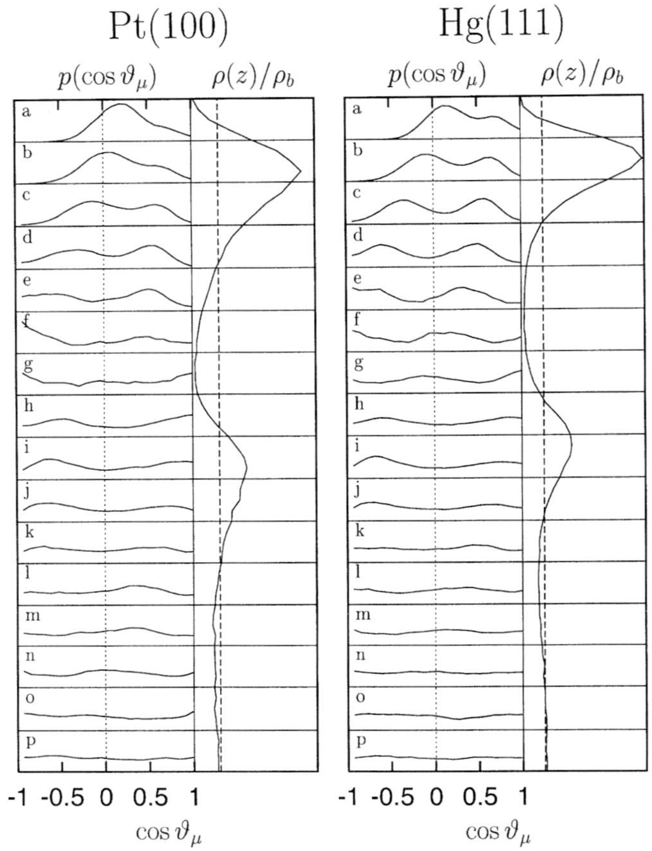

Fig. 16. Orientational distribution of the molecular dipole moment on Pt(100) (left) and Hg(111) (right). $\cos\vartheta_\mu$ is the angle between the water dipole vector and the surface normal that points into the water phase. Panels *a* to *p* on the left are sampled from the distance intervals which are indicated by the cuts through the density profile $\rho(z)/\rho_b$ on the right.

polarization are, according to Lee *et al.* [137], the balance of (i) the packing forces which tend to produce a dense layer in contact with the surface and (ii) the tendency of molecules to maintain a maximal number of hydrogen bonds. These hydrogen bonds are formed mainly within the adsorbate layer but also between the adsorbate and the second layer.

In Ref. 49 the orientational distribution of water near the Pt(100) surface was investigated in great detail. In spite of the preference for adsorption of isolated water molecules through the oxygen atom, which is incorporated into the water-metal interaction potential, relatively few configurations were observed in which the dipole moment of the molecule points into the solution. The analysis will not be repeated here; the interested reader is referred to Ref. 49.

Figure 16 shows the orientational distribution of the molecular dipole moment relative to the surface normal in various distance ranges from the Pt(100) (left) and the Hg(111) surface (right). Additionally, the reduced oxygen density profile is plotted on the right side of each figure. The baselines between distribution functions cut through the density profile. The distribution function in each panel on the left side is for the subset of molecules that are located in the distance range between these lines on the right side. By and large, the orientational distributions are quite similar in both systems. Over the first peak in the density profiles (panels *a* to *d*) there are almost no molecules with the dipole moment perpendicular to the surface. Within the adsorbate layer, there is a transition from the preference for orientations in which the dipoles point more or less into the solution (*a* and *b*) to one where a substantial fraction of the dipoles point more or less towards the surface (*c* and *d*). This behavior is characteristic for the "bilayer" model that has been proposed for the interpretation of the structure of water monolayers adsorbed on metal surfaces under ultrahigh vacuum conditions (see below).

The orientational anisotropy ranges as far into the liquid phase as the density inhomogeneities do (roughly up to panel *m*), with increasingly less pronounced features. Slightly beyond the second maximum in the density profile the orientational distribution is isotropic, as it has to be the case for a bulk-like liquid.

An external electric field changes the orientational distribution and consequently the orientational polarization of the water molecules in the interfacial region. The effect of homogeneous [146–149, 133, 134] and inhomogeneous [100] electric fields on the orientational distribution near smooth model surfaces has been investigated using lattice models [147] and distributed point charge models like the ones discussed in section 3.2.1. Heinzinger and coworkers [150, 151] and later Xia and Berkowitz [135] investigated the changes on the more realistic Pt(100) surface. Because of the absence of free ions in these simulations of pure water, the electric field is only screened by the water dipoles themselves; consequently a net electric field and a concomitant polarization persist through the lamina. At large field strengths, this has been observed to lead to a field-induced phase transition to a crystalline water phase [133–135].

Figure 17 shows the orientational distribution of the dipole moment vectors in the adsorbed water layer for various surface charge densities (corresponding to homogeneous electric fields according to $\sigma = E \cdot \varepsilon_0$). The data are taken from simulations of 700 TIP4P water molecules between two mercury surfaces with (111) surface structure. In the field-free case, a wide bimodal orientational distribution is observed (see also discussion of Fig. 16). The preference of "oxygen down" bonding is manifest in the larger probability to find positive values of $\cos \vartheta$ (corresponding to angles smaller than 90° between dipole vector and the surface normal that points into the liquid phase) than to find negative ones. The tendency of water to form hydrogen bonds leads to the observed wide distribution, where most of the water dipoles are more or less oriented parallel to the surface, similar to the ideal bilayer sketched in Fig. 21 (see below). For positive surface charges the average dipole moment (the first moment of the distribution) shifts with increasing surface charge density towards larger absolute values and the distribution becomes increasingly narrower. For low negative surface charge densities, the dipole orientational distribution becomes more symmetric around the parallel orientation. At larger negative surface charge densities, the orientational distribution changes in such a way that the hydrogen atoms point preferentially to the surface. Even at the highest surface charge densities of

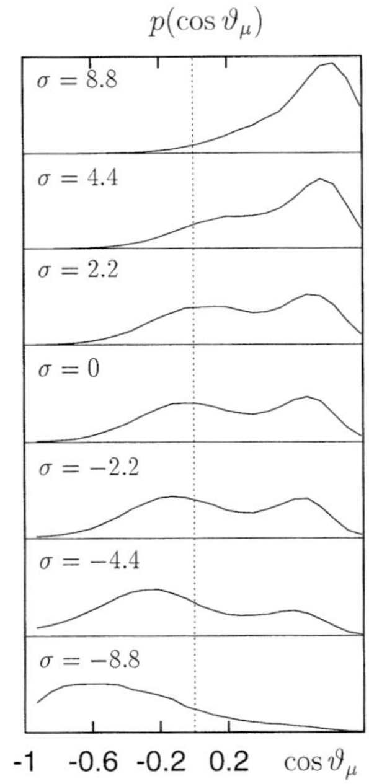

Fig. 17. Distribution of the cosine of the angle ϑ_μ between the dipole moment unit vector of the adsorbed water molecules ($\hat{\mu}$) and the surface normal ($\hat{z}$), which points into the liquid phase, for various homogeneous surface charge densities which are given in units of $\mu C\,cm^{-2}$ on the graphs. Data are from simulations of 700 TIP4P water molecules between Hg(111) surfaces.

$\pm 17.7\,\mu C\,cm^{-2}$ (not shown) no complete ordering is observed. The effect of field-induced crystallization occurs only at even higher surface charges [135].

The behavior near the Pt(100) surface (see Ref. 150, 151) is qualitatively similar to the one on the Hg(111) surface. It is only noted here that with the flexible BJH model used in Ref. 150 the bimodal character of the distribution function is not observed, apparently because the flexibility of the model allows for a wider range of low energy orientations within the adlayer hydrogen bond network.

4.5 Interfacial Potential Drop

The charge density profile $\rho_c(z)$ and the dipole density profile $\rho_\mu(z)$ can be calculated from the atomic density profiles and the orientational distributions. Figures 18b and 18c show the dipole and charge density, respectively, for water near the mercury surface. For both the first and the second water layers, a quadrupolar charge profile with alternating regions of positive, negative, and again positive charge density is observed. The form of the charge profiles originates from the fact that the centers of mass (or the oxygen atom) of the water molecules are mostly well localized while the hydrogen density distribution is broadened due to librational motions and due to the hydrogen bonding between layers. Beyond the second layers the charge density approaches zero within the limits of statistical uncertainty.

The dipole density profile $\rho_\mu(z)$ indicates ordered dipoles in the adsorbate layer. The orientation is largely due to the anisotropy of the water-metal interaction potential, which favors configurations in which the oxygen atom is closer to the

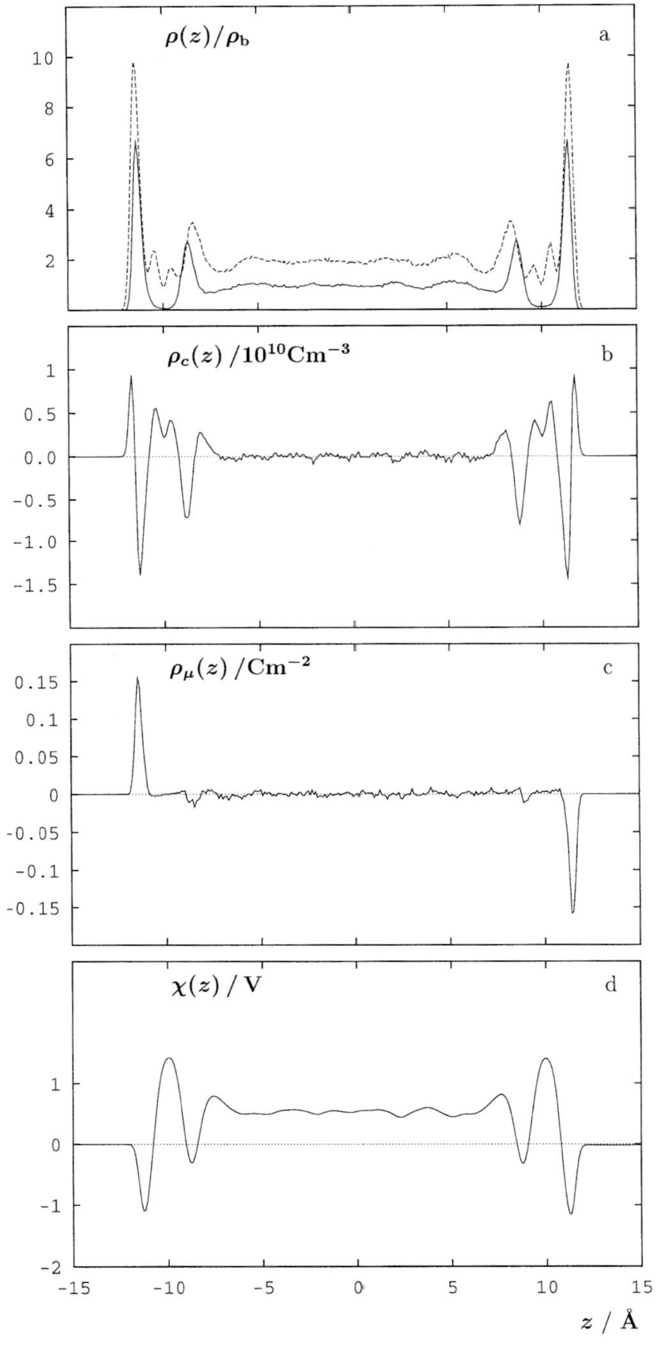

Fig. 18. (a) Oxygen (full) and hydrogen (dashed) density profiles. (b) Charge density ρ_c. (c) Dipole density ρ_μ. (d) Water contribution to the surface potential χ calculated from the charge density ρ_c by means of Eq. (13). All data are taken from a 150 ps simulation of 252 water molecules between two mercury phases with (111) surface structure using Ewald summation in two dimensions for the long-range interactions.

surface. Most quantum chemical calculations of water near metal surfaces to date predict a significant preference of "oxygen-down" configurations over "hydrogen-down" ones at zero electric field (e.g., Ref. 82, 85–88, 152, 153). The difference in sign on the left and right side of the lamina is a consequence of the vector nature of the dipole moment, which points into opposite directions near the two interfaces. The dipole orientation in the second layer is only weakly anisotropic (see also Fig. 16).

Solving the one-dimensional Poisson equation with the charge density profile $\rho_c(z)$ leads to the electrostatic (dipolar) potential drop near the interface according to

$$\chi(z) = -\frac{1}{\varepsilon_0}\int_{-\infty}^{z} \rho_c(z')(z-z')\,\mathrm{d}z' \tag{13}$$

with ε_0 the permittivity of the vacuum (Fig. 18d).

The potential is constant in the center of the lamina. It is demonstrated in Ref. 26, 27 that the use of proper long-range boundary conditions like the Ewald summation method is crucial to obtain the field-free bulk region in the center of the lamina at vanishing surface charge density. In the case of the mercury/water interface, the potential drop across the interface is 0.55 V. Experimentally, no value for χ has been determined that is free from assumptions. However, the most likely value of χ is around 0.2 to 0.3 V [154]. The simulations seem to overestimate the dipolar ordering at the interface. However, there is some indication that the potential drop χ is reduced upon introduction of molecular polarizability (by about 0.1 V) [155]. Results of the liquid water/liquid mercury interface [81] also indicate that the inhomogeneous broadening of the surface leads to a reduction of χ by about 0.2 V. Furthermore, in the model, the electronic response of the metal to the water phase is taken into account only to the extent that it is already included in the quantum chemical cluster calculations with a single water molecule.

4.6 Hydrogen Bonding

Hydrogen bonds are the most characteristic element of liquid water structure. Water models used in computer simulations are able to describe the properties of the hydrogen bond network in a realistic way, contrary to many of the dipolar model fluids used in analytical theories. Much has been learned about bulk water and solutions through an analysis of the hydrogen bond network (e.g., Ref. 156, 157).

The strong changes in the interfacial PCFs discussed in one of the previous sections indicate a substantial perturbation of the hydrogen bond network. These changes have been studied on the basis of a geometrical hydrogen bond definition used previously by Pálinkás and Heinzinger [157]. In what follows, two water molecules will be termed "neighbors" if the O–O distance is smaller than 3.35 Å; they will be termed "hydrogen bonded" if additionally the angle ϑ_{HB} between the intramolecular O–H vector and the intermolecular O–O vector is smaller than 20°.

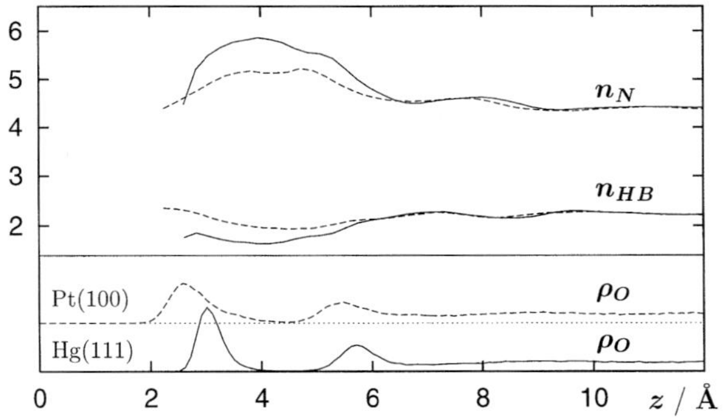

Fig. 19. Top: Number of neighbors n_N and number of hydrogen bonded neighbors n_{HB} (for definition see text) as a function of the distance z from the surface for TIP4P water near Pt(100) (dashed) and near Hg(111) (full). Bottom: the oxygen density profiles near Pt(100) (dashed) and Hz(111) (full).

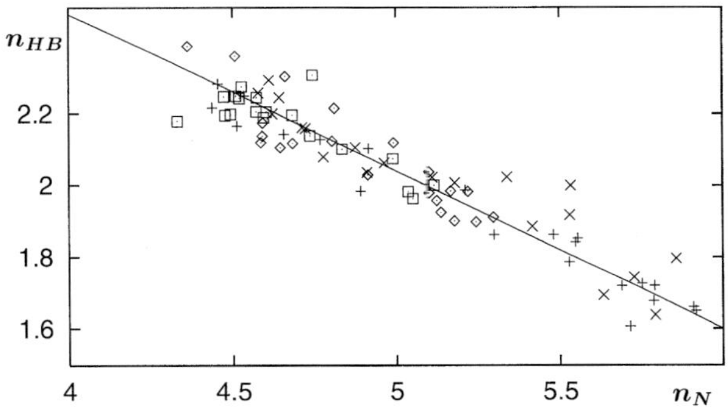

Fig. 20. Number of hydrogen bonds per molecule n_{HB} as a function of the local number of neighbors n_N in various systems: Water near Pt(100) (diamonds), near Hg(111) (+), near a model corrugated surface with adsorption energy of 12 kJ mol^{-1} (squares) and 30 kJ mol^{-1} (×) at 298 K. The line is drawn to guide the eye.

Figure 19 shows the average number of neighbors n_N and the average number of hydrogen bonds of a water molecule, n_{HB}, as a function of distance from the Pt(100) and Hg(111) surfaces. The oxygen density profiles are included in order to identify the length scales. The number of neighbors in the center of the lamina is around 4.4, the same as in bulk TIP4P water using the same definition. In the adsorbed layer and to some extent in the second layer, the number of neighbors increases to a value of approximately 5 in the case of the Pt(100) surface and almost 6 in the case of the mercury surface. The large number of neighbors is due to the fact that most of the adsorption sites are occupied by water molecules.

In Fig. 20 the local number of hydrogen bonds per molecule is plotted versus the local coordination number at all distances and for several simulations. There is a clear correlation between the decrease of the number of hydrogen bonds and the increase in coordination number for all simulations, independent of the interface. The values near the interface (at large n_N) fall on the same line as the bulk values,

$(n_N, n_{HB}) = (4.5, 2.3)$. This demonstrates quite clearly that the local coordination number (which is mostly a function of adsorption energy, curvature of the potential energy surface, and corrugation) plays a more important role than the surface geometry ((100) or (111) surfaces) or the lattice spacing (2.54, 2.77, and 3.00 Å for the model, Pt, and Hg surfaces, respectively). The changes in the hydrogen bond network near the metallic interface thus correlate quite well with the changes in local coordination. The changes are less drastic than one would expect from the geometric changes near the interface. The hydrogen bond network is flexible enough to adapt to the specific geometric situation near the metal surface.

4.7 Water Dynamics

Dynamical properties of water near metallic surfaces have been investigated, among others, by Spohr for flexible water near the Pt(100) surface [158], Heinzinger and coworkers for water near the charged Pt(100) surface [150, 151], Zhu and Robinson for flexible polarizable water between rigid plates [149], Berkowitz and coworkers for water near Pt(100) and Pt(111) [53, 54] and Rose and Benjamin for solutions of Na^+ and Cl^- near the Pt(100) surface. Most groups analyzed the single particle dynamics by computing the self diffusion coefficient (sdc) from mean square displacements and velocity autocorrelation functions. Information about water dynamics near metallic and other interfaces can also be found in Ref. 76, 106, 139, 141, 145, 159–166.

In the interfacial region, the sdc has, in most cases, been calculated for different distance intervals and separately for the directions parallel and perpendicular to the surface. The results of these studies can be summarized by stating that the dynamics is usually different from the one in the bulk only in the first layer of molecules. Beyond the second layer the sdc is very similar to the one in the bulk. If the water-surface interaction energy is large, like in the case of metals, the sdc decreases near the interface by about one order of magnitude, because the increased mobility due to the smaller number of hydrogen bonds is more than compensated by the mobility decrease due to the geometrical confinement (which manifests itself in the increased coordination number; see Fig. 19) and the barriers due to surface corrugation.

The scatter plots shown in Fig. 12 are a very intuitive approach to investigate the mobility parallel to the surface. These plots show that the water molecules next to the (weakly corrugated) mercury surface are obviously more mobile than the molecules on the (strongly corrugated) Pt(100) surface, and that the molecules in the second layers are more mobile than the ones in the adsorbed layers.

Sonnenschein and Heinzinger [163] and later Wallqvist [141] calculated residence times for water in various layers. While the residence times of water near non-metallic surfaces are typically of the order of ten to hundred picoseconds, the values near uncharged metal surfaces are of the order of several hundred picoseconds [167].

The reorientational dynamics of water molecules has been analyzed by calculating the relaxation times of the molecular dipole vector relative to the laboratory fixed

Table 1. Dipole relaxation times, in picoseconds, (calculated from Eq. (14) for water in contact with a corrugated model surface with different adsorption energies. Three regions corresponding to the first (L1; $z < 1.3$ Å in Fig. 8) and second layer (L2; $1.3\,Å < z \leq 3.8\,Å$) and a bulk region with constant density (L3; $6\,Å < z \leq 10\,Å$) are distinguished. Results are given for four different adsorption energies. Also shown are the results for a water slab confined between two mercury surfaces. All simulations were performed using the image charge model. Ewald summation in two dimensions was used to treat the long-range interactions. The numbers in parentheses are obtained from the integration of the autocorrelation function up to 50 ps (Eq. (14)). All other numbers are from a fit to the exponential region of the dipole acf in the time interval from 1 to 7.5 ps.

	parallel			perpendicular		
E_{ads}/kJ mol^{-1}	L1	L2	L3	L1	L2	L3
12	5.5 (4.8)	3.3 (3.0)	3.2 (2.9)	3.7 (2.2)	4.0 (3.4)	3.8 (3.2)
18	6.6 (6.1)	3.6 (3.2)	3.3 (2.9)	3.7 (3.2)	4.0 (3.7)	4.1 (3.6)
24	8.2 (7.6)	3.7 (2.9)	3.6 (3.1)	5.7 (4.4)	4.5 (3.7)	3.9 (3.4)
30	9.4 (8.6)	3.5 (2.8)	3.2 (2.6)	5.2 (4.4)	4.7 (4.9)	3.7 (3.5)
38*	9.5	3 5	3.3	6.3	4.6	3.1

* Water lamina between mercury surfaces

unit vectors in x, y (parallel) and z (perpendicular) directions. The relaxation times are calculated as the integral of the components of the dipole autocorrelation function c_{μ_α} ($\alpha = xy$ or z)

$$\tau = \int_0^\infty c_{\mu_\alpha}(t)\,dt. \tag{14}$$

Table 1 summarizes the results. With increasing adsorption energy the relaxation times in the adsorbate layer increase. In the adsorbed layer, the relaxation of the dipole moment components parallel to the surface is slower than the one for the component perpendicular to the surface. Hydrogen bonding of water molecules to neighboring atoms in the adsorbate layer is obviously more important than hydrogen bonding between adsorbed and second layer. The anisotropy of the relaxation in the second layer increases also slightly when the adsorption energy increases. In this (and to a lesser extent in the third) layer, relaxation parallel to the surface is faster than relaxation perpendicular to the surface. The average bulk values of the relaxation time from the fit (about 3.3 and 3.9 picoseconds for relaxation parallel and perpendicular to the surface, respectively) at temperatures between 297 and 299 K are, within the estimated error limits, equal to the value of pure TIP4P water (3.8 ps at 293 K [168]). It should be noted, however, that the anisotropies of dynamical properties seem to have a slightly larger range than the ones of structural properties. Lee and Rossky [166] and Zhu and Robinson [149] observed a similar slowing down of orientational relaxation relative to the bulk near polar silica surfaces and near a Lennard-Jones wall.

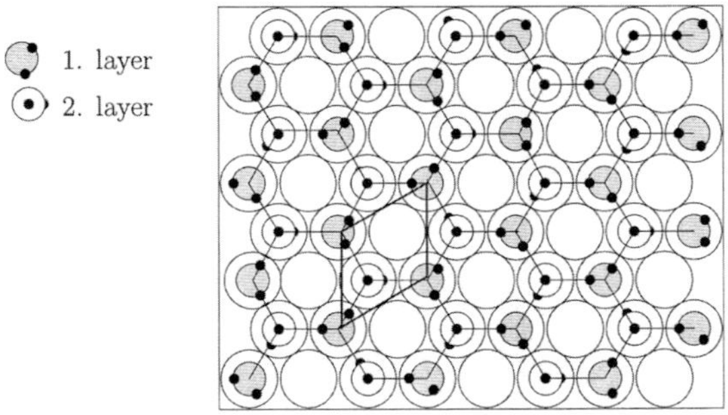

Fig. 21. Schematic diagram of a perfect, infinite water bilayer.

4.8 The Bilayer Model

Low energy electron diffraction (LEED) experiments of water adsorption on metal surfaces in ultra-high vacuum and at temperatures of about 120 to 150 K revealed hexagonal patterns from epitaxial ice grown on Pt(111) and Ag(111) surfaces [169, 170]. These results have been interpreted in terms of the so-called "bilayer" model. In this model, water forms three-dimensional structures, in which a "first layer" of water molecules is bound by direct chemisorption bonds, and molecules in the "second layer" are hydrogen bonded to the first-layer molecules. The "bilayer" can be regarded as 2 layers of molecules cut from a hexagonal ice Ih crystal. Figure 21 illustrates the arrangement of molecules on the hexagonal surface. The LEED pattern indicative of the "bilayer" has later been found on Ru(001) even at low coverages [171]. A $c(2 \times 2)$ LEED pattern on Cu and Ni has been interpreted in terms of a slightly distorted hexagonal ice-like bilayer [172–174]. For further references, see the excellent review by Thiel and Madey [97].

In computer simulation studies of water adsorption on the (100) and (111) faces of Pt, Berkowitz and coworkers [52, 54, 53, 143] interpreted the water dipole and intramolecular oxygen-hydrogen vector orientational distributions near the surface as being consistent with the bilayer structure (Figure 2 in Ref. [143]). They also pointed out that the oscillatory nature of the water motion shows the solid-like character of the adsorbate layer. The orientational distribution functions in Fig. 16 allow the same interpretation: Very close to the surface (panels a and b) there is a predominance for orientations where the dipole moment (i.e., the hydrogen atoms) point into the liquid phase (like the '1. layer' molecules in Fig. 21). Slightly further into the liquid phase (panels c and d) configurations with a dipole moment pointing in an opposite direction become more abundant. Thus, the simulated first layer contains molecular configurations belonging to both of the layers of Fig. 21. Even at elevated temperatures (349 K) remnants of the bilayer structure have been found [49]. However, the distributions are rather broad and substantial deviations from the ideal structure in Fig. 21 are observed. In another computer study, Barnett *et al.* [175] deposited a monolayer of water on Pt(111) at 50 K. They show a snapshot of the resulting equilibrium structure. The hydrogen-bonded bilayer honeycomb superlattice exists with some obvious dislocations. All these studies have been performed with the water-platinum interaction model proposed in Ref. 48, which is discussed in section 3, or a modification of it.

4.9 Summary

A wide variety of different models of the pure water/solid interface have been investigated by Molecular Dynamics or Monte Carlo statistical mechanical simulations. The most realistic models are constructed on the basis of semiempirical or *ab initio* quantum chemical calculations and use an atomic representation of the substrate lattice. Nevertheless, the understanding of the structure of the liquid/metal surface is only at its beginning as (i) the underlying potential energy surfaces are not known very well and (ii) detailed experimental information of the interfacial structure of the *solvent* is not available at the moment (with the notable exception of the controversial study of the water density oscillations near the silver surface by Toney *et al.* [140, 176]).

In this situation computer simulations are especially important, because meaningful results within the framework of classical statistical mechanics can be generated for well-defined models and predictions can be made. A few general results, which are fairly independent of the details of the model assumptions, are summarized below.

- Density inhomogeneities decay rapidly into the liquid phase near uncharged surfaces; usually only two or three, or at most four layers can be discerned.
- Lateral density inhomogeneities are even more short-ranged and are confined almost exclusively to the adsorbate layer; depending on the strength of the potential energy corrugation, the lateral density inhomogeneities mimic the substrate lattice.
- Molecular polarizability is an effect of secondary importance. Effective pair potential models are a reasonable approximation.
- Unscreened electric fields lead to long-range density inhomogeneities.
- Hydrogen bonding in water determines to a large extent the structure near the interface; ice-like arrangements, which are observed in ultrahigh vacuum at low temperatures [97], occur near the interface, but are obscured by large thermal fluctuations.
- The properties of atom-atom pair correlation functions and the hydrogen bond analysis show that the local water structure is affected by the surface only within the range of the density inhomogeneities (a few Ångstrøm); in the constant-density region structural isotropy is observed.
- Orientational distributions near the interface are rather broad; consequently, simple discrete-state models of interfacial structure are unrealistic.
- Electrostatic properties like the surface potential drop χ can be calculated; the simple quantum chemical interaction models lead to qualitatively correct results.
- The mobility of water decreases substantially in the adsorbed layer. The character of the motion becomes more oscillatory than diffusive.
- The reorientation dynamics of water is slowed down and becomes anisotropic near the interface.

5 Electrolyte Solution/Metal Interfaces

5.1 Overview

The behavior of simple and molecular ions at the electrolyte/electrode interface is at the core of many electrochemical processes. The substantial understanding of the structure of the electric double layer has been summarized in various reviews and books (e.g., Ref. 2, 81, 177–183). The complexity of the interactions demands the introduction of simplifying assumptions. In the classical double layer models due to Helmholtz [3], Gouy and Chapman [5, 6], and Stern [7], and in most of the studies cited in the reviews the molecular nature of the solvent has been neglected altogether, or it has been described in a very approximate way, e.g., as a simple dipolar fluid. Computer simulations can overcome this restriction and describe the solvent in a more realistic fashion. They are thus able to paint a detailed picture of the microscopic structure near a metal electrode.

Understanding the structure and dynamics of pure water on a molecular level is only the beginning. Simulations of electrolyte solutions near metallic interface are much more demanding in terms of computer time than those of bulk water, because the relatively small number of ions even in a highly concentrated electrolyte solution mandates the treatment of systems with a much larger total number of particles than in pure water for a longer time span. Furthermore, as was discussed in section 3, much less is known from quantum chemistry about nature and strength of the ion-metal interaction than about the water-metal interactions, so that the interpretation of the results obtained from the simulations is less clear.

Since specific adsorption is an important phenomenon in electrochemistry, the solution/metal interface has nevertheless been studied in various ways. An ion is considered to be adsorbed specifically in the "inner Helmholtz plane" when it is partially dehydrated and in direct contact with the metal surface (see, e.g., Ref. 15). On the other hand, an ion that is adsorbed further away from the electrode with its hydration shell essentially intact is considered to be adsorbed non-specifically in the "outer Helmholtz plane". In the classical treatment of contact adsorption, the balance between the energy of hydration of the ion and the strength of the image interactions determines which ions are specifically adsorbed and which ones are not [15]. The larger alkali and halide ions (Cs^+, Cl^-, Br^-, I^-) are thought to be contact adsorbed at the potential of zero charge, whereas strongly hydrated small ions like Li^+ and F^- do not make contact with the metal surface. Computer simulations are able to test the validity of these predictions; they provide additional insight into the problem since they explicitly incorporate changes in the structure and energetics of water near the electrode, which is completely neglected in the classical treatment.

Two basic simulation approaches can be taken. The most obvious one is the direct 'brute force' simulation of the entire double layer including an electrode with or without surface charge, ions and counterions, which screen this surface charge, and which form a diffuse layer. This approach has become possible only very recently due to the long time scales (in excess of 1 nanosecond) necessary to obtain converged results. Even with present day computing power these calculations are difficult to do.

A more rigorous approach within the framework of statistical mechanics is the calculation of the potential of mean force of the ion as a function of distance from the electrode. It involves a series of simulations of one dissolved ion at different distances or in different distance intervals from the surface. The potential of mean force $W(z)$ is the free energy difference between the ion at position z, $A(z)$, and the bulk, A_{bulk}; it is related to the ion density profile $\rho(z)$ by

$$W(z) = A(z) - A_{bulk} - kT \ln \frac{\rho(z)}{\rho_{bulk}} \tag{15}$$

with Boltzmann's constant k, the absolute temperature T, and the bulk density ρ_{bulk}.

There are two well-established and widely used methods to calculate free energy differences and both have been applied to the problem of ion adsorption. In the umbrella sampling scheme due to Torrie and Valleau [184, 185], an external confining potential is applied to the degree of freedom of interest, in this case the ion-metal distance, in addition to the physical interactions from the environment. During one simulation, the ion can probe only a confined set of distances. By performing several simulations with different confinement regions the whole distance spectrum is sampled (from a non-Boltzmann distribution). After corrections for the confining potential, the distance dependence of the free energy of ion adsorption is obtained. Alternatively, the "constrained MD" simulation method due to Ciccotti *et al.* [186] can be used: Here the ion is fixed at a specified distance from the metal surface (but allowed to move laterally). In a series of simulations in which the ion is kept fixed at different distances from the surface, the mean force acting on the constrained degree of freedom is calculated and the PMF is determined by integrating from a reference state, which is chosen to be the solvated ion far from the metal surface.

Both the brute force approach and the free energy approach have been used. In the following subsections the main results are reviewed and several examples taken from work in my group are discussed.

5.2 Single Ion Studies

Alkali and halide ion adsorption near metal surfaces has been investigated by a variety of groups. A simple (9-3) Lennard-Jones potential, first used by Lee *et al.* [137], together with the image charge model has been employed in a series of direct simulations of single ions by Glosli and Philpott [107–109]. They observed that the larger ions Br^- and I^- are contact adsorbed, and that the small ions Li^+ and F^- are not contact adsorbed, Cl^- being a borderline case. Benjamin and coworkers calculated free energies of adsorption for Na^+ and Cl^- on charged and uncharged platinum surfaces [187, 188] and found Cl^- to be contact adsorbed while Na^+ is not. They used the water-platinum potential function introduced by Spohr and Heinzinger [48] together with a modified image model for the ion-platinum interactions. Seitz-Beywl *et al.* 189 studied the adsorption of Li^+ and I^- on Pt(100) by direct simulation, using the same water-metal interaction potentials and ion-metal interaction potentials derived from *ab initio* calculations at the SCF level [115] of ion-platinum clusters.

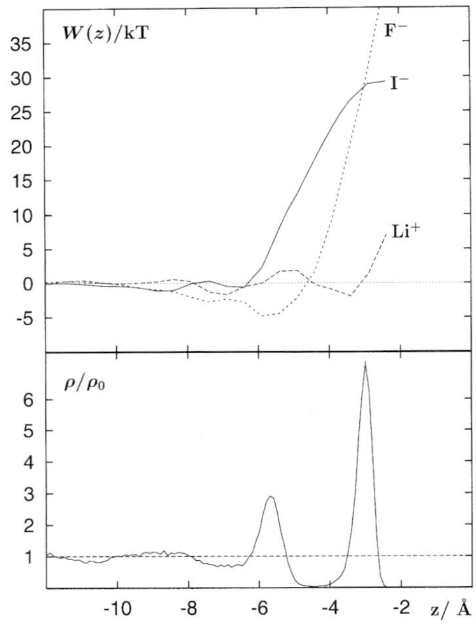

Fig. 22. Ion adsorption near mercury with (111) surface structure. Top: Total potential of mean force for Li^+ (long-dashed), F^- (short-dashed), and I^- (full) ions. Bottom: The reduced oxygen density profile for the same system.

From the positions of the maxima of the ionic density profiles relative to the minima of the ion-metal interaction potentials, they concluded that I^- is contact adsorbed and Li^+ is not. Spohr [190] and later Perera and Berkowitz [191] obtained similar results by means of free energy calculations for I^- and simultaneous Li^+ and I^- adsorption, respectively, on Pt(100), using the same interaction potentials. Eck and Spohr [77, 192] and Tóth and Heinzinger [80] studied the adsorption of Li^+ and several halide ions near the ab initio model of the mercury interface [40]. The liquid/gas interface, contrary to metallic interfaces, is depleted in the interfacial region [193–195]. This is a consequence of the driving force towards fully hydrated ions.

As an example for results obtained by the constrained MD method, we briefly discuss some of the differences between the free energy profiles of Li^+, F^-, and I^- ions on the mercury surface. In this study [192], only the water-metal interactions are described by the SCF interaction energies [40]. The ions interact with the surface exclusively by means of image interactions. Ewald summation in two dimensions is used to properly describe the long range polarization effects near the interface.

Figure 22 shows the PMF in units of the thermal energy at room temperature, $W(z)/kT$, of the three ions. The metal surface is located to the right at $z = 0$. On the left side (at about $z = -25$ Å) the lamina is confined by a smooth (9-3) Lennard-Jones surface like the one in Fig. 7 [137]. At 12 Å from both surfaces $W(z)$ is set equal to zero. $W(z)$ does not change much until about $z = -9$ Å. In this range the forces on the hydrated ion are, on average, isotropic, and, therefore, the hydration shell is on average symmetric. Starting at about -6 Å from the surface, $W(z)$

becomes repulsive for the iodide ion. This behavior is analogous to the one observed on Pt(100) [190]. The repulsion ($W/kT \approx 30$ at $z = -2$ Å) is weaker than in the platinum case. The repulsive region of $W(z)$ originates from the work of hole formation on the surface. Because water molecules are adsorbed on or bound to the surface, work needs to be done to desorb them before the ion can adsorb. The chloride ion behaves similarly to iodide and is not shown; since the ion is smaller, the repulsion of the ion due to the solvent interactions is weaker.

The behavior of the fluoride ion is qualitatively different. The solvent interactions stabilize the ion thermodynamically in the range from −8 to −5 Å, whereas the solvent effect is mostly repulsive for I^- (and Cl^-). For F^- the attraction by the layer of adsorbed water (which are oriented preferentially with the hydrogen atoms pointing into the solution) appears to dominate over the steric repulsion. The distance of the minimum (at about 5.5 Å) corresponds to an adsorbed species $F^-(H_2O)_6$ or $F^-(H_2O)_7$ on top of the adsorbed layer of water molecules. A detailed analysis [196] shows that the very strong repulsion at distances below 5 Å is due to hydration.

The PMF for Li^+ adsorption shows two solvent-induced local minima. In bulk solution, the hydration shell of Li^+ forms a rather rigid octahedral complex. Li^+ and the water molecules in its hydration shell move cooperatively. Consequently, the hydration complex is sensitive to the barriers formed by the two pronounced layers of water molecules around $z = -6$ Å and $z = -3$ Å. This leads to a local free energy minimum on the solution side of each of the two maxima in the oxygen density profiles. There, molecules from the water layers can be part of the hydration complex (see below).

The results of these calculations imply that none of the ions would be contact adsorbed when no specific interactions between ions and metal are taken into account in the model. The Li^+ ion, believed to be nonspecifically adsorbing, would be able to approach the surface more closely than the anions, mostly because of its small size, which allows it to penetrate the surface layers without displacing water molecules. The simulation results thus indirectly demonstrate the importance of specific chemical interactions for realistic models of the electric double layer. Apparently, also some specific features of the hydration shell structure of the ions must be taken into consideration in order to fully understand the adsorption of ions.

5.3 Double Layer Simulations

An early study of an LiI solution in water confined between non-metallic Lennard-Jones surfaces [164] did not last long enough to draw any conclusions about the density distribution. The first simulation studies of full double layers with molecular models of ions and solvent were performed by Philpott and coworkers [102, 101, 103] for the NaCl solution, using the fast multipole method for the calculation of Coulomb interactions. The authors studied the screening of a negative surface charge by the free ions in several highly concentrated NaCl solutions. Figure 23 shows the ionic density profiles and the corresponding predictions of the Gouy-Chapman theory for three different ion concentrations [103]. The results show a large bulk-like electrolyte region with constant ion density, vanishing average electric field, and

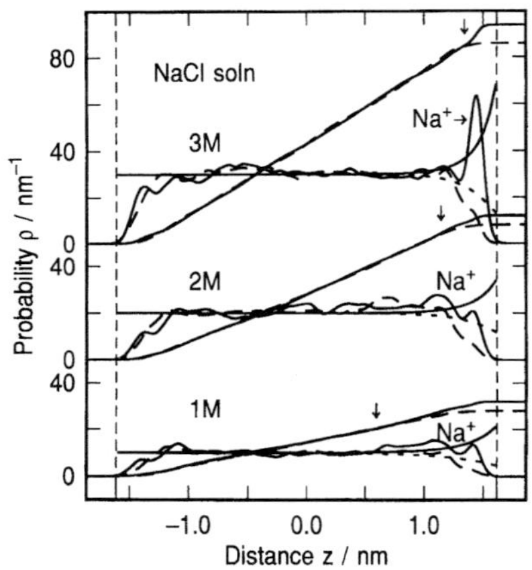

Fig. 23. Ion density profiles for 1 m to 3 m NaCl solutions. Solid line Na^+, broken line Cl^-. 1 m and 2 m solution at 303 K, 3 m solution at 373 K. Vertical dashed lines at $|z| = 1.615$ nm mark where the (9-3) Lennard-Jones wall potentials pass through zero. The metallic surface is to the right of the figure. Inclined curves rising monotonically from left to right are the integrals of the ion density profiles. The profiles from the Gouy-Chapman theory are shown as monotonous curves rising from left to right for Na^+ and falling for Cl^-. (Reproduced with permission from Ref. 103.) The surface charge density is $-4.6\,\mu C\,cm^{-2}$ for the 1 m and 2 m solutions and $-9.2\,\mu C\,cm^{-2}$ for the 3 m solution.

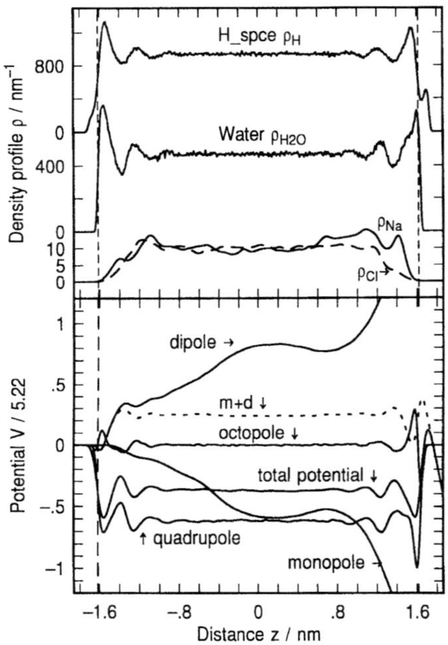

Fig. 24. 1 m NaCl solution at 303 K and surface charge density $\sigma_0 = -4.6\,\mu C\,cm^{-2}$. Top: Atom and ion density profiles. Bottom. Total electrostatic potential and component (monopole, dipole, combined monopole plus dipole, quadrupole and octopole) electric multipole potentials. The total potential was calculated from the total electric charge distribution. Note that the sum of monopole and dipole components is smaller than the quadrupole component. (Reproduced with permission from Ref. 103.)

constant electrostatic potential (Fig. 24, top). The ionic density profiles are clearly nonmonotonic. The Na^+ ions form layers between the first and the second water layer, the height of which increases with increasing concentration. The estimate of the width of the diffuse layer yields consistently larger values than the Debye-Hückel screening length. The authors correctly point out, however, that the Gouy-Chapman theory is not applicable at these high concentrations.

Glosli and Philpott also analyzed the various contributions to the total electrostatic potential. Figure 24 shows, for the 1 m NaCl solution at a surface charge density of $-4.6\,\mu C\,cm^{-2}$, the contribution of the free (ionic) charges and of the dipole, quadrupole, and octupole contributions from the bound (water) charges to

Fig. 25. Snapshot of the simulation with surface charge $\sigma = +9.9\,\mu C\,cm^{-2}$. Na^+: small spheres, Cl^-: large spheres. The liquid slab is confined by the metal surface (bottom) and by a 'free' surface (top).

the electrostatic potential. The variation of the electrostatic potential due to the ionic charge distribution in the central bulk-like region (curve labeled 'monopole') is compensated by appropriate ordering of the water dipoles (curve labeled 'dipole'), as can be seen in the rather small value of the combined ('$m + d$') curve. The quadrupole contribution is important to specify the absolute value of the potential in the liquid phase, while the octupole moments make only minor contributions close to the interfaces.

Quite recently, Spohr [104] investigated the 2.2 molal NaCl solution in the vicinity of a more realistic corrugated surface as a function of surface charge density. In the simulations, water films consisting of 400 water molecules solvate 32 ions in the vicinity of the metal surface, with image charges and a full 2D Ewald treatment for the Coulomb interactions. Like in the studies by Philpott *et al.*, the solution is not necessarily electroneutral; the total charge in the solution is balanced by the image charges which give rise to the surface charge density σ. Surface charge densities $\sigma = -9.9\,\mu C\,cm^{-2}$, $\sigma = 0$, and $\sigma = +9.9\,\mu C\,cm^{-2}$ correspond to simulations with $17\,Na^+ + 15\,Cl^-$, $16\,Na^+ + 16\,Cl^-$, and $15\,Na^+ + 17\,Cl^-$, respectively. Figure 25 shows a snapshot of the simulation cell for surface charge $\sigma = +9.9\,\mu C\,cm^{-2}$. The z-direction is from bottom to top. The large dark spheres are the Cl^- ions, the small

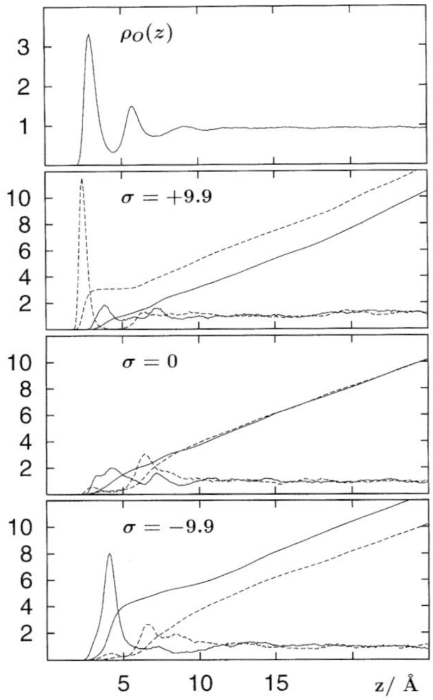

Fig. 26. Density profiles and running integrals $n(z)$ of the ion densities for Na^+ (full lines) and Cl^- (dashed lines) at three different surface charge densities in units of $\mu C\,cm^{-2}$ as indicated.

grey ones the Na^+ ions. In this particular configuration, four anions and one cation are adsorbed on the metal surface.

Figure 26 shows the ion density profiles near the metal surface for all three surface charges investigated. Beyond $z = 20$ Å all ion density profiles are identical within the limits of statistical errors. The oxygen density profile does not change much with surface charge and is repeated here only to provide a geometric reference for the ion positions. The ion densities have been smoothed by a Gaussian filter with half width of 0.25 Å. Together with the ion densities the running integrals, n, of the densities defined by

$$n(z) = L_x L_y \int_0^z \rho(z')\,dz', \tag{16}$$

are plotted for Na^+ (full lines) and Cl^- (dashed lines). $L_x = L_y = 18$ Å are the box dimensions parallel to the interface and ρ is the particle number density.

In the vicinity of the metal surface the ionic densities are not monotonous, as might be expected on the basis of a simple extrapolation of the Gouy-Chapman theory to high concentrations. Rather, the ionic densities oscillate in the region $z < 10$ Å. The ion densities follow to some extent the oscillations of the oxygen density which are the consequence of the 'layering effect' occuring in all liquids near a surface. At finite positive and negative surface charge densities, the counterion density profiles exhibit significantly higher maxima; the counterion positions are correlated much more strongly with the metal surface. This makes possible the definition of planes in which most of the countercharge is included, in analogy to the inner and outer Helmholtz planes of the classical electrochemical models.

At positive surface charge, $\sigma = 9.9\,\mu C\,cm^{-2}$, the chloride ions are attracted directly to the surface. From the running integral it can be concluded that about 3 Cl^- ions are adsorbed on the surface. The three ions overcompensate the surface charge density which corresponds to only two elementary charges for the entire simulation cell. This behavior is reminiscent of the concept of super-equivalent adsorption [15]. Approximately one cation remains in the interlayer region between the adsorbate water and the second layer. Hence, the surface charge of $2e$ is compensated by $-3e$ from the anion density and $+1e$ from the cation density.

At negative surface charge, $\sigma = -9.9\,\mu C\,cm^{-2}$, the maximum of the Na^+ density profile grows substantially but remains more or less at the same position as for the uncharged surface. About 4 Na^+ ions are located in the region of this density maximum. At the same time the maximum of the anion density is shifted to the region between 6 and 9 Å.

When the difference between the running integral of cations and anions becomes equal to the negative value of the surface charge, the entire surface charge can be regarded as screened by the mobile charges. The distance at which this happens changes from about 5.5 Å ($\sigma = +9.9\,\mu C\,cm^{-2}$) to about 8 Å ($\sigma = 0$) to about 10 Å ($\sigma = -9.9\,\mu C\,cm^{-2}$). Hence, the double layer thickness is dependent on the surface charge density. In all cases, the thickness of the double layer is of the order of several Debye lengths ($r_D = 2.1$ Å at the simulated concentration).

5.4 Ion Hydration

Early studies of ion hydration in confined geometries usually found no drastic changes of hydration shell properties. In a study of a 2.2 molal LiI solution between apolar Lennard-Jones walls, Spohr and Heinzinger [164] found the hydration shell of Li^+ unchanged compared to the bulk LiI solution. Seitz-Beywl and Heinzinger [189] studied the hydration of single Li^+ and I^- ions near the platinum (100) surface. The interaction potentials in this study were based on quantum chemical calculations. The rather strong ion-surface interactions lead to adsorption of both ions. For both ions, a significant reduction of the hydration number (from 6 in the bulk to 5 and from 9 to 5 for Li^+ and I^-, respectively) was observed close to the surface. In the Li^+ hydration shell one corner of the octahedron is missing, leading to a pyramidal coordination shell. Because of the large hydration energies, the preferential position of the Li^+ ion is at larger distances from the surface than the position of the energy minimum of the Li^+-surface interaction potential. The hydration shell of the I^- ion near the metal surface does not show any detectable symmetry, but ion-oxygen pair correlation functions show long-range lattice-induced order.

Rose and Benjamin [194, 187] studied the adsorption of Na^+ and Cl^- on the Pt(100) surface and concluded that "the structure of the solvation complex around Na^+ and Cl^- is very similar to the structure in the bulk". In this study, the same model for the water-metal interactions was used as in the preceding study of Li^+ and I^- hydration [189]. The ion-surface interactions, however, were described by a modified image charge potential which was smoothly truncated. The authors discussed in some detail the anisotropy of the hydration complex that is induced by the density inhomogeneities at the interface. In a study of Fe^{2+} and Fe^{3+} hydration [197],

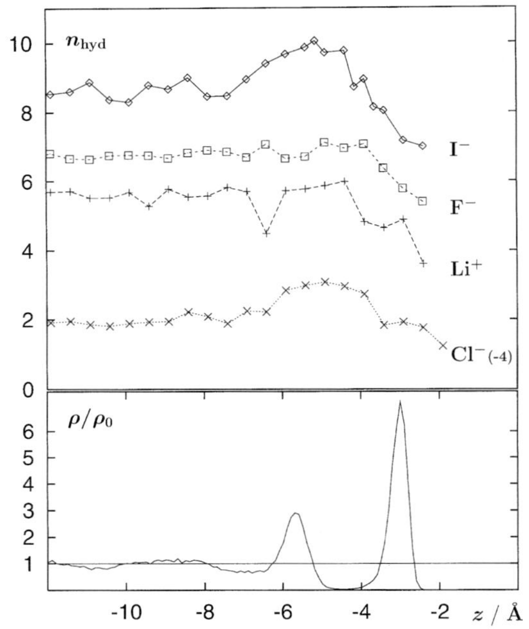

Fig. 27. Top: Average hydration number of alkali and halide ions as a function of their negative distance from the mercury surface (at $z = 0$). The Cl^- hydration number is offset by -4 to improve readability. Data are from four series of simulations of one ion dissolved in 259 water molecules. Bottom: The normalized oxygen density profile for the same system.

the same authors found no disturbance of the octahedral hydration shell in the vicinity of the Pt(100) surface. Because of the strong electrostatic forces holding together the $Fe^{2+}(H_2O)_6$ and $Fe^{3+}(H_2O)_6$ hydration complexes, the ions are not specifically adsorbed.

Spohr [190] studied the adsorption of I^- on the Pt(100) surface. The free energy barrier towards iodide adsorption that is produced by the layers of adsorbed water is associated with a significant intermediate increase in coordination number, before the hydration number decreases at short ion-metal distances for geometrical reasons. Philpott and Glosli [109] observed in a series of MD studies of ion adsorption on charged electrodes that the Li^+ hydration shell structure in the vicinity of a model metal surface does not depend on the halide counterion (F^-, Cl^-, Br^-, I^-). In this study, no specific interactions between the metal surface and water molecules or ions were employed.

As an example of the type of information that can be obtained, the composition and anisotropy of the hydration shells in the vicinity of the mercury surface is analyzed. Figure 27 shows the average coordination number of the ions as a function of position (the mercury surface is located at $z = 0$ Å). Here, the coordination number is defined as the number of water molecules the ion-oxygen distance of which is smaller than 2.5 Å, 3.2 Å, 3.9 Å, and 4.2 Å for Li^+, F^-, Cl^-, and I^-, respectively. This definition is based on the positions of the first minima of the ion-oxygen pair correlation functions.

The hydration number of iodide as a function of its position is roughly constant up to about $z = -6.5$ Å. The penetration of the ion into the layers of adsorbed water leads to an increase up to about $z = -4.5$ Å. At smaller distances from the surface, the coordination number decreases, as expected, due to the geometric effect of the

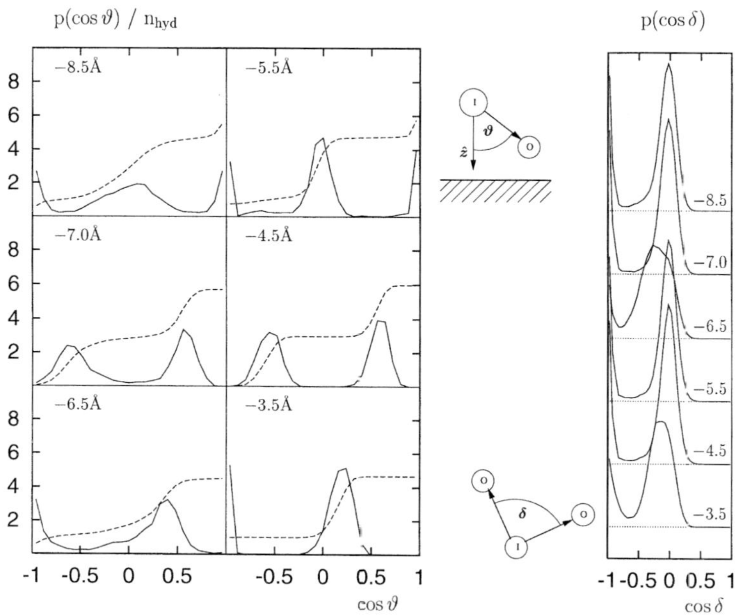

Fig. 28. Left: Distribution of the cosine of the angle between the Li^+-oxygen vector and the inward directed surface normal (full) and running integral of $p(\cos\vartheta)$ (dashed) at various positions relative to the metal surface as indicated. The running integral is normalized to yield the total hydration number at a given distance for $\cos\vartheta = 1$. Right: Distribution of the cosine of the oxygen-Li^+-oxygens angle δ. Only oxygen atoms in the first hydration shell of the Li^+ ion contribute.

surface. The increase in hydration number coincides with the rise of the repulsive solvent free-energy barrier (see Fig. 22). Chloride behaves qualitatively similar to iodide. The hydration number is constant up to the region where the hydration shell penetrates the inhomogeneous water region around $z > -6$ Å and increases significantly in the distance range -6 Å $< z < -4$ Å. Note that in Fig. 27 the hydration number of chloride is offset by -4.

Contrary to the larger halide ions the coordination number of F^- remains constant in the whole region of thermodynamic stability (see Fig. 22). At distances smaller than about 4 Å from the surface, excluded volume effects lead to a similar reduction of the hydration number as for the other anions.

The hydration number of the Li^+ cation is slightly less than 6 up to $z \approx -4.5$ Å, except around $z = -6.5$ Å where there is a significant reduction to a value well below 5. This position coincides with an increase in the free energy (see Fig. 22). Apparently, the penetration of the Li^+ cation into the second layer of water molecules around $z = -6$ Å is associated with partial dehydration (see also the discussion of Fig. 28 below). For small ion-surface distances ($z = -4.5$ Å) the hydration number is reduced once more when the ion penetrates the contact layer of water molecules.

Figure 28 demonstrates some interesting changes in the structure of the hydration shell of the Li^+ ion as a function of its position. It contains the distribution function $p(\cos\vartheta)$, where ϑ is the angle between the ion-oxygen vector and the laboratory z-axis, which points towards the surface in this case. Only water molecules in the first hydration shell of Li^+, i.e., with ion-oxygen distance smaller than 2.5 Å, are taken into account. $\cos \vartheta = -1$ corresponds to an arrangement where the ion is between the hydrating molecule and the surface, whereas $\cos \vartheta = 1$ indicates the situation where the hydrating water molecule is located between the ion and the surface. The dashed lines are the running integral of $p(\cos \vartheta)$, normalized in such a way that the total integral yields the hydration number at the specified position.

The left side of Fig. 28 demonstrates that at z-values of −8.5 and −5.5 Å an octahedral hydration shell complex exists, where one corner points to the surface (which can be deduced from the fact that one water molecule contributes to the coordination number at $\cos \vartheta = 1$). At distances of −7.0 and −4.5 Å from the surface the octahedron remains stable but is rotated in such a way that two triangular planes are now parallel the surface. At −6.5 and −3.5 Å, the first hydration shell consists of only five water molecules in a roughly pyramidal coordination with the basis of the pyramid again parallel to the surface. In all cases, the structure is more pronounced when the ion is closer to the surface.

The sharp maxima at linear and rectangular O-Li^+-O arrangements in the distribution of O-Li^+-O angles (right panel of Fig. 28) at large distances from the surface point towards a regular octahedral hydration complex. At $z = -3.5$ Å and $z = -6.5$ Å, coincident with the reduction in hydration numbers, $p(\cos \vartheta)$ shows significant deviations from the octahedron and the distributions broaden. In between the octahedral shell is reestablished. At $z = -3.5$ Å the hydration number is 5. However, unlike on Pt(100) [189], the hydration shell is distorted from the quadratic pyramid (seen by the rather wide maximum around $\cos\delta \approx -0.2$), because the hexagonal substrate geometry on the Hg(111) surface does not easily support a square arrangement of water molecules.

In short, the structural order observed for the hydration shell of the Li^+ ion at various distances from the mercury surface results from the strong tendency of this ion to form an octahedral arrangement with its six nearest neighbor water molecules in bulk solution. This structure adjusts to the densities and inhomogeneities determined by the metal surface. This example demonstrates how mechanistic information about the process of ion adsorption can be extracted from computer simulations.

5.5 Ion Dynamics

Because of the small number of ions involved in typical simulations, it is difficult to calculate the self diffusion coefficient reliably. However, the differences in mobility between different ions and the trends, when the ion approaches the metal surface, can be qualitatively assessed by a scatter plot like that in Fig. 29. The figure shows the area parallel to the surface probed by I^-, F^-, and Li^+ ions. Each ion is dissolved in a film of adsorbed water on a mercury surface consisting of 150 water molecules and held fixed at the indicated distance from the surface. Trajectories are shown over 60 picoseconds. Both in the bulk (left panels) and closer to the surface (right panels) the mobility decreases in the sequence large anion (I^-) > small anion (F^-) ≥ small cation (Li^+). The mobility close to the surface is reduced quite substantially, especially for Li^+ and F^-.

In simulations of the full electric double layer near the corrugated model surface

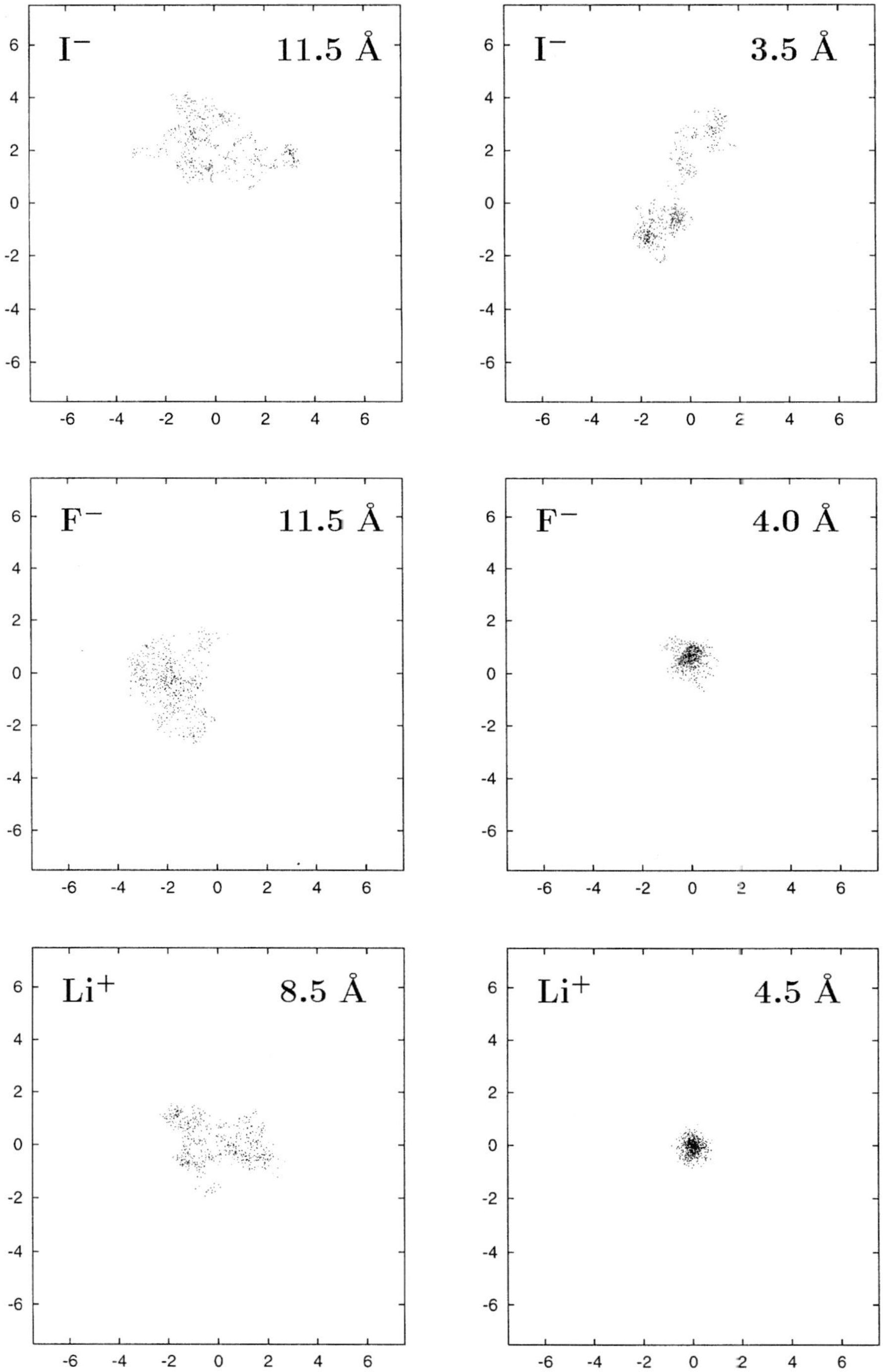

Fig. 29. Lateral density distribution of I^-, F^-, and Li^+ ions far from the electrode surface (left) and close to the electrode. A symbol is plotted every 0.1 ps for a total time of 60 ps during the simulations. The scale on the x and y axes is in Å. The ions are dissolved in a film consisting of 150 water molecules in contact with the mercury surface.

the self diffusion coefficient of ions is reduced slightly at the potential of zero charge for those ions which are close to the electrode. For charged electrodes, the reduction of mobility is more pronounced, and a larger number of ions is affected up to a larger distance from the surface [104].

Benjamin and coworkers studied various aspects of ion dynamics near the liquid/vapor and the liquid/liquid interface. Time-dependent probability distributions of the ion position were studied near the interface between immiscible apolar and polar liquids [198]. The simulation results were in almost quantitative agreement with a one-dimensional diffusion model. Small differences were attributed to the solvent reorganization dynamics. Later this work was extended to the ion and solvent dynamics following charge transfer [199] near the same polar/apolar interface and to the liquid water/vapor interface [200].

Benjamin and Rose [188] also studied the adsorption dynamics of Na^+ and Cl^- on the charged Pt(100) surface. They performed trajectory calculations of a single ion dissolved in water and attracted to the surface by an external electric field. A constant drift velocity proportional to the strength of the electric field is observed until the ion approaches its equilibrium position. The authors conclude that the adsorption proceeds in two steps: the drift of the ion toward the surface and the subsequent displacement of adsorbed water molecules. This mechanism is consistent with the picture emerging from the free energy calculations. During the drift stage the hydration shell of Na^+ remains mostly intact at low field strengths while considerable exchange is observed in the Cl^- hydration shell. At high field, both ions appear to drift independently of their hydration shell.

5.6 Summary

The adsorption of monovalent cations and anions has been investigated using a variety of different methods and interaction models. Electrochemical experience suggests that small ions like Li^+ and F^- are not contact adsorbed to uncharged surfaces whereas larger cations and anions like Cs^+, Cl^-, Br^-, and I^- adsorb specifically on electrodes near the potential of zero charge. In the case of I^- and Cl^- specific adsorption results in the formation of ordered monolayers on platinum and gold surfaces, which can be observed by scanning tunnel microscopy and other techniques (e.g., Ref. 201–205).

The most simple, purely electrostatic model without specific interactions between water molecules/ions and the surface leads to qualitative agreement with the experimental evidence. The use of more sophisticated interaction potentials, however, makes the situation less transparent. Specifically, the following simulation results were obtained by various groups:

- The description of both water-metal and ion-metal interactions by a short-range repulsive potential without significant adsorption energy, augmented by the image charge model of electrostatics, leads to contact adsorption of Br^- and I^- and to no adsorption in the case of Li^+ and F^-. Cl^- is a borderline case.

- With the same ion-metal interactions as above, but a realistic description of water-metal interactions and proper long-range boundary conditions, non-specific adsorption occurs in single ion studies for small and large ions alike. The long-range $1/|z|$ ion-metal attraction is weakened by dipolar screening to such an extent that the solvent repulsion cannot be compensated.
- The description of both water-metal and ion-metal interactions by purely local potentials derived from *ab initio* SCF calculations leads to contact adsorption of the I^- ion on Pt(100). The observed distance spectrum of Li^+ is intermediate between that expected for specific and non-specific adsorption.
- In simulations of the full electric double layer, ionic density profiles are oscillatory in the concentration range between 1 and 3 mol/l. Surface charges are screened by free ions over a distance of several Debye lengths.
- Changes in the free energy of ion adsorption are obviously correlated with changes in orientation, composition, and internal arrangement of the hydration shells. The small, strongly hydrated Li^+ cation is affected as strongly as the anions, in spite of the fact that the PMF shows only relatively low free energy barriers.
- Adsorbed water molecules always form a barrier that must be crossed before the ion is adsorbed. The barrier height is related to the work of hole formation prior to ion adsorption. The small Li^+ ion plays a special role as it can approach the surface without disturbing the water layer significantly; hence the barrier is very small.
- Ion mobility is reduced near the electrode; range and magnitude of the reduction is larger near charged than near uncharged electrodes.
- The simulations can also provide mechanistic insight into the adsorption process. In the case of Li^+ it was shown that the free energy barrier is associated with the temporary loss of hydration shell molecules, while the increase of I^- hydration number is indicative of a mechanism where the steric barrier to adsorption by the adsorbate water layer is most important.

While it is possible to model ion adsorption reasonably well on the basis of these simple models, open questions remain. Experiment, electronic structure calculations and simulations point towards a substantial adsorption energy of water on metal surfaces. Simulations show that the solvent barrier can be strong enough to prohibit ion adsorption. Obviously computer modeling of the adsorption of ions from aqueous solution onto metal surfaces suffers from the present inability to describe the delicate balance between electrostatic, steric, and electronic effects in one (computationally feasible) model. Currently, the biggest problem that awaits solution is the adequate calculation of the ion-metal interactions from quantum mechanics.

6 Reactions at Interfaces

6.1 Outer Sphere Electron Transfer

Electrochemical electron and ion transfer reactions are commonly interpreted by the phenomenological Butler-Volmer law, according to which the rate constant k for the anodic reaction depends on the temperature T and on the electrode potential in the following way:

$$k = A\exp\left(\frac{-E^0_{\mathrm{act}}}{k_BT}\right)\exp\left(\frac{n_e\alpha e\eta}{k_BT}\right) \tag{17}$$

where the pre-exponential factor A is independent of the temperature and the electrode potential, E^0_{act} is the energy of activation at the standard equilibrium potential of the reaction, the overpotential η is the deviation of the electrode potential from the standard equilibrium value. The transfer coefficient α is a phenomenological constant of the order of 1/2, n_e is the number of electrons transferred in the reaction ($n_e = 1$ for electron transfer reactions), e is the elementary charge, and k_B is Boltzmann's constant.

For outer sphere electron transfer reactions the Butler-Volmer law rests on solid experimental and theoretical evidence. An outer sphere electron transfer reaction is the simplest possible case of an electron transfer reaction, a reaction where only an electron is exchanged, no bonds are broken, the reactants are not specifically adsorbed, and catalysts play no role (see, e.g., Ref. 2). Experimental investigations such as those by Curtiss *et al.* [206] have shown that the transfer coefficient of simple electron transfer reactions is independent of temperature. The theoretical basis is given by the theories of Marcus [207] and of Levich and Dogonadze [208]; these theories also predict deviations at high overpotentials which were experimentally confirmed [209, 210].

Several groups have investigated heterogeneous outer-sphere electron transfer reactions in the vicinity of model interfaces by means of computer simulations on the basis of the theory of Marcus. Much of this work is based on the computer simulation of the free energy functions for electron transfer in the bulk liquid by Warshel and other groups [211–216]. According to the Marcus theory, the reaction coordinate for an electron transfer reaction corresponds to a generalized solvent coordinate. There are two potential energy curves, one for the reduced, one for the oxidized state, both of approximately parabolic shape. The oxidation of a molecule occurs in the following way: a thermal fluctuation excites the reduced state and takes the system to the crossing point of the two potential energy surfaces (see Fig. 30), where it gives up an electron. Subsequently the system relaxes to its new equilibrium position on the potential energy curve for the oxidized state. So the electron transfer is caused by a solvent fluctuation and not by the movement of the redox couple. The elementary step is the transfer of the electron to or from the metal; no chemical bonds are broken and the reactants are not specifically adsorbed.

The so-called 'solvent coordinate' is usually chosen as the difference in the solute-solvent energy between the reactant and the product system. For many models, this difference is proportional to the electrostatic potential due to the water molecules at the location of the ion. The approximately parabolic solvent reorganization energy functions can be calculated from computer simulations using umbrella sampling techniques.

Benjamin [217] calculated the solvent reorganization free energy functions for the electron transfer reaction of a monovalent ion near the interface between immiscible

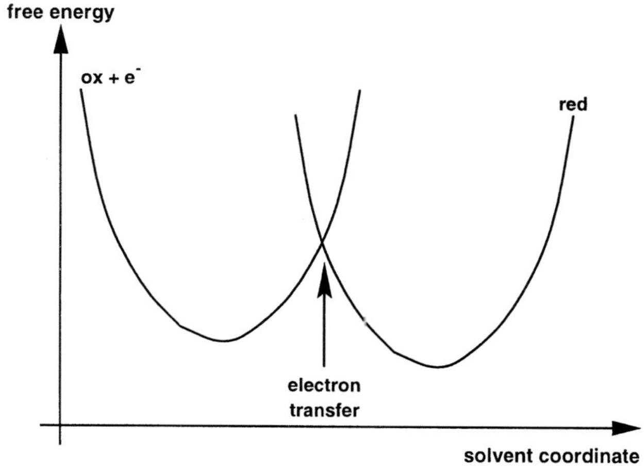

Fig. 30. Free energy curves in the Marcus theory for electron transfer reactions.

polar and nonpolar solvents. Straus and Voth [218] calculated similar functions for the electron transfer between a donor atom in water near the Pt(111) electrode, using model potentials based on those discussed in section 3. The solvent reorganization free energy in the bulk and near the surface exhibits very similar behavior. The authors concluded that the inhomogeneities in the water density near the metal electrode are only of minor importance. They also observed some deviations of the solvent activation free energy from standard Marcus behavior in the inverted region.

Rose and Benjamin [197] calculated the diabatic and adiabatic free energy curves for the reaction $Fe^{3+} + e^- \rightarrow Fe^{2+}$ in aqueous solution near the Pt(100) surface using the ion-water potential functions developed by Kuharski *et al.* [213] for the simulation of electron transfer in bulk aqueous solution. The studies of the Fe^{2+}/Fe^{3+} system near a metal surface were extended by Smith and Halley [219], who studied the free energy curves systematically as a function of the ion-metal distance and the surface charge on the electrode The influence of the magnitude of the ion charge on the electron transfer reaction was investigated by Rose and Benjamin [220]. They conclude that the calculated solvent free energies follow by and large the linear response approximation of the Marcus theory but do observe deviations for the electron transfer involving neutral atoms.

This is illustrated by Fig. 31 which shows the free energy curves for the reaction $Fe^{3+} + e^- \rightarrow Fe^{2+}$ taken from Benjamin's work. The solvent coordinate is given in units of volts and is shifted so that the transition state for the symmetric reaction is at zero. First we note that the calculated free energy curves indeed do not deviate very much from the ones extrapolated from a parabolic fit to the minimum region. Second, the figures indicate that the presence of the metal surface does not change the curvature of the free energy functions by much. Rose and Benjamin also investigated the effect of an external field on the diabatic free energy curves and found in general that the reorganization energy is only mildly affected. Similar results have been reported by Smith and Halley [219] who investigated also the influence of the

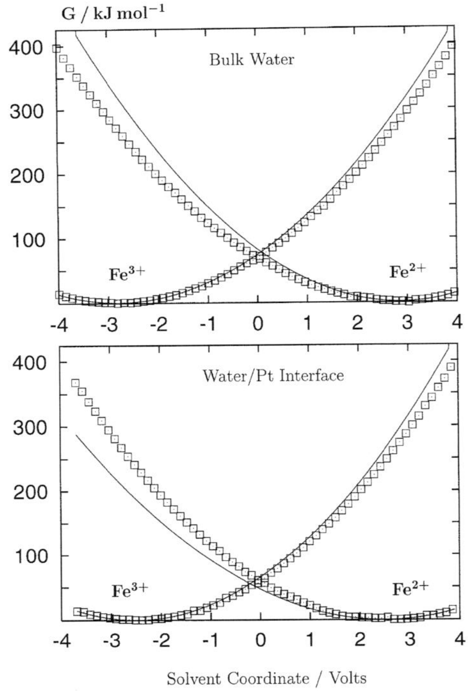

Fig. 31. Solvent free energies for the electron transfer reaction $Fe^{3+} + e^- \rightarrow Fe^{2+}$ in bulk water and at the water/Pt(100) interface at 300 K. Symbols denote the actual simulation data, full lines are obtained by a parabolic fit of the region near the bottom of each well. Data are taken with permission from the work of Rose and Benjamin [197].

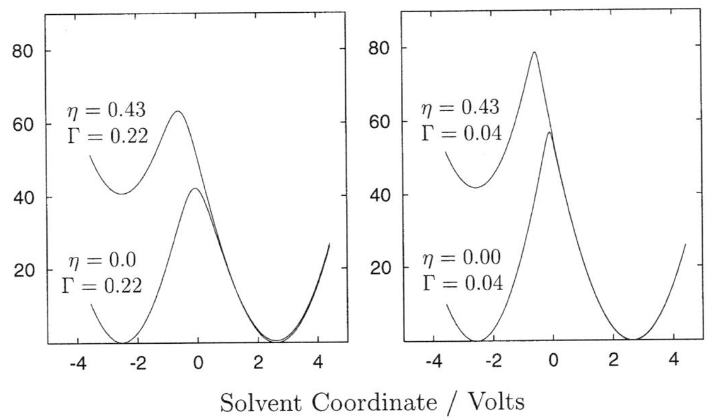

Fig. 32. Adiabatic free energy curves as a function of reaction coordinate for selected overpotential η (in V) and electronic coupling constant Γ (in eV). Data are taken with permission from the work of Rose and Benjamin [197].

distance of the ion from the electrode surface. These authors point out that a transition from diabatic to adiabatic behavior of the electron transfer occurs at short distances.

Rose and Benjamin [197] calculated the adiabatic free energy curves for several choices of overpotential η and electronic coupling parameter Γ (Fig. 32) from the diabatic free energy curves depicted in the bottom of Fig. 31. A change in overpotential not only changes the free energy of the reaction but also the location of the transition state. An increase in the coupling parameter lowers the barrier at constant overpotential, as expected, and makes the barrier less sharp. The activation energy for small values of Γ is found to be in good agreement with the experimental value of 56 kJ/mol [206].

The authors proceed to calculate the reaction rates by the flux correlation method. They find that the molecular dynamics results are well described by the Grote-Hynes theory [221] of activated reactions in solutions, which is based on the generalized Langevin equation, but that the simpler Kramers model [222] is inadequate and overestimates the solvent effect. Quite expectedly, the observed deviations from transition state theory increase with increasing values of Γ.

Straus *et al.* [223] noticed that the quantization of the librational and vibrational modes of water (by means of the Feynman path integral formalism [224]) can significantly influence the solvent free energy barrier and the thermodynamic driving force of the heterogeneous electron transfer process. For more information the reader is referred to the cited publications and the references therein.

6.2 Ion Transfer Reactions

In ion transfer reactions the transfer of an ion or proton from the solution to the surface of an electrode is one elementary step. It is often accompanied by either total discharge (e.g., deposition of a metal ion on a metal electrode of the same composition) or partial discharge (e.g., adsorption of halide ions; see also below). While for outer sphere electron transfer the reaction coordinate describes the solvent reorganization, the reaction coordinate for ion transfer reactions is associated with the motion of the ion. The rate-determining step in an ion transfer reaction is often the adsorption step of the ion on the electrode, which involves the penetration of the barrier formed by the adsorbed solvent (see, e.g., Ref. 2 and section 5.2 for a discussion).

Contrary to outer sphere electron transfer reactions, the validity of the Butler-Volmer law for ion transfer reactions is doubtful. Conway and coworkers [225] have collected data for a number of proton and ion transfer reactions and find a pronounced dependence of the transfer coefficient on temperature in all cases. These findings were supported by experiments conducted in liquid and frozen aqueous electrolytes over a large temperature range [226, 227]. On the other hand, Tsionskii *et al.* [228] have claimed that any apparent dependence of the transfer coefficient on temperature is caused by double layer effects, a statement which is difficult to validate because double layer corrections, in particular their temperature dependence, depend on an exact knowledge of the distribution of the electrostatic potential at the interface, which is not available experimentally. Here, computer simulations may be helpful in the future. Theoretical treatments of ion transfer reactions are few; they are generally based on variants of electron transfer theory, which is surprising in view of the different nature of the elementary act [229].

In the remainder of this subsection the results of recent studies of the diabatic free

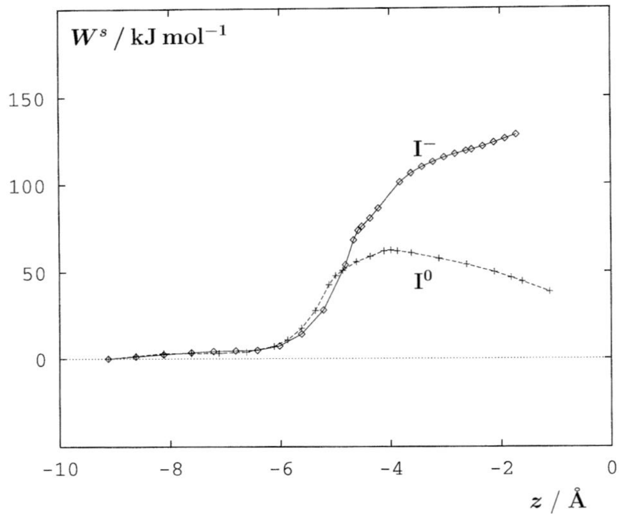

Fig. 33. Adsorption of iodide and iodine from aqueous solution on Pt(100): Solvent contribution to the PMF for I^- (diamonds) and I^0 (crosses).

energy surfaces for adsorption of I^0 and I^- on Pt(100) [230, 231] are summarized. The constancy of the transfer coefficient α is investigated by calculations of the solvent barrier as a function of temperature and external electric field, i.e., surface charge density. Because of strong electronic coupling, the two diabatic free energy curves merge into a single partially charged state close to the metal surface, and partial charge transfer occurs. In subsection 6.3 the role of ion-water interactions on the partial charge transfer near the interface is investigated, again using I^- on Pt(100) as the model system.

The model of iodine is very simple: I^0 interacts with water by the non-electrostatic part of the I^--water interactions [115]. Figure 33 shows the solvent contribution W^s to the potential of mean force (PMF) for the ion and the neutral atom. As for the ion, the solvent PMF is purely repulsive for the atom. In the region from $z = -8$ Å to $z = -5$ Å both curves are almost identical. This is in keeping with the interpretation of this part of the free energy increase being due to "steric" interactions rather than hydration forces (which are absent for the neutral I^0, as the hydration energy becomes almost zero after removal of the electrostatic contributions to the interaction potential).

For distances smaller than about 4 Å from the surface, W^s decreases for the neutral atom. Once the atom has penetrated the compact surface layer and dislocated some of the adsorbed water molecules, the system is stabilized by pushing the atom towards the surface. The behavior is an example of "hydrophobic" interactions where the "insoluble" I^0 atom is pushed out of the aqueous phase. As it cannot be pushed into the solid phase, the contact adsorbed geometry is the more favorable arrangement. In order to keep the model simple, W^s for the ion and the atom has been combined with the same *ab initio* interaction energy obtained for I^--Pt_9 clusters

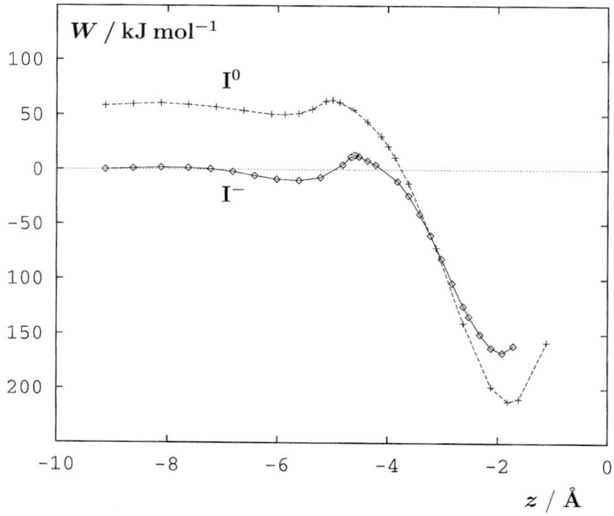

Fig. 34. Adsorption of iodide and iodine from aqueous solution on Pt(100): Total PMF for I^- (diamonds) and I^0 (crosses) on the basis of the quantum chemical ion-metal interactions. The I^0 bulk level has been shifted to account for the differences in bulk free energies (see text).

[115]. The image charge model has not been used, because additional assumptions would have to be made for the I^0-platinum interactions. The total PMFs of the iodine atom and the iodide ion are shown in Fig. 34.

To obtain the total free energy curves experienced by the ion and by the atom the interaction energy of the particles with the metal are added. Since the MD simulations yield only free energy differences relative to the bulk state, the relative energies of the two bulk levels have to be obtained by going through a cycle that decomposes the reaction $I^- \rightarrow I^0 + e^-$ into a series of steps:

$$\begin{aligned}
&\text{I-(solution)} \rightarrow \text{I}^-\text{(above solution)} && \Delta G = -\Delta G_s^r(\text{I}^-) \\
&\text{I}^- \rightarrow \text{I}^0 + \text{e}^- && \Delta G = -A \\
&\text{e}^-\text{(above solution)} \rightarrow \text{e}^-\text{(above metal)} && \Delta G = -e_0(\psi_m - \psi_s) \\
&\text{e}^-\text{(above metal)} \rightarrow \text{e}^-\text{(metal)} && \Delta G = -\Phi \\
&\text{I}^0\text{(above solution)} \rightarrow \text{I}^0\text{(solution)} && \Delta G = -\Delta G_s^r(\text{I}^0)
\end{aligned} \tag{18}$$

In these equations ΔG_s^r denotes the free energy of solvation of the indicated species, A is the electron affinity of the iodine atom, ψ_s and ψ_m denote the outer potentials of the solution and the metal at the potential of zero charge (pzc), and Φ is the work function of the metal; 'above' refers to a position in the vacuum just above the indicated phase. All the quantities in Eq. (18) are measurable, but not all are known exactly. Going through the cycle of Eq. (18) and following the arguments in Ref. 230, a value of 58 kJ mol^{-1} results for the free energy difference between iodine and iodide in the bulk of the solution; due to the uncertainties in the various quantities the error is estimated as $\pm 20\,\text{kJ mol}^{-1}$. As expected, the ionic state is energetically more favorable in the bulk of the solution.

Both particles experience similar free energy curves as they approach the surface from the bulk (see Fig. 34). They pass through a local minimum at a position where they are separated from the surface by one layer of water molecules. This can be interpreted as an adsorption site for a solvated particle. A closer approach to the electrode surface requires an energy of activation as the free energy curves pass through a maximum. This energy barrier contains two contributions: the displacement of water from the metal surface, and the partial loss of the solvation shells. The latter contribution is much larger for the ion, which attracts water molecules through Coulomb forces. Hence the energy barrier, defined as the energy difference between the maximum and the local minimum near -5.5 Å, is higher for the ion (about $22\,\text{kJ mol}^{-1}$) than for the atom (about $12\,\text{kJ mol}^{-1}$). Both free energy curves show pronounced minima right on the metal surface. The calculated minimum is somewhat lower for the atom than for the ion. Because of the electronic coupling to the metal surface, there is only one adsorbed state near the metal surface with a partial charge and a broadened electronic orbital (see below and Ref. 232–234). The curve for the iodine atom has a lower minimum so that the adsorbed state can be expected to carry a partial charge which is substantially lower than the ionic charge.

The rate of an electrochemical ion transfer reaction depends on the electrode potential or on the charge density at the electrode surface. Therefore, simulations in the presence of an external electrostatic field E were performed [230]. The direction of the field was chosen such that the ion is attracted toward the surface. The resulting solvent free energy curves are shown in Fig. 35; they have been adjusted by subtracting the constant force in the center of the lamina that is due to incomplete screening of the external field. Clearly, the solvent barrier decreases with increasing electric field.

Electrochemical ion transfer reactions are usually observed in solutions with a

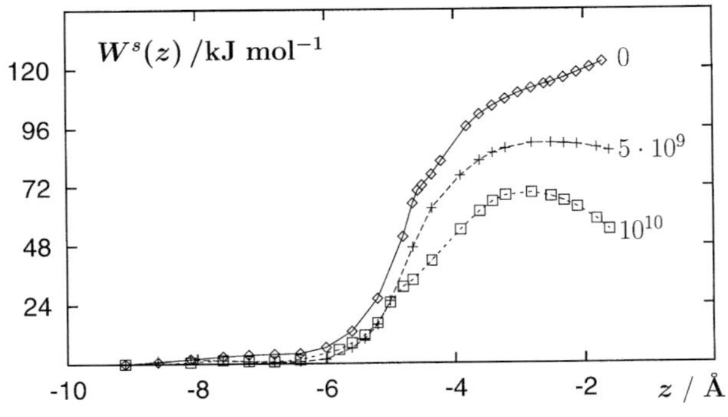

Fig. 35. Solvent contribution, W^s, to the free energy of adsorption of I^- on Pt(100) for different homogeneous electric fields across the lamina. The field strength is indicated in units of Vm^{-1} and the direction of the field is such that the surface carries a positive charge.

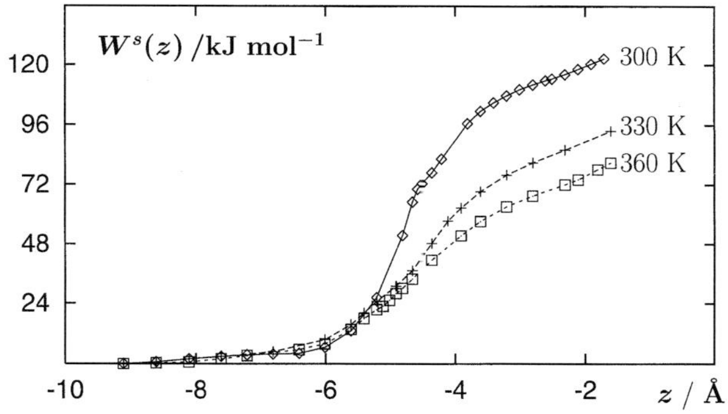

Fig. 36. Solvent contribution W^s to the free energy of adsorption of I^- on Pt(100) at different temperatures as indicated.

high concentration of supporting electrolyte, which shields the external field. The resulting changes in the potential energy profile will affect the free energy curves for the ions. The largest changes will occur in the bulk, where the free energy will attain a constant value, and in the region distant by about one Debye length from the surface. The screening will be less effective directly at the interface; therefore one expects that the variations in the solvent barrier caused by external fields of varying strength will remain qualitatively similar, in this region, to what is observed in Fig. 35.

The solvent free energy barrier for the adsorption of I^- was also calculated at three different temperatures. The results [231] are displayed in Fig. 36. The barrier decreases, as expected, with increasing temperature from about 120 kJ mol^{-1} at room temperature to about 80 kJ mol^{-1} at 360 K.

The simulations suggest the following picture for an ion transfer reaction: before the reaction the ion is located in a weak adsorption site, where it is separated from the electrode by one layer of water molecules. As the ion approaches the electrode surface it displaces water from the surface and partially looses its solvation shell; this requires a substantial energy of activation. Subsequently, the ion moves down the free energy curve towards an adsorption site on the metal surface; simultaneously the electronic interaction with the metal increases, electron exchange becomes adiabatic, and the adsorbed particle carries a partial charge (see next section).

The situation is fundamentally different from that for outer sphere electron transfer reactions where, according to Marcus theory, the solvent reorganization determines the reaction. In contrast, the model calculations discussed in this section indicate that the energy of activation for the ion transfer step is not related to the electron exchange with the electrode, since the crossing between the two diabatic energy states of I^- and I^0 occurs only at such short distances where the ion has already surpassed the solvent barrier. Contrary to the approach discussed here, Xia and Berkowitz [235] assumed the validity of the outer sphere mechanism from the outset. The analysis of the dependence of the solvent barrier on external electric field and temperature indicates that α in Eq. (17) is indeed not constant but depends on temperature.

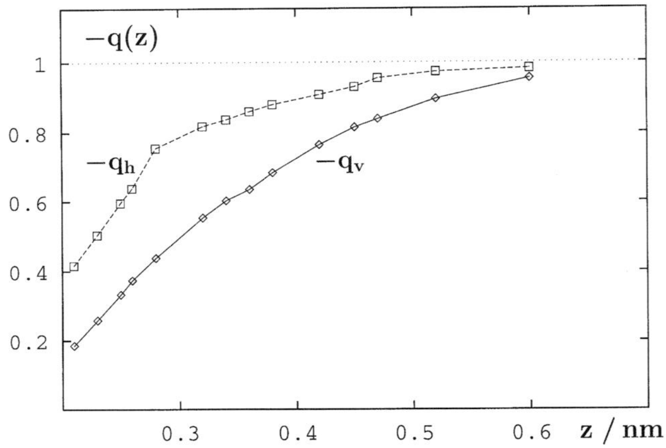

Fig. 37. Charge, in units of the elementary charge e, of the iodide ion above the hollow site of a Pt_9 $(4+5)$ cluster from quantum chemical calculations [115, 239] in vacuum, q_v, (diamonds) and the self-consistently calculated electrochemical charge of the ion, q_h, as a function of the ion-metal distance z. Taken from Ref. 234.

6.3 Partial Charge Transfer

Schmickler [236] recently discussed the iodide ion transfer reaction near the Pt(100) surfaces in some detail based on the Anderson-Newns model of chemisorption [237, 233]. Some of the model parameters used in this work were obtained from computer simulations [190, 230, 234]. Besides the study of Ref. 234, which is outlined in the remainder of this subsection, no other simulations using similar realistic molecular models have been performed.

Figure 37 shows the formal charge of I^- as a function of distance from a Pt_9 cluster, obtained by Mulliken population analysis of *ab initio* calculations of Pt_9I^- clusters [115, 239]. The effect of solvation on the partial charge of the hydrated iodide ion near the Pt(100) surface was investigated [234] by a combination of quantum chemical calculations, the Anderson-Newns model of chemisorption and MD calculations of the free energy of iodide adsorption [190]. On the basis of the approximations discussed in Ref. 234, the charge on the bare ion can also be calculated from the relation

$$q_v = 1 - \frac{2}{\pi}\arctan\left(\frac{\Delta}{\varepsilon_a}\right) \tag{19}$$

ε_a is the electronic energy level of the adsorbate ion relative to the Fermi energy of the metal. The electronic interaction between ion and metal induces the broadening Δ of the adsorbate energy levels. Δ characterizes the strength of the ion-metal interaction and decays rapidly with the distance from the surface. ε_a and q_v are known from the *ab initio* calculations; Δ was then calculated from Eq. (19) for different values of the distance of the ion from the surface, z.

In a polar medium, the partial charge transfer depends on two additional contributions due to the ion-solvent and metal-solvent interactions. Following the theory of "electrochemical" adsorption by Schmickler [232], the adsorbate energy level in solution, ε_a^* is given by

$$\varepsilon_a^* = \varepsilon_a + \varepsilon_S + e\chi \tag{20}$$

where ε_S is the solvation energy of the ion, χ the surface potential drop due to the orientation of the solvent molecules near the interface (see subsection 4.5), and e the elementary charge. Assuming transferability of the parameter Δ from the situation in vacuum to the aqueous solution, the partial charge q_h of the hydrated ion was calculated from

$$q_h = 1 - \frac{2}{\pi}\arctan\left(\frac{\Delta}{\varepsilon_a^*}\right) \tag{21}$$

in analogy to Eq. (19). Clearly, the solvation energy ε_S depends on the charge q_h. This makes a self-consistent evaluation of q_h and ε_S necessary. From the MD simulations in Ref. 190 ε_S was known only for the full ion charge (full line in Fig. 38). ε_S at a given value of q_h was then calculated by a Born-like approximation. (see Ref. 234 for details). Figures 37 and 38 show the partial charge, q_h, of the solvated ion and the associated interpolated hydration energies, ε_S, respectively.

The partial charge on the ion in vacuum, q_v, decreases substantially at small distances from the surface (Fig. 37). Ion hydration stabilizes the ionic charge and consequently the partial charge transfer from the ion to the surface is reduced. Substantial charge transfer in solution occurs only at distances smaller than 4.5 Å from the surface. At larger distances, q_h is almost equal to -1. At 4.5 Å, however, the

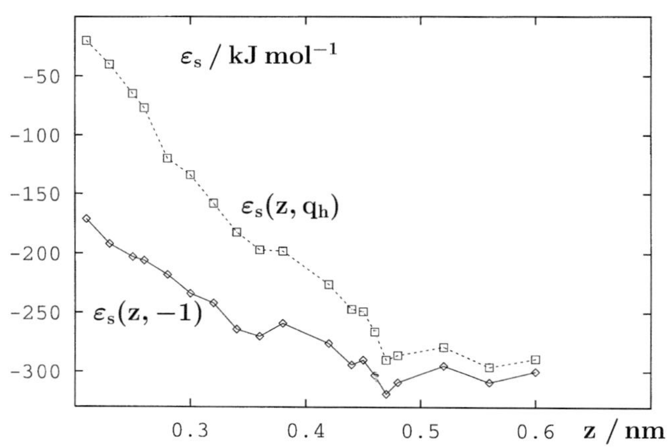

Fig. 38. Distance dependence of the iodide hydration energy from computer simulations, $\varepsilon_s(z, -1)$, (diamonds) and the self-consistent hydration energy at reduced charge, $\varepsilon_s(z, q_h)$ (squares). The hydration energy $\varepsilon_s(z, -1)$ has been shifted to reproduce the experimental value beyond 0.6 nm. Taken from Ref. 234.

layer of adsorbed water exerts already a significant repulsive force on the ion. This supports the conclusions drawn in the previous section and in Ref. 236 that the rate-determining step in the iodide ion transfer reaction near a metal surface is the penetration of the ion into the layer of adsorbed water.

6.4 Proton Transfer

The transport of a proton is involved in many electrochemical reactions like, e.g., the hydrogen evolution reaction. Existing theories have mostly been extensions of the theory of electron transfer reactions, in which the proton tunnels from its initial to its final state. So far, quantum simulation techniques have only been applied to proton transfer in the bulk aqueous phase and clusters (e.g., Ref. 240–246), but not near interfaces.

Quite recently, Pecina and Schmickler [247, 248] proposed a new model for proton transfer, based on the observation that the isotope effects for proton transfer on non-catalytic metals like silver, gold, and mercury are small and similar in magnitude to those in bulk solutions. They suggest that the rate-determining step near these surfaces is similar to the Grotthus mechanism in the bulk. Their model assumes that the rate-determining step is the transfer of a proton from a hydronium ion in the second water layer to the first layer, from where the proton transfer to the metal is assumed to be fast. In a series of Monte Carlo simulations of an aqueous hydronium ion next to a model silver surface, the activation parameters for the reorientation are calculated as a function of surface charge and temperature. Based on the simulation results, the authors offer explanations for the experimentally observed correlation between exchange current density and potential of zero charge and for the experimentally observed entropic component to the transfer coefficient.

Figure 39, taken from their work, shows the decomposition of the activation free energy into enthalpic and entropic contributions. The authors explain the sizable entropic term at small negative surface charges by the strong temperature dependence of the hydrogen bond network in water. At higher temperatures the hydrogen bonds are weakened and it becomes easier to turn the mediating water molecules towards the surface, which is the rate-determining step in their model. Consequently, the free

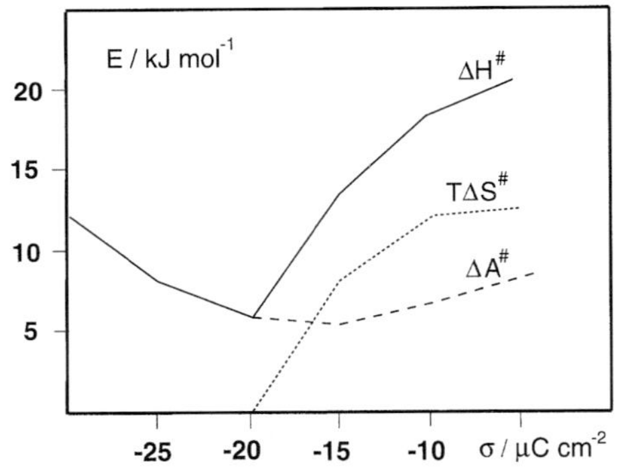

Fig. 39. Activation parameters as a function of surface charge density. Data taken with permission from Ref. 248

energy of activation is expected to decrease with temperature, resulting in the observed positive entropy of activation. At more negative surface charge densities most hydrogen bonds are broken and it is the water dipole that is oriented in the external electrostatic field, resulting in the rise of activation energy and the reduction of the entropic term.

6.5 Other Reactions

Isomerizations are very simple model reactions that are among the most common chemical processes. Solvent effects on these reactions have been studied extensively in the bulk phase. The study of solvent effects near the interface is especially interesting, since the gauche-trans equilibrium can involve two species with significantly different dipole moments; consequently, a substantial interfacial effect can be expected. The isomerization reaction of 1,2-dichloroethane near the water/vacuum interface [249], and of 1,2-dichloroethane and of dialanine near the water/hexane interface [249, 250] has been studied. An increase of the gauche $\rightarrow$ trans isomerization rate is observed near the interface (for a review see Ref. 251).

Computer simulations have also been used by various authors to investigate primary events following electron injection into adsorbed water layers [175] and to calculate tunneling probabilities (e.g., Ref. 252–255).

6.6 Summary

Various simulation approaches to electron and ion transfer reactions are reported in the literature. Outer sphere electron transfer reactions have been investigated by simulation studies within the framework of the Marcus theory. The role of the solvent barrier for ion transfer reactions has been investigated on the basis of free energies of adsorption.

- Simulations of outer sphere electron transfer reactions [197, 217–220, 223] show that the Marcus theory is valid over a wide range. Deviations from the linear response regime were found.
- The diabatic free energy curves for the adsorption of I^- and I^0 near Pt(100) were calculated. I^- is the energetically favorable species in solution, while the more stable species on the surface is I^0. The crossing between the two diabatic curves occurs at short distances, when the ion has already penetrated the adsorbed water layer.
- Due to ionic hydration, partial charge transfer from the iodide ion to the metal is reduced in solution relative to that in vacuum. The partial charge transfer becomes important only at small ion-surface distances.
- Therefore, the rate-determining step for ion transfer reactions near the surface, at least for large ions, appears to be the penetration of the ion into the layer of adsorbed water molecules, rather than the excitation of solvent fluctuations.

7 Adsorbed Organic Molecules

The adsorption of organic molecules from aqueous solutions onto electrodes is quite important for a number of technological processes (see, e.g., Ref. 12). In spite of this fact, the only study on this subject known to the author is about the coadsorption of benzene on an electrode, which was the subject of a simulation study by Philpott and coworkers [102, 256]. The authors observe that adsorption of benzene from aqueous NaCl solution takes place on the uncharged electrode. Charging the electrode negatively leads to desorption while charging it positively does not change the adsorption. The reason for this asymmetry lies apparently in the different behavior of the Na^+ and Cl^- ions: while Cl^- contact adsorbs on the positively charged surface and thus screens the electric field, the Na^+ ions do not adsorb and the benzene molecules leave the high-field region.

Currently, we are studying in our group the adsorption of polyelectrolytes from aqueous solutions on charged and uncharged surfaces. Investigations of thiol adsorption on surfaces (e.g., Ref. 257, 258) conformational equilibria of polymers near surfaces (e.g., Ref. 259), of surfactants with surfaces and micelles [260–263] and others were mostly done without explicit modeling of the solvent.

8 Liquid/Liquid Interfaces

Conformational equilibria [249, 264], electron [217, 265], ion [200, 266, 267], and molecule transfer [268] across liquid/liquid interfaces consisting of water and an organic solvent have recently been studied by several authors. Also, the aqueous/liquid mercury interface has been investigated (see above and Ref. 78, 81). Much of this work has been reviewed recently in two excellent papers by Benjamin [251, 269]. Here, I would like to state that simulations of liquid/liquid interfaces are likely to become increasingly important, for several reasons:

- Much of the interest in the interfaces between two immiscible electrolyte solutions (ITIES) derives from the fact that they can be regarded as models for biological and artificial membranes, which are currently actively studied by computer simulation methods (for an incomplete list, see Ref. 270–275).
- New experimental techniques, foremost the sum frequency generation and other spectroscopic methods, provide a wealth of data on microscopic structure and dynamics that need to be interpretated on the molecular level [276–283].
- Last, but not least, there are less problems to obtain realistic interaction potential surfaces (the most important input to computer simulations) than in the case of liquid/metal interfaces, since *ab initio* calculations can be performed more reliably for molecule-molecule interactions than for molecule-metal interactions.

9 Summary and Outlook

Simulation studies of the interfacial region between aqueous solutions and crystalline or liquid metals focused in recent years increasingly on the realistic description of the distinctive features of these systems. Both Molecular Dynamics and Monte Carlo computer simulations of such realistic models are nowadays a very active area of research.

The field evolved out of the simulation of liquids and solutions and has adopted many of the models from there. Water, ions, and atoms are described by point charge models, augmented by empirical or *ab initio* short-range and dispersion potentials, the use of which has been validated for the bulk phase. Many models of the interface have been employed, ranging from representations of simple, atomically smooth walls to parametrized interaction potentials from *ab initio* quantum chemical calculations that describe the interactions with atomic crystals. The metallic character of the solid phase has in most cases been approximated by the image charge model.

The models used so far in computer simulations are still too unspecific, in spite of the fact that quantum chemical calculations have occasionally been used to generate them. Therefore, the results show trends rather than they reproduce specific experiments. The role of computer simulation in interfacial electrochemistry lies in the visualization of concepts that can lead to simplified pictures, or even cartoons, of reality, and the detailed exploration of the consequences of assumptions made.

A variety of different aspects of the electrochemical interface have been addressed by simulations that have been reviewed and discussed in this chapter:

- The structure of water next to the metal is strongly perturbed only over two layers, and the range of other inhomogeneities and anisotropies appears to be not more than four layers at most. Detailed structural information has been derived. It has been demonstrated that the translational and rotational mobility of water and ions near a metal surface is reduced compared to the bulk.
- Specific and non-specific ion adsorption has been studied and first attempts to study the structure of the diffuse layer have been made. Simulations of the full double layer are becoming possible, at least at ion concentrations of the order of 1 mol/l.
- The interplay between ion adsorption and ion hydration, and the relation between thermodynamic quantities like the free energy of adsorption and microscopic structure, characterized, e.g., by hydration shell properties and hole formation, has been elucidated.
- Electron, ion, and proton transfer reactions have been modeled by several groups as prototypical reactions at the electrochemical interface.

Many open questions remain, however, a challenge for computer simulations of the electrochemical interface. What needs to be done in the future?

- For many electrochemical problems, the models at hand are not adequate. Neither the image charge models nor quantum chemical calculations lead to completely

satisfactory results. More computationally feasible potential energy surfaces need to be developed and extensively tested. This will have to proceed along several paths:
– devise simple empirical one-body potentials that describe efficiently the interaction of water, simple ions and organic molecules with low index faces of noble metals (e.g., copper, silver, gold) and selected transition metals (e.g., nickel, palladium, platinum, ruthenium);
– allow for 'jellium corrections' to the raw image interactions at small distances, which are due to the delocalized electrons in the first atomic layer; work in this direction is currently carried out [72–74]
– perform *ab initio* cluster calculations and density functional calculations on large systems to understand bonding, geometry, and vibrational properties of the adsorbed species on metal surfaces.

- Studies of the structure of the diffuse layer at finite ion concentrations and for different electrode potentials with and without dissolved organic molecules will certainly become very interesting applications of computer simulations. Much larger systems than the ones that have been treated up to now must be simulated.
- A phenomenon so far only touched on by simulations is the dissociation of water near the surface and the water autohydrolysis reaction.
- Many other aspects of electrode reactions like the transport of material to and from the surface and the occurrence of more complex reactions than electron or ion transfer will certainly be investigated. Appropriate simulation techniques have already been tested in bulk liquids.
- Real surfaces have point defects and steps while the interfaces in simulations are usually ideal. Defects play a very important role in electrochemical and catalytic reactions. Models need to be developed that are able to account for the electronic structure near these defects. First steps in this direction have been made [98, 99, 284].
- Realistic simulations should ultimately take into account the electronic structure of reactants, products, the surface, and active sites. The Car-Parrinello simulation scheme seems to be the most appropriate method for the future. First steps into this direction were made by Price and Halley, using a hybrid scheme [285].

Currently, modeling of electrochemical systems and reactions is only in its infancy. However, even today, with state-of-the-art computers, exciting work can be done that could only be dreamt of five years ago. It is certainly not very daring to predict that modeling the electrochemical interface on the basis of realistic yet simple models of solvent, solutes, and reactive species will become increasingly important in the near future.

Acknowledgments. Financial support by the Fonds der Chemischen Industrie is gratefully acknowledged. I thank M. Berkowitz, M. R. Philpott, I. Benjamin, W. Schmickler, and O. Pecina for the permission to reproduce some of their data and figures. I wish to thank these authors and K. Heinzinger, J. C. Shelley and G. Nagy for discussions and preprints, and A. Kohlmeyer for his careful proofreading of the manuscript.

References

1. F. F. Abraham, Rep. Prog. Phys. 45, 1113 (1982).
2. W. Schmickler, *Interfacial Electrochemistry* (Oxford University Press, Oxford, 1995).
3. H. L. F. von Helmholtz, Ann. Physik 89, 211 (1853).
4. H. L. F. von Helmholtz, Ann. Physik 7, 337 (1879).
5. G. Gouy, J. Phys. 9, 457 (1910).
6. D. L. Chapman, Phil. Mag. 25, 475 (1913).
7. O. Stern, Z. Electrochem. 30, 508 (1924).
8. D. C. Grahame, Chem. Rev. 41, 441 (1947).
9. J. O. Bockris, M. A. Devanathan, and K. Müller, Proc. R. Soc. A274, 55 (1963).
10. B. B. Damaskin and A. N. Frumkin, Electrochim. Acta 19, 173 (1974).
11. J. O. Bockris, *The Comprehensive Treatise of Electrochemistry Vol 1* (Plenum Press, New York, 1981).
12. C, M. A. Brett and A. M Oliveira Brett, *Electrochemistry. Principles, Methods, and Applications, Oxford Science Publications* (Oxford University Press, Oxford, 1993).
13. W. Schmickler and D. Henderson, Prog. Surf. Sci. 22, 323 (1986).
14. J. O. Bockris and A. K. N. Reddy, *Modern Electrochemistry, Vol. 1* (Plenum Press, New York, 1977).
15. J. O. Bockris and A. K. N. Reddy, *Modern Electrochemistry, Vol. 2* (Plenum Press, New York, 1977).
16. J. M. Haile, *Molecular Dynamics Simulation. Elementary Methods* (John Wiley & Sons, Inc., New York, 1992).
17. N. A. Metropolis, A. W. Rosenbluth, M. N. Rosenbluth, A. H. Teller, and E. Teller, J. Chem. Phys. 21, 1087 (1953).
18. B. J. Alder and T. E. Wainwright, J. Chem. Phys. 31, 459 (1959).
19. M. P. Allen and D. J. Tildesley, *Computer Simulations of Liquids* (Oxford University Press, New York, 1987).
20. *Monte Carlo Methods in Statistical Physics*, 2nd ed., edited by K. Binder (Springer Verlag, Berlin, 1986).
21. *Applications of the Monte Carlo Method in Statistical Physics*, edited by K. Binder (Springer Verlag, Berlin, 1984).
22. *The Monte Carlo Method in Condensed Matter Physics*, Vol. 71 of *Topics in Applied Physics*, edited by K. Binder (Springer-Verlag, Berlin, 1992).
23. B. Smit and D. Frenkel, *Understanding Molecular Simulations. From Algorithms to Applications* (Academic Press, San Diego, 1996).
24. P. P. Ewald, Ann. Physik 64, 253 (1921).
25. J. H. Petropoulos and J. K. Petrou, J. Chem. Soc. Faraday Trans. 87, 2017 (1991).
26. S. E. Feller, R. W. Pastor, A. Rojnuckarin, S. Bogusz, and B. R. Brooks, J. Phys. Chem. 100, 17011 (1996).
27. E. Spohr, J. Chem. Phys. 107, 6342 (1997).
28. L. F. Greengard, *The Rapid Evaluation of Potential Fields in Particle Systems, ACM Distinguished Dissertations* (MIT Press, Cambridge, Massachusetts, 1988).
29. L. Greengard and V. Rokhlin, J. Comp. Phys 73, 325 (1987).
30. W. G. Hoover, *Computational Statistical Mechanics* (Elsevier, Amsterdam, 1991).
31. W. G. Hoover, *Molecular Dynamics*, No. 258 in *Lecture Notes in Physics* (Springer Verlag, Berlin, 1986).
32. R. Haberlandt, S. Fritzsche, G. Peinel, and K. Heinzinger, *Molekulardynamik. Grundlagen und Anwendungen* (Vieweg, Braunschweig, Wiesbaden, 1995).
33. J. D. Porter and A. S. Zinn, J. Phys. Chem. 97, 1190 (1993).

34. P. Bopp, in *The Physics and Chemistry of Aqueous Ionic Solutions*, edited by M. C. Bellissent-Funel and G. W. Neilson (Reidel, Amsterdam, 1987).
35. H. J. C. Berendsen, J. P. M. Postma, W. F. van Gunsteren, and J. Hermans, in *Intramolecular Forces*, edited by B. Bullman (Reidel, Amsterdam, 1981).
36. W. L. Jorgensen, J. Chandrasekhar, J. D. Madura, R. W. Impey, and M. L. Klein, J. Chem. Phys. 79, 926 (1983).
37. O. Matsuoka, E. Clementi, and M. Yoshimine, J. Chem. Phys. 64, 1351 (1976).
38. P. Bopp, G. Jancsó, and K. Heinzinger, Chem. Phys. Lett. 98, 129 (1983).
39. S. Goldman and P. Backx, J. Chem. Phys. 84, 2761 (1986).
40. J. Böcker, R. R. Nazmutdinov, E. Spohr, and K. Heinzinger, Surf. Sci. 335, 372 (1995).
41. A. Wallqvist and B. J. Berne, J. Phys. Chem. 97, 13841 (1993).
42. S. W. Rick and B. J. Berne, J. Am. Chem. Soc. 116, 3949 (1994).
43. S.-B. Zhu, S. Singh, and G. W. Robinson, J. Chem. Phys. 95, 2791 (1991).
44. A. Kohlmeyer, W. Witschel, and E. Spohr, Chem. Phys. 213, 211 (1996).
45. M. Tuckerman, B. J. Berne, and G. J. Martyna, J. Chem. Phys. 97, 1990 (1992).
46. S. W. Rick, S. J. Stuart, and B. J. Berne, J. Chem. Phys. 101, 6141 (1994).
47. E. Spohr and K. Heinzinger, Chem. Phys. Lett. 123, 218 (1986).
48. E. Spohr and K. Heinzinger, Ber. Bunsenges. Phys. Chem. 92, 1358 (1988).
49. E. Spohr, J. Phys. Chem. 93, 6171 (1989).
50. J. I. Siepmann and M. Sprik, Surf. Sci. Lett. 279, L185 (1992).
51. J. I. Siepmann and M. Sprik, J. Chem. Phys. 102, 511 (1995).
52. K. Foster, K. Raghavan, and M. Berkowitz, Chem. Phys. Lett. 162, 32 (1989).
53. K. Raghavan, K. Foster, and M. Berkowitz, Chem. Phys. Lett. 177, 426 (1991).
54. K. Raghavan, K. Foster, K. Motakabbir, and M. Berkowitz, J. Chem. Phys. 94, 2110 (1991).
55. S.-B. Zhu and M. R. Philpott, J. Chem. Phys. 100, 6961 (1994).
56. E. Spohr, J. Mol. Liquids 64, 91 (1995).
57. J. P. Badiali, Ber. Bunsenges. Phys. Chem. 91, 270 (1987).
58. D. Henderson, E. Leiva, and W. Schmickler, Ber. Bunsenges. Phys. Chem. 91, 280 (1987).
59. W. Schmickler and D. Henderson, J. Chem. Phys. 85, 1650 (1986).
60. V. Russier and M. L. Rosinberg, J. Phys. C: Solid State Phys. 21, L333 (1988).
61. N. D. Lang, Solid State Commun. 7, 1047 (1969).
62. N. D. Lang and W. Kohn, Phys. Rev. B 1, 4555 (1970).
63. J. R. Smith, Phys. Rev. 181, 522 (1969).
64. A. J. Bennett and C. B. Duke, Phys. Rev. 162, 578 (1967).
65. N. D. Lang, Surf. Sci. 299/300, 284 (1994).
66. A. A. Kornyshev, M. B. Partenskii, and W. Schmickler, Z. Naturforsch. 39a, 1122 (1984).
67. M. I. Rojas and E. P. M. Leiva, Surf. Sci. Lett. 227, L121 (1990).
68. W. Schmickler, D. Henderson, and O. R. Melroy, Chem. Phys. Lett. 216, 424 (1993).
69. W. Schmickler, J. Electroanal. Chem. 150, 19 (1983).
70. V. Russier and J. P. Badiali, Phys. Rev. B 39, 13193 (1989).
71. S. Amokrane, V. Russier, and J. P. Badiali, Surface Sci. 217, 425 (1989).
72. W. Schmickler and E. Leiva, Mol. Phys. 86, 737 (1995).
73. J. C. Shelley, G. N. Patey, D. R. Bérard, and G. M. Torrie, J. Chem. Phys. (1997).
74. A. Kohlmeyer and E. Spohr, in preparation.
75. N. Anastasiou, D. Fincham, and K. Singer, J. Chem. Soc. Faraday Trans. II 79, 1639 (1983).
76. K. Heinzinger, Pure & Appl. Chem. 63, 1733 (1991).
77. E. Spohr, Acta Chem. Scand. 49, 189 (1995).
78. J. Böcker, E. Spohr, and K. Heinzinger, Z. Naturforsch. 50a, 611 (1995).

79. E. Spohr, G. Tóth, and K. Heinzinger, Electrochim. Acta 41, 2131 (1996).
80. G. Tóth and K. Heinzinger, Chem. Phys. Lett. 245, 48 (1995).
81. J. Böcker, Z. N. Gurskii, and K. Heinzinger, J. Phys. Chem. 100, 14969 (1996).
82. S. Holloway and K. H. Bennemann, Surf. Sci. 101, 327 (1980).
83. J. E. Müller and J. Harris, Phys. Rev. Lett. 53, 2493 (1984).
84. C. W. Bauschlicher Jr, J. Chem. Phys. 83, 3129 (1985).
85. M. W. Ribarsky, W. D. Luedtke, and U. Landman, Phys. Rev. B 32, 1430 (1985).
86. A. Ignaczak and J. A. N. F. Gomes, J. Electroanal. Chem. 420, 209 (1997).
87. H. Sellers and P. V. Sudhakar, J. Chem. Phys. 97, 6644 (1992).
88. R. R. Nazmutdinov, M. Probst, and K. Heinzinger, J. Electroanal. Chem. 369, 227 (1994).
89. J. Fusy and R. Ducros, Surf. Sci. 176, 157 (1986).
90. G. B. Fisher and J. L. Gland, Surf. Sci. 94, 446 (1980).
91. T. E. Madey and J. T. Yates Jr, Chem. Phys. Lett. 51, 77 (1977).
92. C. Benndorf, C. Nöbl, and F. Thieme, Surf. Sci. 121, 249 (1982).
93. W. A. Steele, Surf. Sci. 36, 317 (1973).
94. E. Spohr, M. Wolfsberg, and P. Bopp, Z. Naturforsch. 46a, 174 (1991).
95. E. Spohr and M. Wolfsberg, Surf. Sci. 253, L395 (1991).
96. F. H. Stillinger and T. A. Weber, Phys. Rev. B 31, 5262 (1985).
97. P. A. Thiel and T. E. Madey, Surf. Sci. Reports 7, 211 (1987).
98. G. Nagy and G. Denuault, J. Electroanal. Chem. 433, 153 (1997).
99. G. Nagy and G. Denuault, J. Electroanal. Chem. 433, 161 (1997).
100. A. A. Gardner and J. P. Valleau, J. Chem. Phys. 86, 4171 (1987).
101. M. R. Philpott and J. N. Glosli, J. Electrochem. Soc. 142, L25 (1995).
102. M. R. Philpott, J. N. Glosli, and S. B. Zhu, Surf. Sci. 335, 422 (1995).
103. M. R. Philpott and J. N. Glosli, in *Solid-Liquid Electrochemical Interfaces*, Vol. 656 of *ACS Symposium Series*, edited by G. Jerkiewicz, M. P. Soriaga, K. Uosaki, and A. Wieckowski (ACS, Washington, 1997), Chap. 2. Molecular Dynamics Simulation of Interfacial Electrochemical Processes: Electric Double Layer Screening, pp. 13–30.
104. E. Spohr, Electrochim. Acta submitted (1997).
105. K. Heinzinger, Physica 131 B, 196 (1985).
106. K. Heinzinger, in *Computer Modelling Of Fluids Polymers and Solids*, edited by C. R. A. Catlow et al. (Kluwer Academic Publishers, Dordrecht, 1990), Chap. Molecular Dynamics Simulations of Aqueous Systems, pp. 357–394.
107. J. N. Glosli and M. R. Philpott, J. Chem. Phys. 96, 6962 (1992).
108. J. N. Glosli and M. R. Philpott, in *Microscopic Models of Electrode-Electrolyte Interfaces*, edited by J. W. Halley and L. Blum (Electrochemical Society Inc., Pennington, 1993), No. 93-5, pp. 90–103.
109. J. N. Glosli and M. R. Philpott, J. Chem. Phys. 98, 9995 (1993).
110. Y. Tamura, K. Tanaka, E. Spohr, and K. Heinzinger, Z. Naturforsch. 43a, 1103 (1988).
111. K. Heinzinger, P. Bopp, and G. Jancsó, Acta Chimica Hungarica 121, 27 (1986).
112. M. M. Probst, E. Spohr, and K. Heinzinger, Chem. Phys. Lett. 161, 405 (1989).
113. M. M. Probst, E. Spohr, K. Heinzinger, and P. Bopp, Molec. Simulation 7, 43 (1991).
114. R. P. Messmer, in *The Nature of the Surface Chemical Bond*, edited by T. N. Rhodin and G. Ertl (North-Holland, Amsterdam, 1979), Chap. 2. Cluster Model Theory and its Application to Metal Surface-Adsorbate Systems, p. 51.
115. J. Seitz-Beywl, M. Poxleitner, M. M. Probst, and K. Heinzinger, Int. J. Quant. Chem. 42, 1141 (1992).
116. G. Pacchioni, P. S. Bagus, M. R. Philpott, and C. J. Nelin, Int. J. Quant. Chem. 38, 675 (1990).
117. P. S. Bagus, G. Pacchioni, and M. R. Philpott, J. Chem. Phys. 90, 4287 (1989).
118. G. Pacchioni, P. S. Bagus, and M. R. Philpott, Z. Phys. D 12, 543 (1989).

119. M. Blanco, J. M. Ricart, J. Rubio, and F. Illas, J. Electroanal. Chem. 267, 243 (1989).
120. M. Blanco, J. Rubio, and F. Illas, J. Electroanal. Chem. 261, 39 (1989).
121. G. Tóth, E. Spohr, and K. Heinzinger, Chem. Phys. 200, 347 (1995).
122. A. Ignaczak and J. A. N. F. Gomes, J. Electroanal. Chem. 420, 71 (1997).
123. T. Kramar, D. Vogtenhuber, R. Podloucky, and A. Neckel, Electrochim. Acta 40, 43 (1995).
124. J. C. Shelley and G. N. Patey, Mol. Phys. 88, 385 (1996).
125. D. Henderson, F. F. Abraham, and J. A. Barker, Mol. Phys. 31, 1291 (1976).
126. D. Henderson and M. Plischke, J. Phys. Chem. 92, 7177 (1988).
127. J. Quintana, D. Henderson, and A. D. J. Haymet, J. Chem. Phys 98, 1486 (1993).
128. J. K. Percus, J. Stat. Phys. 15, 1772 (1976).
129. M. J. Booth, D. M. Duh, and A. D. J. Haymet, J. Chem. Phys. 101, 7925 (1994).
130. J. G. Harris, J. Gryko, and S. A. Rice, J. Chem. Phys. 86, 1067 (1987).
131. M. P. d'Evelyn and S. A. Rice, J. Chem. Phys. 78, 5081 (1983).
132. L. Bosio, R. Cortes, and C. Segaud, J. Chem. Phys. 71, 3595 (1979).
133. A. M. Brodsky, M. Watanabe, and W. P. Reinhardt, Electrochim. Acta 36, 1695 (1991).
134. M. Watanabe, A. M. Brodsky, and W. P. Reinhardt, J. Chem. Phys. 95, 4593 (1991).
135. X. Xia and M. L. Berkowitz, Phys. Rev. Lett. 74, 3193 (1995).
136. K. J. Schweighofer, X. Xia, and M. L. Berkowitz, Langmuir 12, 3747 (1996).
137. C. Y. Lee, J. A. McCammon, and P. J. Rossky, J. Chem. Phys. 80, 4448 (1984).
138. E. Spohr, J. Chem. Phys. 106, 388 (1997).
139. E. Spohr and K. Heinzinger, Electrochim. Acta 33, 1211 (1988).
140. M. F. Toney, J. N. Howard, J. Richter, G. L. Borges, J. G. Gordon, O. R. Melroy, D. G. Wiesler, D. Yee, and L. B. Sorensen, Nature 368, 444 (1994).
141. A. Wallqvist, Chem. Phys. Lett. 165, 437 (1990).
142. K. A. Motakabbir and M. L. Berkowitz, Chem. Phys. Lett. 176, 61 (1991).
143. M. L. Berkowitz and L. Perera, in *Theoretical and Computational Approaches to Interface Phenomena*, edited by H. L. Seller and J. T. Golab (Plenum Press, New York, 1994).
144. K. Heinzinger, Fluid Phase Equilibria 104, 277 (1995).
145. K. Heinzinger, in *Structure of Electrified Interfaces, Frontiers of Electrochemistry*, edited by J. Lipkowski and P. N. Ross (VCH, New York, 1993), Chap. 7. Molecular Dynamics of Water at Interfaces, p. 239.
146. G. Aloisi, M. L. Foresti, R. Guidelli, and P. Barnes, J. Chem. Phys. 91, 5592 (1989).
147. G. Aloisi and R. Guidelli, J. Chem. Phys. 95, 3679 (1991).
148. J. Hautman, J. W. Halley, and Y.-J. Rhee, J. Chem. Phys. 91, 467 (1989).
149. S. B. Zhu and G. W. Robinson, J. Chem. Phys. 94, 1403 (1991).
150. G. Nagy and K. Heinzinger, J. Electroanal. Chem. 296, 549 (1990).
151. G. Nagy, K. Heinzinger, and E. Spohr, Faraday Discuss. 94, 307 (1992).
152. H. Yang and J. L. Whitten, Surf. Sci. 223, 131 (1989).
153. M. Rosi and C. W. Bauschlicher Jr, J. Chem. Phys. 90, 7264 (1989).
154. S. Trasatti, personal communication.
155. A. Kohlmeyer, Computersimulation von Wasserensembles unter Berücksichtigung der Polarisierbarkeit, Diplomarbeit, Universität Ulm, 1995.
156. A. G. Kalinichev and K. Heinzinger, Geochim. Cosmochim. Acta 59, 641 (1995).
157. K. Heinzinger and G. Pálinkás, in *Interactions of Water in Ionic and Nonionic Hydrates*, edited by H. Kleeberg (Springer, Heidelberg, 1987), p. 1.
158. E. Spohr, Chem. Phys. 141, 87 (1990).
159. M. Marchesi, Chem. Phys. Lett. 97, 224 (1983).
160. G. Barabino, C. Gavotti, and M. Marchesi, Chem. Phys. Lett 104, 478 (1984).
161. R. M. Townsend and S. A. Rice, J. Chem. Phys. 94, 2207 (1991).
162. P. J. Rossky and S. H. Lee, Chemica Scripta 29a, 1 (1989).
163. R. Sonnenschein and K. Heinzinger, Chem. Phys. Lett. 102, 550 (1983).

164. E. Spohr and K. Heinzinger, J. Chem. Phys. 84, 2304 (1986).
165. O. A. Karim and A. D. J. Haymet, Chem. Phys. Lett. 138, 531 (1987).
166. S. H. Lee and P. J. Rossky, J. Chem. Phys. 100, 3334 (1994).
167. E. Spohr, Computer Modeling of Aqueous/Metallic Interfaces, Habilitationsschrift, Universität Ulm, 1995.
168. M. Neumann, J. Chem. Phys. 85, 1567 (1986).
169. L. E. Firment and G. A. Somorjai, J. Chem. Phys. 63, 1037 (1975).
170. L. E. Firment and G. A. Somorjai, Surf. Sci. 84, 275 (1979).
171. P. A. Thiel, F. M. Hoffmann, and W. H. Weinberg, J. Chem. Phys. 75, 5556 (1981).
172. C. Noebl, C. Benndorf, and T. E. Madey, Surface Sci. 157, 29 (1985).
173. A. Spitzer and H. Lueth, Surf. Sci. 120, 376 (1982).
174. K. Bange, D. E. Grider, T. E. Madey, and J. K. Sass, Surf. Sci. 136, 38 (1984).
175. R. N. Barnett, U. Landman, and A. Nitzan, J. Chem. Phys. 93, 6535 (1990).
176. J. G. Gordon, O. R. Melroy, and M. F. Toney, Electrochim. Acta 40, 3 (1995).
177. K. Nicklas, P. Bopp, and J. Brickmann, J. Chem. Phys. 101, 3157 (1994).
178. D. Henderson, *Fundamentals of Inhomogeneous Fluids* (Marcel Dekker, New York, 1992).
179. S. L. Carnie, Ber. Bunsenges. Phys. Chem. 91, 262 (1987).
180. S. L. Carnie and G. M. Torrie, Adv. Chem. Phys. 56, 141 (1984).
181. G. M. Torrie, P. G. Kusalik, and G. N. Patey, J. Chem. Phys. 91, 6367 (1989).
182. L. Blum, Adv. Chem. Phys. 78, 171 (1990).
183. P. Attard, Adv. Chem. Phys. 92, 1 (1996).
184. G. M. Torrie and J. P. Valleau, Chem. Phys. Lett. 28, 578 (1974).
185. G. M. Torrie and J. P. Valleau, J. Comput. Phys. 23, 187 (1977).
186. G. Ciccotti, M. Ferrario, J. T. Hynes, and R. Kapral, Chem. Phys. 129, 241 (1989).
187. D. A. Rose and I. Benjamin, J. Chem. Phys. 95, 6956 (1991).
188. D. A. Rose and I. Benjamin, J. Chem. Phys. 98, 2283 (1993).
189. J. Seitz-Beywl, M. Poxleitner, and K. Heinzinger, Z. Naturforsch. 46a, 876 (1991).
190. E. Spohr, Chem. Phys. Lett. 207, 214 (1993).
191. L. Perera and M. L. Berkowitz, J. Phys. Chem. 97, 13803 (1993).
192. B. Eck and E. Spohr, Electrochim. Acta 42, 2779 (1997).
193. M. A. Wilson, A. Pohorille, and L. R. Pratt, Chem. Phys. 129, 209 (1989).
194. I. Benjamin, J. Chem. Phys. 95, 3698 (1991).
195. M. A. Wilson and A. Pohorille, J. Chem. Phys. 95, 6005 (1991).
196. B. Eck, Computersimulation von Grenzflächen zwischen wäßrigen Elektrolytlosungen und einer Metalloberfläche, Diplomarbeit, Universität Ulm, 1994.
197. D. A. Rose and I. Benjamin, J. Chem. Phys. 100, 3545 (1994).
198. I. Benjamin, J. Chem. Phys. 96, 577 (1992).
199. I. Benjamin, Chem. Phys. 180, 287 (1994).
200. K. J. Schweighofer and I. Benjamin, Chem. Phys. Lett. 202, 379 (1993).
201. J. L. Stickney, S. D. Rosasco, G. N. Salaita, and A. T. Hubbard, Langmuir 1, 66 (1985).
202. A. T. Hubbard, Chem. Rev. 99, 633 (1988).
203. R. Vogel, I. Kamphausen, and H. Baltruschat, Ber. Bunsenges. Phys. Chem. 96, 525 (1992).
204. B. M. Ocko, G. M. Watson, and J. Wang, J. Phys. Chem. 98, 897 (1994).
205. O. M. Magnussen, B. M. Ocko, R. R. Adzic, and J. X. Wang, Phys. Rev. B 51, 5510 (1995).
206. L. A. Curtiss, J. W. Halley, J. Hautman, N. C. Hung, Z. Nagy, Y.-J. Rhee, and R. M. Yonco, J. Electrochem. Soc. 138, 2032 (1991).
207. R. Marcus, J. Chem. Phys. 24, 966 (1965).
208. V. G. Levich, *Physical Chemistry: An Advanced Treatise*, edited by H. Eyring, D. Henderson, and W. Jost (Academic Press Inc, New York, 1970), No. IX b.

209. C. Miller and M. Grätzel, J. Phys. Chem. 95, 5225 (1991).
210. C. E. D. Chidsey, Science 251, 919 (1991).
211. A. Warshel, J. Phys. Chem. 86, 2218 (1982).
212. J. K. Hwang and A. Warshel, J. Am. Chem. Soc. 109, 715 (1987).
213. R. A. Kuharski, J. S. Bader, D. Chandler, M. Sprik, M. L. Klein, and R. W. Impey, J. Chem. Phys. 89, 3248 (1988).
214. Y. Hatano, M. Saito, T. Kakitani, and N. Mataga, J. Phys. Chem. 92, 1008 (1988).
215. E. A. Carter and J. T. Hynes, J. Phys. Chem. 93, 2184 (1989).
216. G. King and A. Warshel, J. Chem. Phys. 93, 8682 (1990).
217. I. Benjamin, J. Phys. Chem. 95, 6675 (1991).
218. J. B. Straus and G. A. Voth, J. Phys. Chem. 97, 7388 (1993).
219. B. B. Smith and J. W. Halley, J. Chem. Phys. 101, 10915 (1994).
220. D. A. Rose and I. Benjamin, Chem. Phys. Lett. 234, 209 (1995).
221. R. F. Grote and J. T. Hynes, J. Chem. Phys. 73, 2715 (1980).
222. H. A. Kramers, Physica 7, 284 (1940).
223. J. B. Straus, A. Calhoun, and G. A. Voth, J. Chem. Phys. 102, 529 (1995).
224. R. P. Feynman, *Statistical Mechanics* (Addison-Wesley, Reading, MA, 1982).
225. B. E. Conway, *Modern Aspects of Electrochemistry*, edited by B. E. Conway, R. E. White, and J. O. Bockris (Plenum Press, New York, 1985), Vol. 16.
226. U. Frese and W. Schmickler, Ber. Bunsenges. Phys. Chem. 92, 1413 (1988).
227. U. Frese and U. Stimming, J. Electroanal. Chem. 198, 409 (1986).
228. V. M. Tsionskii, L. I. Krishtalik, and L. B. Kriksunov, Electrochim. Acta 33, 623 (1988).
229. W. Schmickler, J. Electroanal. Chem. 284, 269 (1990).
230. O. Pecina, W. Schmickler, and E. Spohr, J. Electroanal. Chem. 394, 29 (1995).
231. O. Pecina, W. Schmickler, and E. Spohr, J. Electroanal. Chem. 405, 239 (1995).
232. W. Schmickler, J. Electroanal. Chem. 100, 533 (1979).
233. A. A. Kornyshev and W. Schmickler, J. Electroanal. Chem. 185, 253 (1985).
234. R. R. Nazmutdinov and E. Spohr, J. Phys. Chem. 98, 5956 (1994).
235. X. Xia and M. L. Berkowitz, Chem. Phys. Lett. 227, 561 (1994).
236. W. Schmickler, Chem. Phys. Lett. 237, 152 (1995).
237. P. W. Anderson, Phys. Rev. 124, 41 (1961).
238. M. D. Newns, Phys. Rev. 178, 1123 (1969).
239. R. Nazmutdinov, unpublished results.
240. D. Borgis and J. Hynes, J. Phys. Chem. 100, 1118 (1996).
241. A. Staib, D. Borgis, and J. Hynes, J. Chem. Phys. 102, 2487 (1995).
242. D. Borgis and J. Hynes, Chem. Phys. 170, 315 (1993).
243. J. Timoneda and J. Hynes, J. Phys. Chem. 95, 10431 (1991).
244. D. Borgis and J. Hynes, J. Chem. Phys. 94, 3619 (1991).
245. D. Borgis, S. Lee, and J. Hynes, Chem. Phys. Lett. 162, 19 (1989).
246. M. Tuckerman, D. Marx, M. Klein, and M. Parrínello, Science 275, 817 (1997).
247. O. Pecina, Ph.D. thesis, Universität Ulm, Ulm, 1996.
248. O. Pecina and W. Schmickler, J. Electroanal. Chem. 431, 47 (1997).
249. I. Benjamin and A. Pohorille, J. Chem. Phys. 98, 236 (1993).
250. A. Pohorille and M. A. Wilson, in *Reaction Dynamics in Clusters and Condensed Phases*, edited by J. Jorner, R. D. Levine, and B. Pullman (Kluwer, Dordrecht, 1994).
251. I. Benjamin, Chem. Rev. 96, 1449 (1996).
252. I. Benjamin, D. Evans, and A. Nitzan, J. Chem. Phys. 106, 6647 (1997).
253. A. Mosyak, P. Graf, I. Benjamin, and A. Nitzan, J. Phys. Chem. A 101, 429 (1997).
254. I. Benjamin, D. Evans, and A. Nitzan, J. Chem. Phys. 106, 1291 (1997).
255. G. Nagy and G. Denuault, J. Electroanal. Chem. 433, 37 (1997).
256. M. R. Philpott and J. N. Glosli, Chem. Phys. 198, 53 (1995).
257. J. Hautman and M. Klein, J. Chem. Phys. 91, 4994 (1989).

258. S. H. Lee and H. Kim, Bulletin of the Korean Chemical Society 17, 700 (1996).
259. E. Kramarenko, R. Winkler, P. Khalatur, and A. Kohkhlov, J. Chem. Phys. 104, 4806 (1996).
260. C. Wijmans and P. Linse, J. Chem. Phys. 106, 328 (1997).
261. T. Wallin and P. Linse, J. Phys. Chem. 100, 17873 (1996).
262. C. Wijmans and P. Linse, J. Phys. Chem. 100, 12583 (1996).
263. P. Linse, Macromolecules 29, 326 (1996).
264. A. Pohorille and I. Benjamin, J. Phys. Chem. 97, 2665 (1993).
265. D. A. Rose and I. Benjamin, J. Phys. Chem. 96, 9561 (1992).
266. I. Benjamin, Science 261, 1558 (1993).
267. K. J. Schweighofer and I. Benjamin, J. Electroanal. Chem. 391, 1 (1995).
268. T.-M. Chang and L. X. Dang, Chem. Phys. Lett. 263, 39 (1996).
269. I. Benjamin, Mod. Aspects of Electrochemistry 31, 115 (1997).
270. L. Perera, U. Essmann, and M. L. Berkowitz, Langmuir 12, 2625 (1996).
271. K. Raghavan, M. R. Reddy, and M. L. Berkowitz, Langmuir 8, 233 (1992).
272. S.-J. Marrink, M. Berkowitz, and H. J. C. Berendsen, Langmuir 9, 3122 (1993).
273. J. Böcker, J. Brickmann, and P. Bopp, J. Phys. Chem. 98, 712 (1994).
274. J. Böcker, M. Schlenkrich, P. Bopp, and J. Brickmann, J. Phys. Chem. 96, 9915 (1992).
275. J. Böcker, M. Schlenkrich, K. Nicklas, J. Brickmann, and P. Bopp, J. Chim. Phys. 88, 2535 (1991).
276. R. Dryfe, Z. Ding, R. Wellington, P. F. Brevet, A. M. Kuznetzov, and H. H. Girault, J. Phys. Chem. A 101, 2519 (1997).
277. A. Tamburelloluca, P. Hebert, P. Brevet, and H. Girault, J. Electroanal. Chem. 409, 123 (1996).
278. Z. Ding, R. Wellington, P. Brevet, and H. Girault, J. Phys. Chem. 100, 10658 (1996).
279. H. Girault, J. Electroanal. Chem. 388, 93 (1995).
280. H. Wang, E. Borguet, and K. Eisenthal, J. Phys. Chem. A 101, 713 (1997).
281. K. Eisenthal, J. Phys. Chem. 100, 12997 (1996).
282. K. Eisenthal, Chem. Rev. 96, 1343 (1996).
283. X. Shi, E. Borguet, A. Tarnovsky, and K. Eisenthal, Chem. Phys. 205, 167 (1996).
284. G. Nagy and G. Denuault, J. Electroanal. Chem. 450, 159 (1997).
285. D. L. Price and J. W. Halley, J. Chem. Phys. 102, 6603 (1995).

Time and Frequency Resolved Studies of Photoelectrochemical Kinetics

L. M. Peter[1] and D. Vanmaekelbergh[2]

[1] Department of Chemistry, University of Bath, Bath BA2 7AY UK
[2] Debye Institute, University of Utrecht, PO Box 80000; 3508 TA Utrecht NL

Contents

List of Symbols

A	electrode area
A_{direct}	constant for direct optical transitions

$A_{indirect}$	constant for indirect optical transitions
C	effective capacitance $= C_{sc}C_H/(C_{sc}+C_H)$
C_H	Helmholtz capacitance
C_{HF}	limiting high frequency capacitance in PEIS
C_{LF}	limiting low frequency capacitance in PEIS
C_{sc}	space charge capacitance
d_{sc}	width of the space charge region at the semiconductor electrolyte junction
D	diffusion coefficient
D_{eff}	effective diffusion coefficient in presence of trapping
E_a	activation energy for electron injection during photocurrent multiplication
E_C	energy of the conduction band edge
E_n	demarcation level for electrons
E_p	demarcation level for holes
E_V	energy of the valence band edge
E_F	Fermi energy
$E_{F,n}$	quasi Fermi energy for electrons
$E_{F,p}$	quasi Fermi energy for holes
$E_{F,redox}$	redox Fermi energy
E_g	bandgap energy
f	ac frequency in Hz
$f(E)$	Fermi occupancy factor
g	minority carrier flux predicted by the Gärtner equation (Eq. 20)
g_{ac}	modulated component of the Gärtner flux
$G(\omega)$	dimensionless system transfer function
h	Planck constant
i	$\sqrt{-1}$
$I(0)$	incident photon flux corrected for reflection loss
j_H	current density for electron transfer through the Helmholtz layer
j_{photo}	photocurrent density measured in external circuit
j_{sc}	current density for to minority carrier transport through the depletion layer
I_{ac}	modulated photon flux
I_{dc}	constant photon flux
J_n	electron flux
J_p	hole flux
$k(E)$	energy dependent electron transfer rate constant
k_{inj}	rate constant for electron injection during photocurrent multiplication
k_{tr}	phenomenological rate first order constant for electron transfer
k_{rec}	phenomenological first order rate constant for recombination
k_V	rate constant for electron transfer from the valence band
k_B	Boltzmann constant
L	diffusion length of minority carriers
n_0	electron density in the dark at equilibrium
$n(0)$	electron density at surface (cm^{-3})
n_{bulk}	bulk volume electron density
n_s	surface electron concentration (cm^{-2})
N_C	effective density of states in the conduction band
N_d	donor density
N_V	effective density of states in the valence band
Ox	oxidised species in solution
$p(0)$	hole concentration at the surface (cm^{-3})
p_{bulk}	bulk volume hole concentration
p_s	surface hole concentration (cm^{-2})
p_{free}	concentration of free holes
p_{trap}	concentration of trapped holes
q	elementary charge

Q_s	surface charge density (C cm^{-2})
R	resistance
R_{el}	electrolyte resistance
R_{HF}	limiting high frequency resistance in PEIS
R_{LF}	limiting low frequency resistance in PEIS
R_m	series measuring resistance
Red	reduced species in solution
S	sensitivity factor in light modulated microwave reflectivity
U	electrode potential
U_{FB}	flat band potential
v_n, v_p	drift velocity of electrons or holes
X	filled surface state
X^+	vacant surface state
α	absorption coefficient of semiconductor
β_p	hole capture rate constant
δ	mean tunnelling distance for electron transfer
$\Delta\phi_H$	potential drop across the Helmholtz layer
$\Delta\phi_{sc}$	potential drop across the space charge region
ΔI_0	modulated incident photon flux corrected for reflection loss
ΔV	modulated voltage
Δj_{photo}	modulated photocurrent density
ΔV_{photo}	modulated photovoltage
ΔR_M	modulated microwave reflectivity
ΔQ_s	modulated surface charge density
$\Delta\phi_{sc}$	potential difference across the space charge (depletion) layer
ε	relative permittivity
ε_0	permittivity of free space
ϕ	potential
Φ	photocurrent conversion efficiency j_{photo}/qI_0
γ	dimensionless term in IMPS solution for diffusion controlled electron transport
η	efficiency of electron injection in dye sensitised cells
λ	wavelength of light
μ_n, μ_p	mobility of electrons and holes
v	pre-exponential frequency factor
θ	phase shift in ac response
σ_n, σ_p	capture cross section for electrons and holes
τ_d	transit time for transport across a nanocrystalline film
τ_n, τ_p	minority electron and hole lifetimes
τ_{RC}	RC time constant
ω	radial frequency $2\pi f$
ω_{max}	radial frequency corresponding to maximum in complex plane plot
ω_{min}	radial frequency corresponding to minimum in complex plane plot

1 Introduction

1.1 Photoelectrochemistry

As its name suggests, photoelectrochemistry is a hybrid of electrochemistry and photochemistry, and it also incorporates aspects of solid state physics, photophysics and surface science [1–6]. The common feature of all photoelectrochemical systems is that they involve electron transfer induced by light absorption. An explosion of interest in photoelectrochemistry in the 1980s was stimulated by the exciting possibility that semiconductor/electrolyte junctions could be utilised for solar energy conversion [7–16]. The extraordinary level of research activity during this period is reflected in the impressive conversion efficiencies that were achieved for liquid junction photovoltaic cells. In spite of their initial promise, however, fundamental problems with stability [17], and ultimately with cost, have so far prevented practical large scale implementation of these devices. The hydrogen economy is an important goal for the next century, and photoelectrolysis systems aimed at generating hydrogen and other chemical fuels by light driven reactions at semiconductor electrodes were also developed during the 1980s [8, 18, 19]. Unfortunately, the solar conversion efficiencies of photoelectrolysis cells have remained stubbornly low. Failure to develop low cost electrolyte-based systems to challenge established photovoltaic technologies led inevitably to a decline in interest (and funding) in photoelectrochemistry at the end of the last decade. However, the development of the dye sensitised TiO_2 cell by Grätzel and co-workers [20–26] in the early 1990s triggered a resurgence of interest in the practical potential of photoelectrochemical solar energy conversion. At the same time, the fundamental properties of nanocrystalline semiconductor electrodes have become the subject of widespread research activity.

In fact, the scope of photoelectrochemistry is more general than the preceding brief historical sketch might suggest. Light driven electron transfer lies at the heart of many important processes. Photosynthesis and colour photography both involve light stimulated charge carrier separation, and mundane but practically important processes such as the photodegradation of paint layers containing TiO_2 are also essentially photoelectrochemical in nature. Photoelectrochemical reactions may in principle also involve light absorption by solution species rather than by the electrode. These photogalvanic reactions are also part of photoelectrochemistry, although they are rarely considered alongside semiconductor photoelectrochemistry. This chapter restricts its focus to photoelectrochemical systems in which light absorption takes place either in a semiconducting solid or in a monolayer of dye adsorbed on the surface of a wide bandgap semiconductor like TiO_2. However, the conceptual and experimental approaches outlined here may well be applicable to other systems.

1.2 A Generalised Approach to Dynamic Processes

Conventional kinetics is largely concerned with the description of dynamic processes in the time domain, and in consequence few conceptual problems are encountered in understanding time resolved experiments. By contrast, frequency resolved measurements often pose more of a challenge to understanding, in spite of the obvious correspondence between the time and frequency domains. This conceptual difficulty may explain why the only frequency resolved method to achieve universal acceptance in electrochemistry is electrochemical impedance spectroscopy (EIS) [27–29], which analyses the response of electrochemical systems to periodic (sinusoidal) perturbations of voltage or current. It is clear that EIS is a very powerful method, and there are relatively few cases where small amplitude transient experiments performed on the same systems and analysed in the time domain are preferable. Since in EIS both the perturbation and the response are electrical, the analysis of experimental data is normally based on linear equivalent circuit elements (resistors and capacitors). Unfortunately, the equivalent circuit approach can easily obscure the physical and chemical origin of the electrochemical response, and it can be argued that it introduces an unnecessary conceptual barrier that hinders more widespread application of frequency resolved techniques in modern electrochemistry.

Recent progress in photoelectrochemistry has shown clearly that frequency resolved methods offer unmatched diagnostic power and resolution, particularly for processes taking place on a time-scale of microseconds to seconds. EIS can be used to study semiconductor electrodes not only in the dark, but also under conditions of steady illumination, in which case it has been termed photoelectrochemical impedance spectroscopy (PEIS) [30–35]. Another widely applied method is intensity modulated photocurrent spectroscopy (IMPS) [36–90] which involves modulation of the intensity of illumination incident on a photoelectrode and measurement of the magnitude and phase shift of the resulting photocurrent relative to the periodic illumination. IMPS has been used to study a wide range of single crystal semiconductors as well as anodic films on metals [54–56, 69, 70, 71] and porous or nanocrystalline systems [78–80, 87, 90]. Methods closely related to IMPS include intensity modulated photovoltage spectroscopy [91–93] and light modulated microwave reflectance [73, 94–96].

Intensity modulated methods were originally developed using conventional lamps [36], but these were replaced by cw laser sources coupled to acousto-optic modulators [38]. More recently, the availability of efficient light emitting diodes with emission wavelengths ranging from the blue to the near IR has greatly simplified the optical arrangement; laser sources are now only necessary for measurements at high intensities or short wavelengths. IMPS measurements have also been carried out using stepped or scanning laser microscope systems in order to relate information about photochemical kinetics to surface topography and composition [67, 72, 76].

Many of the techniques used to study electrochemical kinetics involve perturbation and measurement of electrical variables such as voltage and current. However, an electrochemical system in some initial steady state condition can also be perturbed by a suitable periodic non-electrical stimulus, and the monitored response may also be non-electrical: examples are periodic modulation of mass (electrochemical quartz crystal microbalance [97] and of optical transmission (electrochromic systems) [98–100]. In general, the relationship between input and response is described by the transfer function, G, which contains information about the system under study [101], and analysis can be performed either in the time or frequency domain. In the latter case, the frequency dependent transfer function $G(\omega)$ is defined as

$$G(\omega) = \frac{output(\omega)}{input(\omega)} \tag{1}$$

where the input and output are characterised by their magnitude, phase and appropriate physical dimensions. Examples of some of the types of perturbation and response that are relevant to electrochemical systems are listed in Table 1.

Table 1

Perturbation	Response	Type of transfer function (units)
voltage ΔV	current ΔI	admittance $\Delta I/\Delta V$ ($AV^{-1} = \Omega^{-1}$)
current ΔI	voltage ΔV	impedance $\Delta V/\Delta I$ ($VA^{-1} = \Omega$)
voltage ΔV	charge ΔQ	integrated admittance $\Delta Q/\Delta V$ ($CV^{-1} = F$)
voltage ΔV	absorbance ΔA	optical transfer function $\Delta A/\Delta V$ (V^{-1})
voltage ΔV	mass change (Δm)	gravimetric transfer function $\Delta m/\Delta V$ (kg V^{-1})
light flux ΔI_o	photocurrent Δj_{photo}	quantum efficiency $\Phi = \Delta j_{photo}/q\Delta I_o$ (dimensionless)
light flux ΔI_o	photovoltage ΔV_{photo}	photovoltage impedance $\Delta V_{photo}/q\Delta I_o$ ($VA^{-1} = \Omega$)

It is clear that different transfer functions are often related to each other. For example, changes in optical absorbance of an electrochromic system due to a Faradaic reaction are proportional to the charge passed. Inspection of the definitions in Table 1 confirms that the optical transfer function is therefore also proportional to the integrated Faradaic admittance [99]. The same relationship also holds for the gravimetric transfer function [97, 101]. The advantage of measuring absorbance or mass changes is that they depend only on the Faradaic charge and not on the charge associated with double layer charging. This opens the way to deconvolution of the overall response of electrochemical systems in a way that cannot be achieved by EIS alone.

It can be seen from this discussion that a more broadly based approach to the characterisation of the dynamic properties of electrochemical systems is desirable. One of the objectives of this review is to provide a cohesive and unifying approach to the application of time and frequency resolved methods in photoelectrochemistry. The input functions are generally either light or voltage, and the output functions may be current, charge, voltage, optical absorbance/reflectance or microwave reflectivity.

The terminology used for the transfer function requires some care. If the input function is an intensive quantity such as voltage (also called an 'across' function), and the output function is an extensive quantity such as current (also called a 'through' function), the ratio of output to input can be referred to as an admittance. If the type of input and output functions are reversed, then the ratio becomes an impedance. If the input and output functions are both of the same type, the ratio of output to input can be referred to as a gain function. However, these conventions, which are usually employed in network analysis, have not always been followed consistently in the development of new photoelectrochemical techniques. For example, the frequency dependent photocurrent efficiency, Φ, shown in Table 1 is often referred to as the opto-electrical admittance and its inverse as the opto-electrical impedance [62, 64–66], in spite of the fact that it is the ratio of two through functions. It would be preferable to use the term opto-electrical transfer function. The inverse of Φ has also been called the photoelectrochemical impedance [53, 70]. To avoid confusion, the use

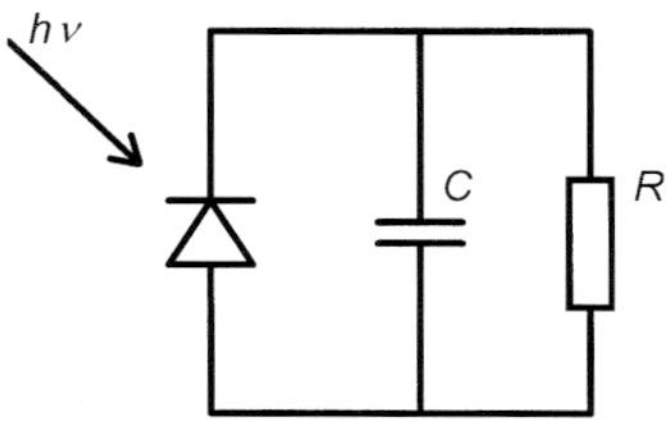

Fig. 1. Simple photodiode circuit. R is the current measuring resistor and C is the junction capacitance of the diode. The photodiode signal is attenuated at frequencies above $f = 1/2\,\pi RC$.

of the term 'photoelectrochemical impedance' is restricted in this review to the description of the electrochemical impedance of an electrode measured under conditions of steady rather than modulated illumination, in other words the periodically perturbed variable is the voltage rather than the photon flux.

A simple example of how the frequency response of an opto-electronic system is defined by its transfer function is provided by the equivalent circuit for a conventional photodiode shown in Fig. 1.

The photodiode generates a current in response to illumination, and the current is measured as a voltage difference across an external measuring resistor. The space charge capacitance of the diode appears in parallel with the measuring resistor, leading to attenuation of the response at high frequencies. The transfer function of the photodiode is determined by the product of the transfer function $\Phi(\omega)$ for the process of conversion of light to current in the semiconductor junction and the dimensionless complex transfer function $A(\omega) = 1/(1 + i\omega RC)$ due to RC attenuation:

$$G(\omega) = \Phi(\omega)\frac{1}{1 + i\omega RC} \tag{2}$$

Here $i = \sqrt{-1}$. $\Phi(\omega)$ corresponds to the quantum efficiency of the photodiode, and except at very high frequencies where electron transport effects become important, it is effectively constant so that the frequency response is determined only by $A(\omega)$. The product RC is referred to as the RC time constant, τ_{RC}. The complex attenuation factor $A(\omega) = Real[A(\omega)] + i\ Imaginary[A(\omega)]$ can be plotted in the complex plane as shown in Fig. 2a, noting that it has real and imaginary components given by

$$Re[A(\omega)] = \frac{1}{1 + \omega^2 R^2 C} \tag{3a}$$

$$Im[A(\omega)] = \frac{-\omega RC}{1 + \omega^2 R^2 C^2} \tag{3b}$$

It can be seen the plot of $A(\omega)$ describes a semicircle in the lower complex plane with the lowest point at $\omega = 1/RC = 1/\tau_{RC}$. The corresponding Bode plots of the magnitude and phase of $A(\omega)$ are shown in Fig. 2b.

Until recently, most experimental photoelectrochemical studies have focused on single crystal semiconductors, but an upsurge of interest in nanocrystalline semiconductors has been stimulated by the development by Grätzel and co-workers of

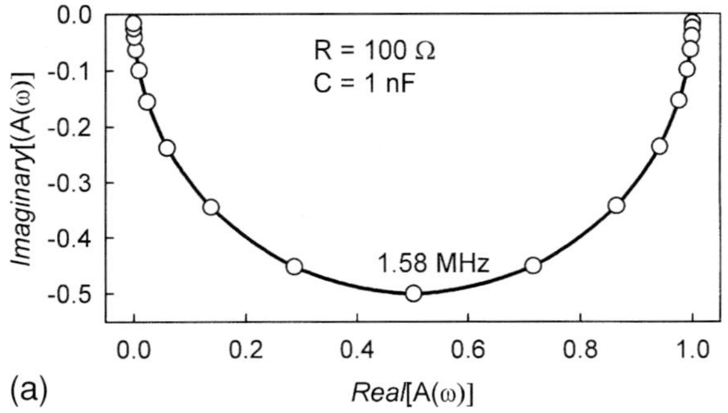

Fig. 2a. Complex plane plot of the attenuation factor $A(\omega)$ calculated for the circuit in Fig. 1 with $R = 100\ \Omega$ and $C = 1$ nF. Note that the minimum of the semicircle is located at a frequency $f_{min} = 1/2\ \pi RC = 1.58$ MHz.

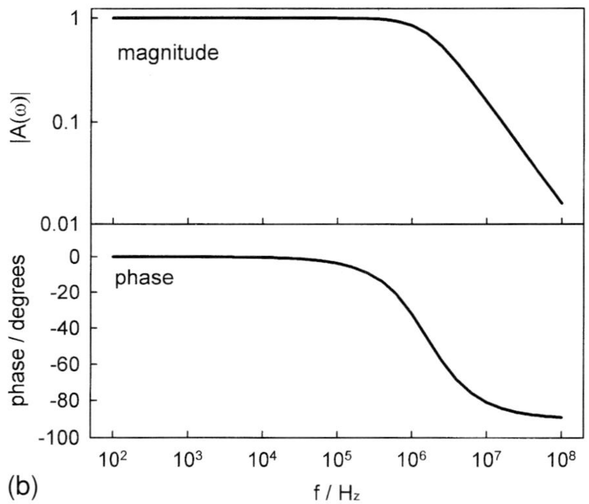

Fig. 2b. Bode plot of magnitude $|A(\omega)|$ and phase shift of the attenuation factor $A(\omega)$. The plot is an alternative way of representing the data shown in Fig. 1. Note that the magnitude of $A(\omega)$ is 0.5 and the phase shift is 45° when $f = 1/2\pi RC$. This corresponds to the minimum of the semicircle in Fig. 2a.

efficient solar cells based on dye sensitised titanium dioxide [20–26]. The porous electrodes in these cells consist of interpenetrating solid and liquid phases, and they exhibit a number of unique properties, some of which are not well understood. Time and frequency resolved methods are being used to study charge generation and collection in these interesting systems, and this review therefore encompasses not only studies on well-characterised (generally single crystal) semiconductors but also more recent work on nanocrystalline electrodes. The main emphasis is placed on small amplitude periodic perturbation with analysis of the response in the frequency domain, but time resolved measurements are also included in order to establish a clear conceptual bridge between the two domains.

1.3 Photoelectrochemical Systems: Processes and Time-scales

Processes in photoelectrochemical systems take place over an enormous range of time and distance scales. Interfacial electron transfer over tunnelling distances of atomic dimensions is coupled to the transport of electrons, holes and ions over macroscopic distances in the solid and liquid phases. The shortest time scales are associated with the process of dye sensitisation of TiO_2 by adsorbed ruthenium dyes (the basis of the Grätzel cell), where spectroscopic studies [102, 103] have established that the excited state of the dye injects an electron into the conduction band of the oxide on a femtosecond timescale. Electron and hole transport times can vary considerably from system to system. Section 2 of this chapter discusses field driven collection of photo-excited minority carriers in the space charge region of semiconductor junctions: this process is also very fast – generally on the picosecond time-scale. At the other end of the timescale, electron diffusion in nanocrystalline semiconductor electrodes is many orders of magnitude slower; as discussed in section 5, the collection of electrons in a dye sensitised TiO_2 cell can take several tens of milliseconds. A similarly wide range of time-scales is also encountered when considering trapping and recombination processes. Trapping of electrons and holes can be fast if the trap density is high, taking place in nanoseconds. By contrast, bulk electron-hole recombination in very pure silicon occurs on a millisecond time scale, whereas in most II–VI and III–V compound semiconductors it is much faster. Electron-hole recombination at semiconductor surfaces can also be slow, particularly when the density of majority carriers is reduced by the presence of a space charge region.

Heterogeneous electron transfer reactions have been studied extensively [104–106], mostly at metal electrodes. The electron density in the metal can be considered to be invariant, and theoretical considerations suggest that the upper limit for the first order heterogeneous rate constant is of the order of 10^4 cm s^{-1}. The electronic density of states in semi-metallic highly oriented pyrolytic graphite (HOPG) is considerably lower than that of metals, and it has been observed that the rate constants for electron transfer at HOPG are correspondingly lower by 3–5 orders of magnitude [107]. In the case of metal electrodes, experimental limitations such as mass transfer and the RC time constant of the electrochemical cell usually restrict the experimentally accessible upper limit of rate constants to 10 cm s^{-1} or less. The situation is different in the case of majority carrier electron transfer at semiconductor electrodes because the rates of electron transfer depend not only on the concentration of electron donor (or acceptor) species in the solution but also on the density of carriers in the semiconductor. The simplest case arises where measurements are performed in the dark and the reaction involves majority carriers from the semiconductor. Under weak depletion, the density of majority carriers at the semiconductor surface is much lower than in the case of a metal, and in the absence of surface charging effects it varies exponentially with potential according to the Boltzmann equation. This means that high values of the electron transfer rate constant can be determined directly from the dark current-voltage characteristics by appropriate selection of the potential range. For example if we consider an n-type semiconductor where the bulk density of electrons is 10^{16} cm^{-3}, the surface density of electrons is reduced to 10^{14} cm^{-3} at a potential 120 mV positive of the flatband potential. Consequently, the electron transfer rate can be determined readily since it is six to seven orders of magnitude lower than at a metal electrode where the electron density approaches 10^{21} cm^{-3}. The experimental access to high values of the electron transfer rate constant provided by semiconductor electrodes has led to recent controversy about the upper limit to the value of rate constants. Lewis and co-workers [108–111] have argued that the upper limit to the second order heterogeneous rate constants of outer sphere redox reactions taking place at semiconductor electrodes lies around 10^{-17} cm^4 s^{-1}, and their measurements on well defined silicon surfaces and on InP appear to confirm this conclusion. However, values of rate constants as high as 10^{-10} cm^4 s^{-1} have been derived by Memming, Nozik and co-workers [112–114] from measurements on GaAs and GaInP. To place this value in context, it can be translated into an equivalent rate constant for a metal electrode, where for an electron density of 10^{21} cm^{-3}, the corresponding rate constant would be 10^{11} cm s^{-1}. This is more than seven orders of magnitude higher than the generally accepted upper limit of 10^4 cm s^{-1}.

In the case of photoelectrochemical reactions, it is the minority carriers that take part in the interfacial electron transfer. Carriers are generated by absorption of light in the bulk of the semiconductor and collected at the interface by diffusion and migration. Unless the illumination level is very high, the rate of photoelectrochemical reaction is constrained by the supply of carriers to the surface. This means that the density of minority carriers at the surface is not known a priori because it depends on the rate of arrival from the bulk and on the rate of removal by electron transfer and surface recombination. An idea of the shortest time scale associated with minority carrier reactions can be obtained by considering a fast redox reaction with a rate constant at the theoretical upper limit for outer sphere reactions of 10^{-17} cm^4 s^{-1}. A typical value of the concentration of redox species in solution is 10^{19} cm^{-3} (corresponding to 1.6×10^{-2} mol dm^{-3}). The pseudo-first order rate constant is then given by $10^{-17}\,\text{cm}^4\,\text{s}^{-1} \times 10^{19}\,\text{cm}^{-3}/\delta\,\text{cm} = 10^2/\delta\ \text{s}^{-1}$ where δ is a reaction length of the order of the tunnelling distance. For $\delta = 10^{-7}$ cm, the upper limit to the pseudo-first order rate constant for the reaction of minority carriers is 10^9 s^{-1}, corresponding to a nanosecond time-scale. As shown in section 3, the time required for a carrier to reach the surface is even shorter. Such rapid processes lie outside the range of conventional frequency response analysis, which for electrochemical systems is generally restricted by experimental problems to frequencies below 1 MHz. However, it has been found that many multistep photoelectrochemical reactions of interest, such as hydrogen evolution on InP and Si [35, 73], oxygen evolution on TiO_2 [84], CdS photodecomposition [59] or photocurrent multiplication during photoanodic dissolution of Si [46, 49], occur on a time-scale of microseconds to milliseconds, so that they can easily be studied by frequency response analysis.

Surface recombination of electrons and holes is also relatively slow. Usually, the recombination lifetime depends inversely on the majority carrier concentration at the surface. For a n-type semiconductor, the lifetime of a hole trapped at a surface state is given by $\tau_p = 1/\beta n_{surf}$, where β is the rate constant for electron capture For example if $n_{bulk} = 10^{17}$ cm^{-3} and the band-bending $q\Delta\phi_{SC} = 10k_BT$, $n_{surf} = 5.10^{12}$ cm^{-3}. Taking a typical value of $\beta = 10^{-10}$ cm^{-3} s^{-1}, the recombination lifetime is obtained as $\tau_p = 2.10^{-3}$ s. Under flat-band conditions where $n_{surf} = n_{bulk}$, τ_p is reduced to 100 ns. Surface recombination rates can therefore also be measured under depletion conditions using frequency resolved methods.

The accessibility of different time domains under well-defined conditions is subject to instrumental limitations. Intensity modulated photocurrent measurements are commonly performed under potentiostatic conditions using a fast potentiostat, and roll-off and phase shift normally restrict the upper frequency limit for reliable data to 30 kHz or less. Data acquired at higher frequencies must be corrected for the combined transfer function of the potentiostat and electrochemical cell. An alternative approach is to eliminate the potentiostat and to use a simple circuit with a small measuring resistor. Since the measuring resistor is in series with the electrolyte and contact resistances, it increases the *RC* time constant of the cell (the influence of the *RC* time constant of the cell on the measured response is analogous to the attenuation of the photodiode response illustrated in section 1.2, and it is discussed in more detail in section 3.2). The *RC* time constant for a moderately doped semiconductor electrode under depletion conditions is two to three orders of magnitude lower than the value for a metal electrode. The difference arises from the fact that the space charge capacitance of the semiconductor electrode is typically 100 nF cm^{-2} or less, whereas the Helmholtz capacitance is around 10 μF cm^{-2}. For a series resistance of 10 Ω cm^2, the corresponding *RC* time constants are 10^{-6} s and 10^{-4} s respectively. As shown in section 3.2, at times shorter than the *RC* time constant, the photogenerated charge is stored in the electrode capacitance, so that for example an experiment involving a very short (ns) light pulse corresponds essentially to coulostatic conditions, and the measured quantity is a photopotential rather than a photocurrent.

2 Review of Fundamental Processes

2.1 Electron-hole Pair Generation

An electron hole pair is created in a semiconductor when a photon of sufficient energy is absorbed, resulting in excitation of an electron from the valence band to the conduction band [115]. In the context of semiconductor photoelectrochemistry, it is useful to distinguish between direct and indirect optical transitions. If the top of the valence band and the bottom of the conduction band are both situated at $\boldsymbol{k} = 0$ ($\boldsymbol{k}$ being the electron wave vector), one-step optical processes between delocalised states in the valence and conduction band can occur. The absorption coefficient for direct absorption of photons of energy $h\nu$, in a semiconductor with bandgap E_g is given by

$$\alpha_{direct}(h\nu) = A_{direct} \times \sqrt{h\nu - E_g} \tag{4}$$

The proportionality constant A_{direct} is determined by the effective masses of the electron and hole and by the index of refraction of the semiconductor. Typically, the absorption depth $1/\alpha_{direct}(h\nu)$ of photons absorbed in a direct transition is 100–1000 nm.

If the bottom of the conduction band is not located at $\boldsymbol{k} = 0$, an optical transition from the top of the valence band to the bottom of the conduction band requires absorption or emission of one (or more) quantised energies of lattice vibrations (phonons) to conserve the momentum of the electron (also denoted as crystal momentum). The absorption probability depends on the phonon density and hence on the temperature. At a given temperature (>0 K), the absorption coefficient can be approximated by

$$\alpha_{indirect}(hv) = A_{indirect}(hv - E_g)^2 \tag{5}$$

Indirect optical transitions are generally characterised by an absorption depth $1/\alpha_{indirect}(hv)$ that is orders of magnitude larger than for direct transitions. For instance, the absorption depth of green light (wavelength 515 nm) in GaP is about 10^4 nm at room temperature because its energy falls below the direct bandgap but above the indirect bandgap of the material [116, 117].

In direct and indirect optical transitions, a delocalised electron in the valence band is excited into a delocalised state in the conduction band. Absorption of a photon can also result in an electron transition from the valence band to a localised state in the band gap or from a localised state in the band gap to a delocalised state in the conduction band. Such transitions have been studied less extensively than band to band transitions. (Transient) light absorption, photoconductivity and photovoltage studies have been used to probe transitions with photons of energy considerably below the band gap [118–121]. In a few cases, photocurrents induced by absorption of sub-bandgap light in bulk single crystalline semiconductor electrodes have been studied [122–125]. The mechanisms of free carrier and photocurrent generation in semiconductors induced by absorption of a sub-band gap photon have been discussed recently [126]. The work was motivated by the growing interest in nanostructured nanoporous semiconductors which exhibit a huge volume density of surface atoms and in consequence a strongly enhanced probability for light absorption mediated by localised band gap states [127, 128].

Fig. 3 summarises the possible sub-band gap transitions between localised and delocalised states.

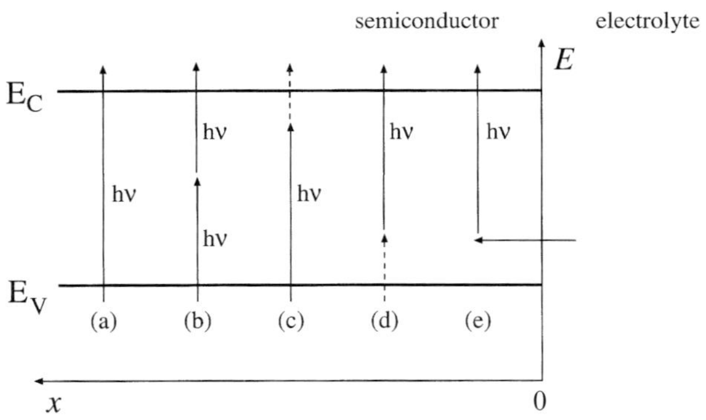

Fig. 3. Mechanisms that generate free electrons and holes (and photocurrent) by absorption of sub-bandgap photons. Continuous vertical arrows represent optical transitions and broken vertical arrows thermal transitions. Interfacial electron transfer is indicated by the horizontal arrow.

It is important to recognise that a sub-band gap optical transition leads to a delocalised carrier of one type and a localised carrier of opposite type. Steady-state photocurrent flow requires that the localised carrier is excited subsequently to the valence or conduction band, either by absorption of a second photon (process (b) in Fig. 3) or by thermal excitation (processes (c, d)). Bandgap states localised at the semiconductor surface may be of special importance for sub-band gap photocurrent flow. In process (e), an electron (majority carrier) is optically excited into the conduction band, and the resulting empty surface state is refilled by an interfacial electron transfer process. The latter process is similar to the process of dye sensitised electron injection in the nanocrystalline TiO_2 solar cell [20–26, 129].

2.2 Electron-hole Pair Separation in a Semiconductor Electrode

2.2.1 The Efficiency of Electron-hole Pair Separation

The simplest photoelectrochemical cells consist of a semiconductor working electrode and a metal counter electrode, both of which are in contact with a redox electrolyte. In the dark, the potential difference between the two electrodes is zero. The open circuit potential difference between the two electrodes that arises from illumination of the semiconductor electrode is referred to as the photovoltage. When the semiconductor and counter electrode are short circuited, a light induced photocurrent can be measured in the external circuit. These phenomena originate from the effective separation of photogenerated electron-hole pairs in the semiconductor. In conventional photoelectrochemical studies, the interface between the flat surface of a bulk single crystalline semiconductor and the electrolyte is two dimensional, and the electrode is illuminated from the electrolyte side. However, in the last decade, research into the properties of nanoporous semiconductor electrodes interpenetrated by an electrolyte solution has expanded substantially. If a nanocrystalline electrode is prepared as a film on a transparent conducting substrate, it can be illuminated from either side. The obvious differences between a flat (two dimensional) semiconductor/electrolyte junction and the (three dimensional) interface in a nanoporous electrode justify a separate treatment of the two cases.

Fig. 4 illustrates the energetics (in the dark) of a single crystalline bulk n-type electrode in equilibrium with a redox system.

The equilibrium Nernst-potential of the redox electrolyte determines the redox Fermi level $E_{F,redox}$. In the case illustrated here, the redox Fermi level lies below the Fermi level in the semiconductor before contact is made. On contact, electronic equilibrium is established by the transfer of electrons from the semiconductor to the oxidising agent in solution until the Fermi energies of the two phases are equal. As a result of the removal (depletion) of electrons, a space charge region consisting of immobile positively charged donor ions is formed in the semiconductor. The positive charge of the ionised donors in the depleted region of the semiconductor is compensated by an excess negative charge in the Helmholtz-layer on the electrolyte side. From Poisson's equation, and applying Fermi-statistics for the free electrons, it is found that the width, of the depletion region, d_{sc}, and the potential drop between the

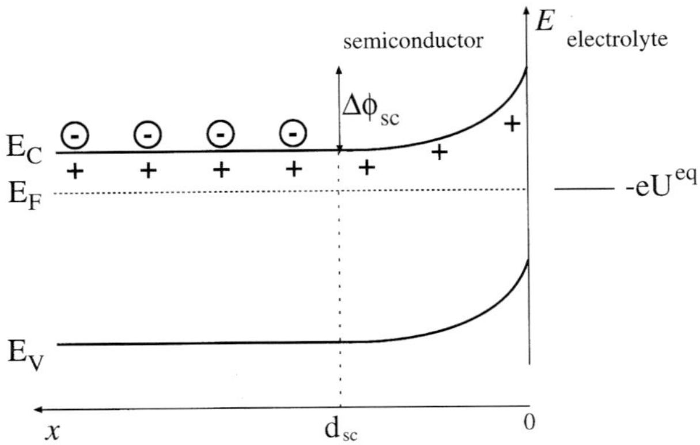

Fig. 4. Energy diagram for a bulk monocrystalline n-type semiconductor electrode in the dark, in equilibrium with a redox system that has an equilibrium potential U^{eq}. The Fermi-level (E_F) and the energy of the band edges are shown as a function of distance, x, perpendicular to the surface. The electrode is depleted of majority carriers at its surface. d_{SC} is the width of the depletion layer and $\Delta\phi_{SC}$ is the potential drop over the depletion layer.

surface and the bulk of the semiconductor, $\Delta\phi_{sc}$, are related by

$$\Delta\phi_{sc} = (qn_{bulk}/2\varepsilon_0\varepsilon)\, d_{sc}^2 \tag{6}$$

Here n_{bulk} is the free electron concentration in the bulk of the semiconductor and ε is the relative permittivity of the solid. For $\varepsilon = 10$ and $\Delta\phi_{sc} = 1$ V, the width of the depletion layer is 10, 100, and 1000 nm for free carrier concentrations of 10^{19}, 10^{17}, and $10^{15}\,\mathrm{cm}^{-3}$, respectively. The capacitance C_{SC} of the depletion layer can be calculated from

$$C_{sc} = \varepsilon\varepsilon_o/d_{sc} \tag{7}$$

For densities of free carriers between 10^{15} and $10^{19}\,\mathrm{cm}^{-3}$, C_{SC} ranges from 10^{-9} to $10^{-7}\,\mathrm{F\,cm}^{-2}$.

In the absence of electronic equilibrium established by a redox electrolyte, the potential difference across the semiconductor/electrolyte junction can be controlled in a three electrode cell with a reference electrode. The variation of the depletion layer capacitance with potential, U, is described by the Mott-Schottky equation [1]:

$$C_{SC}^{-2} = \frac{2}{\varepsilon\varepsilon_0 q N_d}\left(U - U_{FB} - \frac{k_B T}{q}\right) \tag{8}$$

Plots of C_{SC}^{-2} vs. potential are used to determine the flatband potential, U_{FB}, and the doping density N_d.

The situation sketched in Fig. 4 pertains to equilibrium conditions in the dark, and consequently the spatial derivative of the electron Fermi energy, dE_F/dx, is zero. If the n-type semiconductor electrode is illuminated from the electrolyte side with

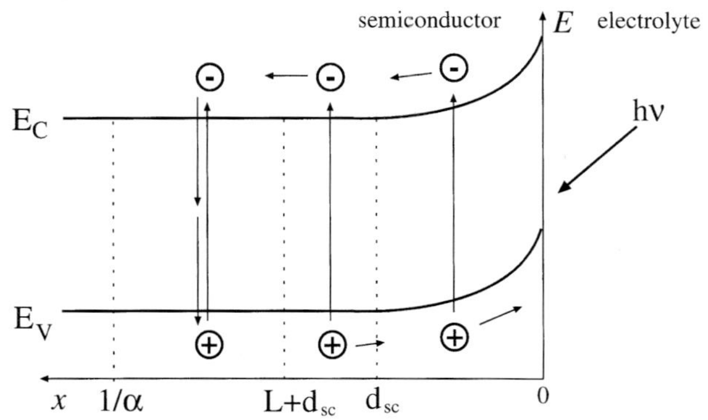

Fig. 5. An *n*-type semiconductor electrode under depletion conditions, illuminated from the electrolyte side. Holes generated in the depletion layer and in the diffusion layer (width L) reach the surface and can contribute to the photocurrent. Holes generated deeper than $d_{SC} + L$ are lost by bulk recombination.

monochromatic light ($h_\nu > E_g$) with an intensity I ($x = 0$) (corrected for reflection losses), electron-hole pairs are generated in the semiconductor at a rate $\alpha I(x)$. The holes are minority carriers, and so they have a finite lifetime in the bulk of the semiconductor that is determined by recombination with electrons. Holes generated in the depletion layer migrate towards the surface and escape bulk recombination (see Fig. 5). If the holes react sufficiently rapidly with the reducing agent in solution, the electrostatics of the semiconductor/electrolyte interface under short circuit conditions can still approximated by Fig. 4. In general, light absorption increases the hole concentration substantially above the dark value, tending to drive the electrochemical cell away from equilibrium. The departure from equilibrium can be described by introducing the concept of quasi-Fermi levels for electrons and holes, denoted as $E_{F,n}$ and $E_{F,p}$ respectively (see section 2.3.3).

If the density of photogenerated electrons is negligible compared with the electron density in the dark (i.e. $E_{F,n}(x) = E_F$), the recombination rate is first order in the hole concentration and characterised by a (minority carrier) life time for holes denoted as τ_p. This is usually the case in the bulk semiconductor unless the doping is low or the illumination intensity is very high. Holes generated adjacent to the depletion layer may reach the edge of the depletion layer by diffusion. This is accounted for by assuming a diffusion length L, in which electron-hole pair separation is fully effective:

$$L = \sqrt{D\tau_p} = \sqrt{k_B T \mu_p \tau_p} \tag{9}$$

Here D is the diffusion coefficient and μ_p is the corresponding mobility of the hole. Electron-hole pairs generated in the neutral bulk region deeper than the diffusion layer (i.e. at $x > d_{SC} + L$) will recombine.

It is clear that the photocurrent quantum yield of a flat semiconductor electrode will depend on the relative magnitudes of the width of the retrieval layer, $d_{SC} + L$, and the penetration depth of the light, $(1/\alpha)$. An expression for the flux of photogenerated minority carriers arriving at the surface, g, was derived originally by

Gärtner [130] for the boundary condition that the concentration of minority carriers tends to zero in the depletion layer. The Gärtner equation can be written in the normalised form

$$\frac{g}{I(0)} = 1 - \frac{e^{-\alpha d_{sc}}}{1 + \alpha L} = \Phi \tag{10}$$

The ratio $g/I(0)$ defines the photocurrent efficiency Φ. In the absence of surface recombination, qg corresponds to the photocurrent density j_{photo} measured in the external circuit. The Gärtner equation has been used successfully to explain the photocurrent-potential characteristics of many semiconductor electrodes under conditions where surface recombination is absent. Plots of $ln(1 - \Phi)$ against d_{SC} (which according to the Mott-Schottky relationship is proportional to $(U - U_{FB})^{1/2}$) have been used to determine the absorption coefficient of the light and the diffusion length L of minority carriers. The method is most accurate for cases where $1/\alpha \gg d_{sc} + L$ [44, 125, 131–134].

The preceding discussion is based on a simple one-dimensional model of the semiconductor/electrolyte junction. This model is not appropriate for porous network electrodes. Remarkably high values of the photocurrent quantum yield are found with nanostructured porous semiconductor electrodes when an efficient scavenger for the minority carrier is present in the interpenetrating electrolyte solution. Such electrodes have been prepared from single crystals by (photo) anodic etching leading to a single crystalline semiconductor network with interconnected structural units [135–141]. Nanocrystalline porous electrodes can also be fabricated by coating a layer of semiconductor colloid (size 1–100 nm) on a conducting substrate and slight sintering [20–26] or by electrodeposition [142].

A comparison between electron-hole pair separation in bulk single crystalline and nanostructured electrodes can be made using n-GaP as an example, since both single crystalline and nanostructured GaP electrodes have been studied. n-type GaP can be made porous by anodic etching under conditions of dielectric breakdown. The porous network formed (Fig. 6) has interconnected structural units in the 100 nm range.

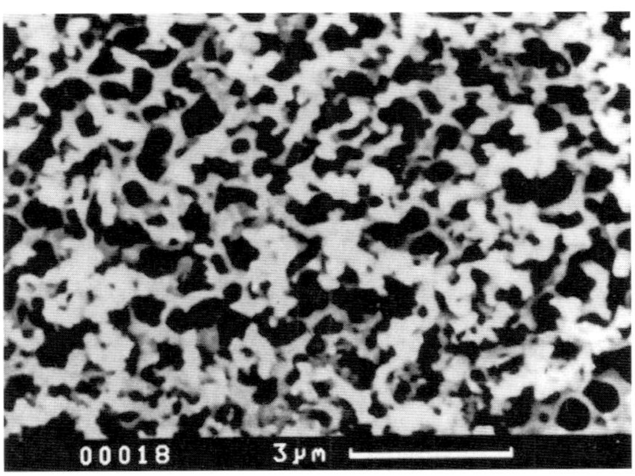

Fig. 6. SEM photograph of a nanoporous GaP network obtained by anodic etching of single crystalline n-GaP. The white areas are GaP and the dark areas pores. Note that the GaP retains its monocrystalline character after etching.

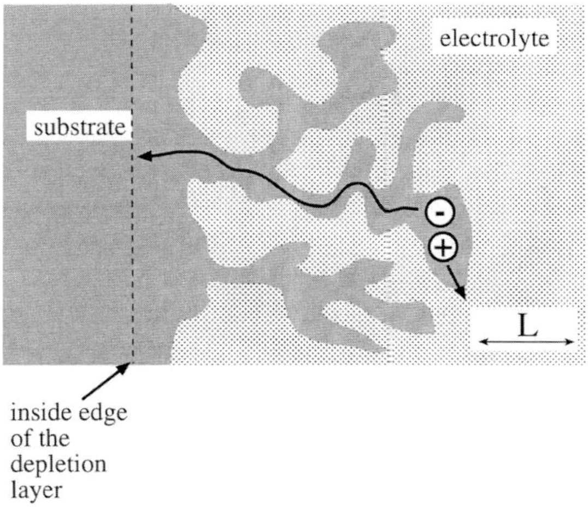

Fig. 7. Electron-hole pair separation in a nanostructured semiconductor electrode (the darker area represents the nanocrystalline material). The charge carriers are generated within a distance from the internal surface that is comparable with the minority carrier diffusion length L. This leads to very efficient electron hole separation.

Optical transitions in GaP are indirect in the photon energy range $2.25\,\mathrm{eV} < h\nu < 2.76\,\mathrm{eV}$, so that the penetration depth for green light is large ($1/\alpha \approx 10^4\,\mathrm{nm}$). For the doping densities of the samples used, the width of the depletion layer and the diffusion length of holes were both much smaller (around 50 nm). For the unetched electrode, the photocurrent efficiency, Φ, for green light was found to be 0.01 in the absence of surface recombination. This value agrees with the prediction based on the Gärtner equation (Eq. 10). For the porous electrode, on the other hand, it was found that Φ tends to unity for all light of energy greater than the indirect bandgap of 2.25 eV. The reason for the remarkable difference between the flat and porous electrodes is that in the latter all photogenerated holes can easily reach the semiconductor/electrolyte interface at a pore wall before they recombine because the characteristic dimensions of the network are comparable with the diffusion length, L (see Fig. 7). Since the interface extends into a third dimension, the photocurrent quantum efficiency depends only on the relative magnitudes of the penetration depth of the light and the film thickness.

High photocurrent quantum yields have also been obtained with porous, nanocrystalline CdS, CdSe, TiO_2, and ZnO electrodes [142–144]. In these systems, bulk recombination is negligible due to the small size of the structural units ($L >$ size). Surface recombination is prevented by effective scavenging of the photogenerated holes by a reducing agent interpenetrated in the nanoporous system.

Nanostructured particulate semiconductor electrodes are a topic of current interest since they offer the possibility of cheap and efficient dye sensitised solar cells [20–26]. These cells rely on efficient light harvesting by a dye adsorbed on the high internal surface area of the electrode. Rapid electron injection from the excited dye into a TiO_2 particle and subsequent regeneration of the dye from its oxidised state by electron donation from I^- make the primary electron-hole pair separation very efficient (see Fig. 3, process e).

2.2.2 Dynamic Aspects of Electron-hole Pair Separation

It is useful to estimate the characteristic time required to achieve primary separation of the electron-hole pair in a bulk single crystalline semiconductor electrode with a flat interface with the electrolyte. If there is a depletion layer at the interface in which an electron-hole pair is generated (see Fig. 5), the electron will move to the interior and the hole to the surface. Provided that the mean free path length for electron-lattice scattering is much smaller than the width of the depletion layer, the drift velocities of the electron and hole are described by the relationships

$$v_n(x) = \mu_n \frac{dE_{F,n}}{q\,dx} \tag{11a}$$

$$v_p(x) = \mu_p \frac{dE_{F,p}}{q\,dx} \tag{11b}$$

It is difficult to estimate the driving force $dE_{F,n}/q\,dx$ for the movement of the majority carriers, since it is necessary to account for diffusion and migration. However, for minority carriers, it can be assumed that the driving force $dE_{F,p}/q\,dx$ is entirely due to the electrical field in the depletion layer. Since the mean electric field is $\Delta\phi_{sc}/d_{sc}$, the drift velocity of a hole is of the order of $\mu_p\Delta\phi_{SC}/d_{SC}$. The transit time of the minority carrier through the depletion layer, $\tau(d_{SC})$, can be expressed as equivalent to an RC time constant using Eq. 6–7:

$$\tau(d_{sc}) = \frac{d_{sc}}{v_p} = \frac{d_{sc}^2}{\mu_p \Delta\phi_{sc}} = \frac{2\varepsilon\varepsilon_o}{q\mu_p n_{bulk}} = 2C_{sc}\left(\frac{d_{sc}}{q\mu_p n_{bulk}}\right) \tag{12}$$

For $\varepsilon = 10$ and $\mu_p = 10^{-2}\,\text{m}^2\,\text{V}^{-1}\,\text{s}^{-1}$ (for simplicity, the dependence of mobility on doping density is neglected), the transit time through the depletion layer is about 10^{-10}, 10^{-12}, and 10^{-14} s for doping densities of 10^{15}, 10^{17}, and $10^{19}\,\text{cm}^{-3}$ respectively. The equivalent resistance term, $d_{SC}/q\mu_p n_{bulk}$, is many orders of magnitude smaller than the electrolyte resistance R_{el}, and correspondingly the time required for photogenerated charge carriers to move through the space charge region is orders of magnitude smaller than the time constant, $\tau_{cell} = R_{el}C_{SC}$, of the electrochemical cell. This conclusion is important with respect to time resolved measurements with flat semiconductor electrodes (see section 3). The time needed for minority carriers photogenerated in the neutral bulk region to reach the edge of the depletion layer is, by definition, the bulk recombination lifetime τ_p (see Eq. 8). Experimental results indicate that photo-induced electrical transients can be considerably influenced by minority carrier diffusion from the neutral bulk region to the edge of the depletion layer [145–146].

It is reasonable to assume that in a nanoporous electrode (Fig. 7), the time required for a photogenerated minority carrier to reach the semiconductor surface by drift or diffusion is not very different from that estimated for a flat semiconductor/electrolyte interface. However, the photogenerated majority carrier must travel through the porous structure to reach the back contact. Much work has been devoted recently to the study of electronic transport through nanostructured electrodes (cf. sections 3 and 5). It has been found that the characteristic time required for a majority carrier to travel through the porous network can be extremely long (in the ms to s range).

2.3 Interfacial Processes

2.3.1 Introduction

For a flat semiconductor/electrolyte contact, the flux, g, of photogenerated minority carriers close to the semiconductor surface can be predicted (see Eq. 10) if it is assumed that minority carriers are consumed sufficiently rapidly by interfacial electron transfer processes to ensure that their concentration in the depletion layer and at the surface can be neglected. The Gärtner model has been used successfully to explain the photocurrent-potential characteristics of many single crystalline semiconductor electrodes, especially in the potential region where the band-bending, $q\Delta\phi_{SC} = q(U - U_{FB})$, is relatively large ($>10k_BT$). However, for lower values of band bending, it is often observed that the photocurrent quantum yield decreases more steeply with decreasing potential than predicted by Eq. 6 and 10. In fact, the photocurrent often falls to zero while there is still a considerable depletion layer. Since this implies that the flux of minority carriers taking part in electrochemical reactions is smaller than the flux, g, arriving at the surface from the interior of the electrode, it is assumed that minority carriers recombine at the surface. The rate of surface mediated recombination depends on the local concentration of majority carriers. Here, a n-type photoanode is taken as an example, in which electrons in the conduction band are the majority carriers. By use of Fermi-statistics, it is found that the surface concentration of electrons is given by

$$n(x = 0) = n_{bulk} e^{-(q\Delta\phi_{sc}/k_BT)} \tag{13}$$

The occurrence of surface recombination can be taken to indicate that interfacial electron transfer proceeds rather slowly, leading to a non-negligible density of free and surface trapped minority carriers. Moreover, the density of recombination-mediating bandgap states is evidently much higher at the surface and in the near surface region [125] than in the bulk. Interfacial bandgap states may originate from bonds to surface atoms, from chemisorbed ions and molecules or from surface and near surface defects.

2.3.2 Competition for Minority Carriers between Direct Transfer and Trapping

Holes generated by illumination of a flat *n*-type single crystalline semiconductor electrode in contact with a simple redox system may be consumed at the surface by electrochemical electron transfer, i.e. oxidation of a reduced species in the electrolyte:

$$h^+ + Red \overset{k_V}{\rightarrow} O \tag{14}$$

Several reviews of the Marcus and Gerischer theories of electrochemical electron transfer (EET) are available in the literature [104–106, 147–150]. It is assumed that EET involves iso-energetic tunnelling between electronic states in the electrode and in the (activated) redox-system. The oxidation of the reduced species involves tunnelling of an electron from a thermally activated state of *Red* to a vacant level (hole) at an energy close to the valence band edge E_V. The fact that iso-energetic tunnelling generally requires a non-equilibrium state of the reducing agent means that electrochemical electron transfer is not necessarily very fast (maximum values for electrochemical rate constants are discussed in section 1.3.) For instance, electrochemical electron transfer is usually much slower than electron tunnelling from a metal to a hole in the valence band (which occurs, for instance, in a semiconductor/(insulator)/metal Schottky barrier). An important reason for this difference is the fact that the density of states involved in tunnelling is much higher in a metal than in a redox electrolyte.

Kinetic investigations with semiconductor electrodes have shown that the localisation of holes (trapping of holes) in interfacial bandgap states may compete effectively with direct electron transfer (comprehensive reviews of this work are given in [48, 131, 151, 152]). Trapping of a hole in an interfacial state in the gap can be written as:

$$h^+ + X \overset{\beta_p}{\rightarrow} X^+ \tag{15}$$

X and X^+ represent bandgap states occupied and unoccupied by an electron respectively. The detailed mechanism of process (15) is not very clear. One might think that hole trapping in an interfacial state is very similar to trapping of a hole in a localised bulk state. The rate constant for hole trapping, β_p, should then be defined as the product of the thermal velocity of free electrons and the capture cross section. (Detailed reviews of electronic processes in bulk semiconductors can be found in [153–157]). Alternatively, the interfacial state X^+/X might be considered to be similar to a redox system chemisorbed at the electrode surface. Hole trapping should then be seen as an electrochemical reaction, which might require a finite solvent reorganisation energy.

Formally, the rate of hole trapping per state can be written as $\beta_p p(0) f(E)$ with $p(0)$ being the concentration of free holes at the surface and $f(E)$ the electron occupation factor of the bandgap state. Trapped holes may be thermally excited from an interfacial state into the valence band. The rate of this process depends exponentially on the energy difference between the state and the valence band. In the framework of the theory developed for electronic transitions in bulk semiconductors, the rate per state at an energy E is given by $\beta_p N_v \, exp[-(E_V - E_F)/k_B T]$. Thermal release of holes from band gap states is usually neglected in photoelectrochemical kinetics (a simplified phenomenological treatment of the competition between direct transfer and hole trapping based on this approximation is given in section 4). Neglect of the thermally activated reverse process implies that only those states are considered for which the rate of some other competing processes (e.g. capture of an electron from the conduction band or from a reduced species by interfacial electron transfer) is faster than the thermal release process. The competition between direct transfer and trapping in interfacial bandgap states is expressed as

$$\delta \frac{dp(0)}{dt} = g - k_V p(0)[Red] - \beta_p p(0) \int_{E_{VB}}^{E_{CB}} [X(e)] \left\{ f(E) - [1 - f(E)] \frac{N_V}{p(0)} e^{-(E - E_V)/(k_B T)} \right\} dE \quad (16)$$

Here g, the flux of photogenerated minority carriers arriving at the surface, is given by Eq. 10, $[X(E)]$ is the concentration of interfacial bandgap states at a given energy E (in $cm^{-2}\,eV^{-1}$), δ is the distance from the surface ($x = 0$) over which the potential varies with an amount $k_B T$; and $[Red]$ is the concentration of the reducing agent within tunnelling distance from the surface.

2.3.3 Competition between Interfacial Electron Transfer and Recombination

The vacant interfacial bandgap state X^+ formed by hole trapping can be refilled with an electron in two ways: by electron donation from a reducing agent in solution, or by capture of an electron from the conduction band. The first process involves iso-energetic electron tunnelling from a suitable non-equilibrium state of the reducing agent at a given energy E.

$$X^+ + Red \xrightarrow{k(E)} X + Ox \quad (17)$$

Alternatively, X^+ can capture an electron (majority carrier) from the conduction band:

$$X^+ + e^- \xrightarrow{\beta_n} X \quad (18)$$

At a given energy, the electron capture rate per state is given by $\beta_n n(0)[1 - f(E)]$, where the concentration of majority carriers at the surface, $n(0)$, is given by Eq. (13). Electron release from X, leading to a free electron in the conduction band, (i.e. the reverse of process 18) occurs with a rate which increases exponentially with decreasing energy difference between the state at an energy E and the conduction band edge E_C. The rate of thermal electron excitation per state can be written as the product $\beta_n N_C \, exp[-(E_C - E)/k_B T] f(E)$.

The electron occupation factor $f(E)$ for the interfacial bandgap states is determined by the rates of electron capture and removal:

$$\frac{df(E)}{dt} = k(E)(Red)[1 - f(E)] - \beta_p p(0)\left\{f(E) - \frac{N_V}{p(0)} e^{-(E-E_V)/(k_BT)}[1 - f(E)]\right\}$$
$$- \beta_n n(0)\left\{[1 - f(E)] - \frac{N_C}{n(0)} e^{(E-E_C)/(k_BT)} f(E)\right\} \quad (19)$$

From Eq. 19, it follows that the electron occupation factor under steady-state conditions $(df(E)/dt = 0)$ is given by

$$f(E) = \frac{k(E)(Red) + \beta_n n(0) + \beta_p N_V e^{-(E-E_V)/(k_BT)}}{k(E)(Red) + \beta_n n(0) + \beta_p N_V e^{-(E-E_V)/(k_BT)} + \beta_p p(0) + \beta_n N_C e^{(E-E_C)/(k_BT)}} \quad (20)$$

It is useful to consider the energy dependence of the electron occupation factor in more detail. A comprehensive treatment dealing with localised bandgap states in the bulk of a semiconductor can be found in [157]. In the energy range close to the valence and conduction band edges, thermal excitation processes predominate, and according to Eq. 20, the electron occupancy factor $f(E)$ becomes unity near to the valence band edge and zero near to the conduction band edge. More precisely, the states are fully occupied below a demarcation energy, E_p, defined by

$$E_p = E_V - k_BT\, ln\left[\frac{k(E)(Red) + \beta_n n(0) + \beta_p p(0)}{\beta_p N_V}\right] \quad (21)$$

The density of holes at the surface can be described conveniently by the quasi-Fermi level for holes, $E_{F,p}(x = 0)$, given by $E_V - k_BT\, ln[p(0)/N_V]$. The states are empty above a demarcation energy, E_n, defined by:

$$E_n = E_C + k_BT\, ln\left[\frac{k(E)(Red) + \beta_n n(0) + \beta_p p(0)}{\beta_n N_C}\right] \quad (22)$$

Between the demarcation levels E_n and E_p, the occupation factor is determined by the rates of hole and electron capture and interfacial electron transfer between the states and the reducing agent:

$$f(E) = \frac{k(E)(Red) + \beta_n n(0)}{k(E)(Red) + \beta_n n(0) + \beta_p p(0)} \quad (23)$$

The effective density of recombination states used in the phenomenological treatment of photoelectrochemical kinetics in section 4 corresponds to the density of states $X(E)$ integrated over the energy range between the demarcation levels. The electrical flux of holes lost by surface recombination is given by

$$J_r = \beta_n n(0) \int_{E_p}^{E_n} [X(E)][1 - f(E)]\, dE$$

$$= \beta_n n(0) \int_{E_p}^{E_n} [X(E)] \left[\frac{\beta_p p(0)}{k(E)(Red) + \beta_n n(0) + \beta_p p(0)} \right] dE \tag{24a}$$

The recombination flux can be approximated by

$$J_r = \frac{\beta_p p(0) \beta_n n(0)}{k(E)(Red) + \beta_n n(0) + \beta_p p(0)} \int_{E_p}^{E_n} X(E)\, dE \tag{24b}$$

if the occupation factor between E_n and E_p does not depend strongly on the energy, which is the case if the interfacial rate constant $k(E)$ is approximately constant between the demarcation levels.

When losses due to surface recombination are taken into account, the photocurrent efficiency, Φ, is given by

$$\Phi = \frac{j_{photo}}{qI(0)} = \frac{g - J_r}{I(0)} \tag{25}$$

The photocurrent is due to direct and surface state mediated transfer of holes (processes 14 and 17 respectively). The competitive processes which determine the photocurrent quantum yield are summarised in Fig. 8.

In a number of systems, it has been observed that the photocurrent quantum yield exceeds the Gärtner limit given by Eq. 10. Examples are GaP and GaAs photocathodes in a solution containing H_2O_2 as an oxidant, and ZnO, TiO_2, CdS and CdSe photoanodes in aqueous solutions containing two equivalent reducing agents

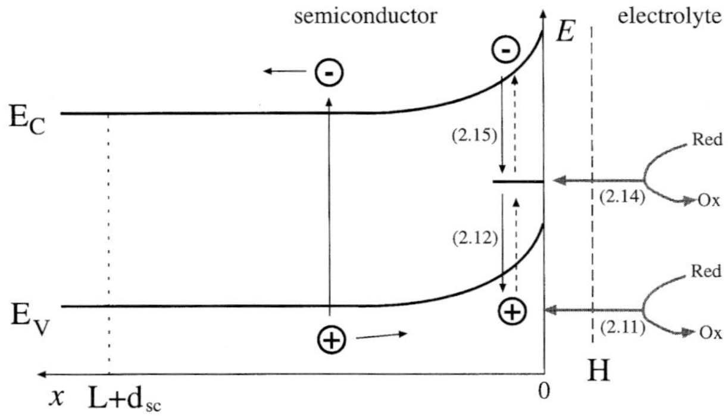

Fig. 8. Overview of the interfacial processes induced by photogenerated minority carriers. The numbers in parenthesis refer to the processes described in text.

such as alcohols, aldehydes, and carboxylic acids. The phenomenon, referred to as photocurrent multiplication, is attributed to majority carrier injection in an electrochemical process initiated by photogenerated minority carriers (for details and references see section 4).

3 Time and Frequency Windows in a Photoelectrochemical Cell

3.1 Introduction

Reliable interpretation of time or frequency resolved photocurrent measurements requires a clear understanding of the relationship between the time resolved current or voltage measured in the external circuit and the photoinduced processes in the semiconductor electrode. The dynamics of photogenerated carriers in a bulk single crystalline semiconductor electrode involve primary electron-hole pair separation in the semiconductor solid and interfacial processes induced by photogenerated minority carriers. If the semiconductor electrode consists of a nanoporous network, transport of photogenerated majority carriers through the network in the direction of the substrate is involved too. It is important to realise that there are fundamental and practical limitations to the way in which the dynamics of photogenerated charge carriers in a semiconductor electrode can be probed by measurement of a photo-induced electrical signal (current or potential) in the external circuit. These limitations are illustrated here for an *n*-type bulk electrode as an example. The semiconductor/electrolyte interface is considered to be a series connection of two parallel plate capacitors, corresponding to the depletion and Helmholtz layers respectively. In principle, the movement of a charge carrier through the depletion and Helmholtz-layers is related to current in the external circuit. However, since the capacitance of the depletion layer, C_{sc}, is usually much smaller than that of the Helmholtz-layer, C_H, it is the movement of a charge carrier through the depletion layer that will be detected in the external circuit. In other words, with a semiconductor electrode, interfacial electron transfer (i.e. electron tunnelling through the Helmholtz-layer) cannot be probed directly by time resolved current measurements. Information about the kinetics of interfacial electron transfer at semiconductor/electrolyte junctions can, however, be derived from time or frequency resolved measurement of the interfacial charge using microwave reflectance. This method is discussed in section 4.

After the absorption of a photon sufficiently close to the semiconductor/electrolyte interface, primary separation of the electron and hole will occur. The minority carrier (hole) may diffuse to the inner edge of the depletion layer, and migrate through the depletion layer towards the surface. Diffusion occurs in a time equal to the life time of the minority carrier in the bulk, τ_p, which for compound semiconductors is often in the microsecond or nanosecond range (in very pure Si, the minority carrier life time is much larger (ms) [158]). In section 2.2, it was shown that the transit time of a charge carrier through the depletion layer lies in the range 10^{-14} to 10^{-10} s. When the hole arrives at the semiconductor surface, it can be involved directly in interfacial electron transfer or it may become trapped in an interfacial state before being consumed by interfacial electron transfer or recombination (capture of an electron from the conduction band). The time-scales for electron transfer and surface recombination are discussed in section 1: they range from nanoseconds to milliseconds. In general, interfacial processes in which majority carriers are consumed (electron transfer and recombination) or injected (in photocurrent doubling, see section 4) occur on a time scale that is much longer than the transit time through the depletion layer. It follows that when an electron is captured by a surface trapped hole, equilibration in the depletion layer (corresponding to the movement of an electron through the depletion layer) is essentially instantaneous. This means that the kinetics of surface recombination and photocurrent doubling processes can be probed by time resolved (photo) current measurement in the external circuit.

The timescales of the processes occurring after photo-generation of an electron hole pair can be shorter or longer than the *RC* time constant of the photoelectrochemical cell, τ_{cell}. In many cases, the time scale for interfacial processes at a semiconductor/electrolyte junction is longer than τ_{cell}, whereas electron-hole separation and electron transit through the depletion layer are much faster. The apparatus used to probe electron-hole separation is therefore different from that used to study the kinetics of photoinduced interfacial processes.

3.2 Two Time Windows, Separated by τ_{cell}

Fig. 9 is an electrical representation of an electrochemical cell with a semiconductor working electrode and metal counter electrode.

The system is perturbed by time dependent illumination that generates electron-hole pairs in the semiconductor electrode. The time resolved voltage, $j_{photo}AR_m$, is measured in the external circuit. (*A* is surface area of working electrode). The elec-

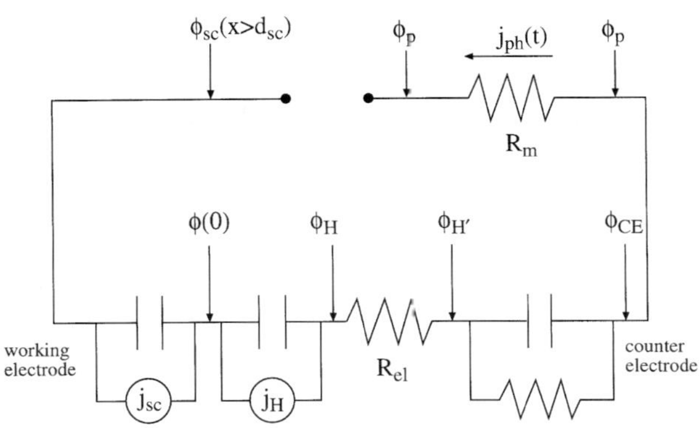

Fig. 9. Electrical representation of a photoelectrochemical cell with an illuminated semiconductor working electrode and a metal counter electrode (see text for details and definition of terms).

trodes are connected to a power source that maintains a constant potential difference, $\phi_{SC}\ (x > d_{SC}) - \phi_p$ over the cell at all times. Here $\phi_{SC}\ (x > d_{SC})$ is the potential of the semiconductor electrode in the neutral bulk region, and ϕ_p is the potential on one side of the measuring resistance R_m, which is connected to the counter electrode at the other side. The potential at the semiconductor surface is $\phi(0)$ and the potential at the outside of the Helmholtz-layer is ϕ_H. The potential of the outer Helmholtz-plane of the counter electrode is denoted as ϕ'_H, and the potential of the lead connected to the counter electrode as ϕ_{CE}. Electron-hole pairs generated by illumination of the semiconductor electrode are separated rapidly, and minority carriers participate in interfacial processes such as recombination and electrochemical electron transfer (electron tunnelling across the Helmholtz-layer). These processes are accounted for by the photocurrent densities $j_{sc}(t)$ and $j_H(t)$ that flow in parallel with the capacitance of the depletion (C_{sc}) and Helmholtz-layer (C_H) respectively. For the circuit of Fig. 9

$$[\phi_{sc}(x > d_{sc}) - \phi(0)] + [\phi(0) - \phi_H] + [\phi_H - \phi'_H] + [\phi'_H - \phi_{CE}] + [\phi_{CE} - \phi_p] + [\phi_p - \phi_{sc}(x > d_{sc})] = 0 \tag{26}$$

In Eq. 26, $\phi_{SC}(x > d_{SC}) - \phi(0)$ is the potential drop over the depletion layer $\Delta\phi_{SC}$, $\phi(0) - \phi_H$ is the potential drop over the Helmholtz-layer at the semiconductor electrode $\Delta\phi_H$, $\phi_H - \phi'_H$ and $\phi_{CE} - \phi_p$ are equal to $j_{photo}(t)AR_{el}$ and $j_{photo}(t)AR_m$ respectively. It is assumed that the impedance of the counter electrode is so small that the potential drop $\phi'_H - \phi_{CE}$ is constant and independent of the current through the cell. It follows from Eq. 26 that

$$\frac{d\Delta\phi_{sc}}{dt} + \frac{d\Delta\phi_H}{dt} = -A(R_m + R_{el})\frac{dj_{photo}(t)}{dt} \tag{27}$$

Application of Kirchoffs laws to the electrical circuit of Fig. 9 gives

$$C_{sc}\frac{d\Delta\phi_{sc}}{dt} + j_{sc}(t) = C_H\frac{d\Delta\phi_H}{dt} + j_H(t) \tag{28}$$

$$C_{sc}\frac{d\Delta\phi_{sc}}{dt} + j_{sc}(t) = j_{photo}(t) \tag{29}$$

If $d\Delta\phi_H/dt$ is negligible with respect to $d\Delta\phi_{SC}/dt$, the relationship between internal current flow through the depletion layer, $j_{sc}(t)$, and the measured voltage over R_m can be obtained directly from Eq. 27 and 29. If $d\Delta\phi_H/dt$ is not negligible, a somewhat more elaborate treatment is necessary. Eq. 28 allows $d\Delta\phi_H/dt$ to be expressed as a function of $d\Delta\phi_{SC}/dt$, and using Eq. 27 and 29, it is found that

$$\frac{C_H}{C_H + C_{sc}}j_{sc}(t) + \frac{C_{sc}}{C_H + C_{sc}}j_H(t) = j_{photo}(t) + \tau_{RC}\frac{dj_{photo}(t)}{dt} \tag{30}$$

The RC time constant of the electrochemical cell, τ_{cell}, is given by

$$\tau_{cell} = \frac{C_{sc}C_H}{C_{sc} + C_H} \times (R_m + R_{el}) \tag{31}$$

The value of τ_{cell} can be determined independently from electrical impedance measurements. Eq. 30 gives the relationship between the photoinduced current flow through the depletion and Helmholtz-layers and the transient voltage measured in the external circuit. Solving Eq. 30 makes it clear that the relationship between the externally measured signal and the photoinduced current flow through the semiconductor electrode (closely related to the rate of photoelectrochemical processes) depends on the time, t, relative to τ_{cell}. Willig et al. [159] have solved Eq. 30 for perturbation by a short light pulse to obtain an analytical relationship between the currents $j_{SC}(t)$ and $j_H(t)$ and the voltage $j_{photo}R_m$ measured in the external circuit. Peter [48] has also shown that definition of the RC attenuated photocurrent transfer function in the Laplace domain leads to solutions for a range of perturbation functions (delta pulse, step function and sinusoidal modulation). To simplify the discussion here, it assumed that $C_{SC} \ll C_H$; this is usually the case under depletion conditions. The analysis shows that two time windows must be considered, separated by the time constant, τ_{cell}, of the photoelectrochemical cell. In the time window $0 < t \ll \tau_{cell}$, the measured voltage $j_{photo}(t)R_m$ is proportional to the integral of $j_{SC}(t)$ over time. This can be rationalised by considering the depletion layer of the semiconductor electrode as a parallel plate capacitor, electron-hole pair separation leads to charging of the capacitor plates and a photovoltage arises between the plates. This photovoltage is detected in the external circuit since $\phi_p - \phi_{SC}(x > d_{SC})$ is constant. In the time region close to τ_{cell}, the photoinduced charge on the capacitor plates is discharged through R_m, and the photovoltage decays exponentially with a time constant τ_{cell} (it is worth noting at this point that Peter [48] has shown that the apparent time constant can in fact be shorter than τ_{cell} if fast surface recombination occurs giving rise to an additional flux of majority carriers into the surface). In the time window $t \gg \tau_{cell}$, the photocurrent density $j_{photo}(t)$ measured in the external circuit is equal to the photoinduced current density through the depletion layer $j_{SC}(t)$. It follows that photoelectrochemical processes in this time window that lead to the flow of charge carriers through the depletion layer can be followed directly by measurement of $j_{photo}(t)$ in the external circuit.

A similar analysis can be performed for the sinusoidal intensity modulation used in intensity modulated photocurrent spectroscopy (IMPS). Under conditions of harmonic light modulation, $dj_{photo}(t)/dt = -i\omega j_{photo}(t)$, and Eq. 30 becomes

$$j_{photo}(\omega) = \frac{\dfrac{C_H}{C_H + C_{sc}} j_{sc}(\omega) + \dfrac{C_{sc}}{C_H + C_{sc}} j_H(\omega)}{1 + i\omega\tau_{cell}} \tag{32}$$

It is clear that two frequency windows must be considered, corresponding to the time windows defined above. In the high frequency window $\omega \gg 1/\tau_{cell}, \omega\tau_{cell} j_{photo}(\omega)$

is proportional to the photocurrent flow, $j_{SC}(\omega)$, in the depletion layer and $R_m\omega\tau_{cell}j_{photo}(\omega)$ is a measure of the modulated photopotential caused by photoinduced charging of the depletion layer capacitance. In the low frequency window $\omega \ll 1/\tau_{cell}$, the ac photocurrent density through the depletion layer, $j_{SC}(\omega)$, is measured directly in the external circuit as the photocurrent density response $j_{photo}(\omega)$. IMPS studies of the mechanisms and kinetics of relatively slow photoinduced interfacial processes at single crystalline semiconductor/electrolyte interfaces are usually performed in the low frequency window $\omega \leq 1/\tau_{cell}$. In principle, the method can also be used in the high frequency window $\omega \gg 1/\tau_{cell}$ to study primary electron-hole pair separation and faster interfacial processes. Intensity modulated photovoltage measurements at modulation frequencies higher than $1/\tau_{cell}$ have been used to determine depletion layer capacitances [91, 92].

3.3 Experimental Results

This section presents results that show how the rates of photoelectrochemical processes can be derived from time resolved measurement of the photoinduced current or potential in the external circuit of a photoelectrochemical cell. The capacitance of the Helmholtz-double layer is of the order of 10^{-5} F cm^{-2}, the depletion layer capacitance of an extrinsic semiconductor junction is typically 10^{-8}–10^{-9} F cm^{-2}, while the capacitance of an insulator is orders of magnitude lower. With a value of 100 Ohm for the resistance $R_{el} + R_m$ of the cell, the time constant of photoelectrochemical cells is 10^{-3} s for metallic electrodes, 10^{-6}–10^{-5} s for semiconductor electrodes and much lower for insulator electrodes. The rates of photoelectrochemical processes also span a wide range. This makes photoelectrochemical kinetics a rich, albeit demanding, area for research.

An excellent example of time resolved photoelectrochemistry is provided by a study by Willig and co-workers [159] of primary electron-hole pair separation at the single crystalline *n*-type GaAs/electrolyte interface. Primary electron-hole pair separation occurs in the sub-nanosecond range in the early time window $0 < t < \tau_{cell}$, and short light pulses with time resolution of at least 10^{-10} s are required in order to investigate the process. Fig. 10 illustrates a photocurrent transient recorded for a *n*-GaAs semiconductor electrode in the nanosecond regime [159].

The 2.14 eV laser pulse is also shown; its pulse width is about 1 ns. The photoinduced potential over the measuring resistance, $R_mAj_{photo}(t)$, rises and reaches a plateau after 4 ns. The shape of the transient voltage corresponds to integration over time of the laser pulse, and this indicates that the electron-hole pair separation in the depletion layer of the GaAs electrode is faster than the time resolution of the laser pulse so that the internal photocurrent charging the depletion layer capacitance simply follows the shape of the laser pulse. The plateau in the photovoltage is followed by an exponential decay in the microsecond time range determined by the value of τ_{cell}. In some experimental systems it has been noted that the photopotential decays faster than expected from τ_{cell}. This has been explained by Peter [48] in terms of the discharging of the depletion capacitance by the flow of majority carriers into surface states brought about by fast surface recombination.

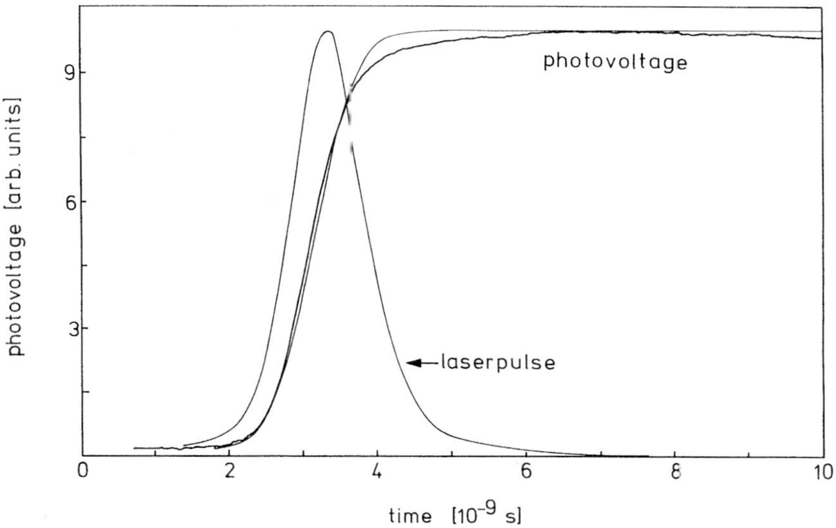

Fig. 10. Photocurrent transient recorded for a *n*-GaAs semiconductor electrode in the nanosecond regime [159]. The 2.14 eV laser pulse is also shown. Note that the initial shape of the transient voltage corresponds to integration over time of the laser pulse. The subsequent exponential decay is determined by τ_{cell} (taken from ref [159]).

It is clear that laser pulses shorter than 1 ns are required to deconvolute the dynamics of electron-hole pair separation in semiconductor electrodes. Bitterling et al. [160] have reported transient measurements with an GaAs electrode illuminated with 20 ps light pulses. The rise-time of the electrical response $R_m A j_{photo}(t)$ was attributed to diffusion of (screened) minority carriers generated in the neutral region adjacent to the edge of the depletion layer (see section 2.2.2), allowing determination of the bulk recombination life time of the minority carriers. Other groups have reported transient photopotential measurements in the nanosecond range with TiO_2 electrodes [146, 161] and with InP electrodes [61, 162]. The photopotential transients were attributed to electron-hole pair separation in the depletion layer.

Insulator electrodes have much lower capacitances than semiconductor electrodes, so that τ_{cell} is moved to times shorter than those in which charge carrier transport can be characterised. For example, the photoelectrochemical cell with a dye sensitised anthracene electrode studied by Willig and co-workers [159] is characterised by a cell time constant of 1 ns. Since the mobility of the photoinjected holes in anthracene is lower than in most semiconductors, their transit time τ_d through the anthracene crystal can be measured because it is larger than τ_{cell}. In this case, the externally measured transient voltage $R_m A j_{photo}(t)$ is directly related to the movement of charge carriers through the anthracene crystal. The average velocity of the photogenerated holes and hence the hole mobility (1 cm^{-2} V^{-1} s^{-1}) could therefore be derived from the photocurrent transient. Similar results have been reported earlier by Kepler [163].

In semiconductor device physics, transient photocurrent measurements are known as time-of-flight measurements and have been used to measure the transit time of electrons or holes through a semiconducting or insulating crystal (see, for instance [164–166]). Much work has been devoted to the study of the dynamics of photogenerated charge carriers in amorphous hydrogenated silicon, which forms the active layer in p-i-n α-Si:H solar cells (see, for example, [167–173]). In these devices, the insulating layer is relatively thick (typically 500 nm), and transport of charge carriers is attenuated by an enhanced scattering of the carriers with the lattice and by multiple trapping of the carriers in bulk defects. Generally, transit times are considerably larger than the time constant of the device.

More recently, the transport characteristics of photogenerated electrons through nanoporous semiconductor networks interpenetrated with an electrolyte have become a matter of interest. This is due to novel and remarkable properties of such electrodes and their potential use in photoelectrochemical solar cells (see section 5). Nanoporous electrodes have a huge interfacial area which, under accumulation conditions at least, may increase τ_{cell} up to about 1 ms or more. However, the transit of electrons through a nanoporous network is also very slow (in the ms – s range) and most of the work performed until now pertained to the long time window ($t \gg \tau_{cell}$ or low frequency regime $\omega < 1/\tau_{cell}$). A clear example of how τ_{cell} can influence IMPS results has been observed with nanoporous GaP networks prepared on single crystalline GaP substrates by anodic etching [78] (see section 5). Interpretation of the IMPS results obtained with these electrodes was made possible by independent determination of the potential dependence of τ_{cell} by electrical impedance measurements.

4 Frequency Resolved Kinetic Studies at Single Crystal Photoelectrodes

4.1 IMPS: Competition between Electron Transfer and Recombination

In the absence of surface recombination, all minority carriers that are collected by diffusion and migration in the semiconductor/electrolyte junction will eventually either transfer to redox species in the solution or react with the semiconductor itself leading to anodic or cathodic photodecomposition. Slow interfacial kinetics will result in the build up of photogenerated carriers at the interface, but unless photocurrent multiplication occurs, the saturation photocurrent will simply be determined by the light intensity, and the quantum efficiency will be unity. This means that the photocurrent contains no information about interfacial kinetics. In reality, most semiconductor/electrolyte interfaces are non-ideal, and a substantial fraction of the photogenerated electrons or holes do not take part in interfacial redox reactions because they recombine via surface states (see section 2.3.3). It is this competition between interfacial electron transfer and surface recombination that opens the way to obtain information about the rates of interfacial processes.

The competition between charge transfer and recombination via surface states can be treated exactly by Hall-Shockley-Read statistics, taking proper account of

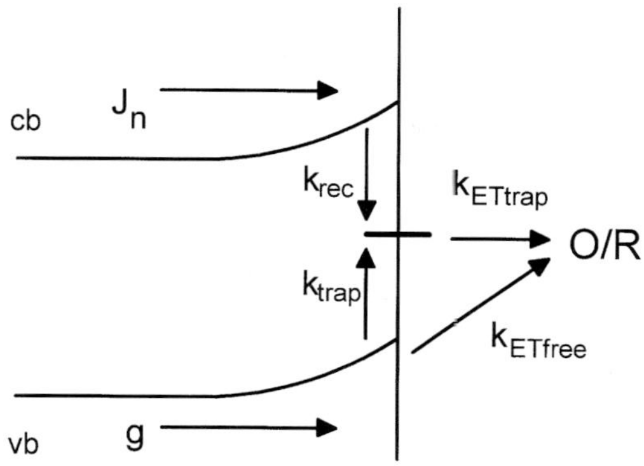

Fig. 11. Simplified scheme showing reaction routes for photogenerated holes in an n-type semiconductor electrode. The rate constants are defined in the text. g is the hole flux and J_n is the electron flux arising from recombination via surface energy levels.

the capture and thermal release of carriers (see sections 2.3.2 and 2.3.3 and [174]). The present section uses a simplified phenomenological approach developed by Ponomarev and Peter [74] because it leads to less cumbersome expressions for the steady state and dynamic photocurrent responses that allow IMPS to be compared with PEIS and light modulated microwave reflectivity measurements.

Fig. 11 illustrates the simplified scheme of minority carrier generation, collection and reaction for an n-type semiconductor.

Note that it is assumed that trapping and charge transfer processes are effectively irreversible. Holes generated in the bulk of the semiconductor can diffuse into the space charge region where they are swept rapidly to the surface. The flux, g, of holes generated by these processes is given by the Gärtner equation (Eq. 10). In the absence of photocorrosion reactions, holes in the valence band can either react with an electron donor in solution (rate constant $k'_{ET\,free}$) or become trapped at the surface (rate constant k'_{trap}). The trapped holes can either react with the electron donor (rate constant k'_{ETtrap}) or recombine with an electron from the conduction band (rate constant k'_{rec})

The minority carrier density can be expressed in terms of an equivalent surface concentration, p_s (cm^{-2}), since this allows a convenient formulation of the kinetic equations. The surface concentrations can be converted to equivalent volume densities, $p(0)$ (cm^{-3}) by dividing by a nominal reaction length δ. Further simplification is achieved by considering the concentrations of redox species and majority carriers to be time invariant.

Consideration of the rates of arrival, charge transfer, trapping and recombination of minority carriers leads to expressions for the time dependent surface concentrations (cm^{-2}) of free $p_{s,free}$ and trapped $p_{s,trap}$ holes:

$$\frac{dp_{s,free}}{dt} = g - k'_{ET\,free}[Red]p_{s,free} - k'_{trap}N_{trap}p_{s,free} \tag{33}$$

and

$$\frac{dp_{s,trap}}{dt} = k'_{trap}N_{trap}p_{s,free} - k'_{ET\,trap}p_{s,trap}[Red] - k'_{rec}n_s p_{s,trap} \tag{34}$$

Here g is the Gärtner flux of holes into the surface generated by illumination, $[Red]$ is the bulk concentration of reduced species in solution and n_s is the surface electron density. Two limiting cases have been considered by Peat et al. [43]. In one case, hole transfer to redox species occurs entirely via the valence band; trapped holes can only recombine. This limit is appropriate for so-called 'near-surface' states, which arise from defects close to the surface that interact strongly with the conduction and valence bands but not with electrolyte energy levels [125]. In the second case, holes are trapped very rapidly at surface states, but the resulting vacant surface electron energy level can accept an electron from a redox species as well as from the conduction band (recombination). For the present discussion, it is assumed that the second case applies, i.e. electron transfer and recombination involve a common intermediate state. The corresponding simplification of Eq. 33 and 34 is obtained by assumed $k'_{trap}N_{trap} \gg k'_{ET\,free}[Red]$) and the concentration of free holes and its time derivative are negligible. In this case, Eq. 34 simplifies to

$$\frac{dp_s}{dt} = g - k'_{ET\,trap}[Red]p_s - k'_{rec}n_s p_s \tag{35}$$

Here $p_{s,trap}$ has been replaced by p_s, since essentially all holes are trapped. If $[Red]$ is considered constant (the pseudo first order situation), $k'_{ET2}[Red]$ can be replaced simply by k_{tr} (units s^{-1}). Similarly, if the density of electrons is assumed to be constant, i.e. there is no change in band bending under illumination, $k'_{rec}n_s$ can be replaced by k_{rec} (units s^{-1}). Eq. 35 then assumes the very simple phenomenological form

$$\frac{dp(0)}{dt} = g - k_{tr}p_s - k_{rec}p_s \tag{36}$$

where p_s is the density (cm^{-2}) of trapped holes at the surface.

The surface concentration of electrons depends on the potential drop (band bending) in the semiconductor, and in the absence of complications due to surface state charging (Fermi level pinning) it is given by

$$\frac{n_s}{\delta} = n_b e^{-q((U-U_{fb})/(kT))} \tag{37}$$

where U is the electrode potential and U_{fb} is the flatband potential. Consequently the first order rate constant for surface recombination $k_{rec} = k'_{rec}n_s$ is expected to depend exponentially on potential:

$$k_{rec} = k^0_{rec}e^{-q((U-U_{fb})/(kT))} \tag{38}$$

The recombination rate constant k^0_{rec} depends on the doping density N_d, the thermal velocity v_n of majority carriers and the majority carrier capture cross section σ_n of recombination centres:

$$k^0_{rec} = N_d \sigma_n v_n \tag{39}$$

For $N_d = 10^{17}\,\mathrm{cm}^{-3}$, $\sigma_n = 10^{-16}\,\mathrm{cm}^2$, $v_n = 10^5\,\mathrm{cm\,s}^{-1}$, $k^0_{rec} = 10^6\,\mathrm{s}^{-1}$.

Under conditions of steady state illumination, $dp_s/dt = 0$ in Eq. 33, and it follows that the surface hole concentration is given by

$$p_s = \frac{g}{k_{tr} + k_{rec}} \tag{40}$$

The photocurrent density is therefore

$$j_{photo} = qk_{tr}p_s = qg\frac{k_{tr}}{k_{tr} + k_{rec}} \tag{41}$$

where q is the elementary charge. The dimensionless ratio $k_{tr}/(k_{tr} + k_{rec})$ represents the fraction of minority carriers that undergo electron transfer. This simple treatment suffices to show that the rate constants for electron transfer and recombination cannot be separated by steady state measurements.

In order to separate k_{tr} and k_{rec} it is necessary to carry out a non steady state measurement. The simplest non steady state measurement that can be envisaged is switching the illumination on and off. The photogenerated flux of holes towards the surface can be coupled to three processes: surface charge storage, interfacial electron transfer and recombination. The corresponding components of the total current density can be written in terms of the surface charge Q_s due to minority carriers:

$$j_{ch} = \frac{dQ_s}{dt} \tag{42a}$$

$$j_{tr} = k_{tr}Q_s \tag{42b}$$

$$j_{rec} = k_{rec}Q_s \tag{42c}$$

The measured photocurrent response is the sum of the charging and transfer components, since holes that recombine are not measured in the external circuit:

$$j_{photo} = j_{ch} + j_{tr} = qg - j_{rec} \tag{43}$$

It follows that the time dependent behaviour is described by the differential equation

$$qg(t) = \frac{dQ_s(t)}{dt} + (k_{tr} + k_{rec})Q_s(t) \tag{44}$$

It is useful to consider the case where the illumination is switched on at $t = 0$. The initial condition is $Q_s(0) = 0$. The transfer, recombination and charging components of the time dependent photocurrent are

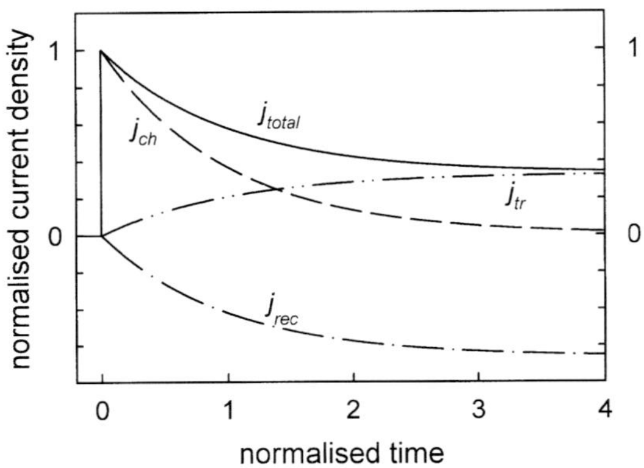

Fig. 12. Components of the current response of a *n*-type semiconductor electrode to an illumination step. j_{ch} is the charging current, j_{tr} is the current due to interfacial electron transfer and j_{rec} is the current due to electrons recombining with holes via surface states. The total current, given by the sum of j_{ch} and j_{tr}, is equal to $qg - j_{rec}$, where g is the flux of minority carriers given by the Gärtner equation (Eq. 10). The dimensionless normalised time axis is $(k_{tr} + k_{rec})t$. The dimensionless normalised current axis is j/qg.

$$j_{tr}(t) = \frac{qgk_{tr}[1 - e^{-(k_{tr}+k_{rec})t}]}{k_{tr} + k_{rec}} \tag{45a}$$

$$j_{rec}(t) = \frac{qgk_{rec}[1 - e^{-(k_{tr}+k_{rec})t}]}{k_{tr} + k_{rec}} \tag{45b}$$

$$j_{ch}(t) = qge^{-(k_{tr}+k_{rec})t} \tag{45c}$$

The total time dependent photocurrent can therefore be written is dimensionless form as

$$\Phi(t) = \frac{j_{photo}(t)}{qg} = \frac{(k_{tr} + k_{rec}e^{-(k_{tr}+k_{rec})t})}{k_{tr} + k_{rec}} \tag{46}$$

Fig. 12 illustrates the way in which the time dependence of the photocurrent is related to the charging and transfer components.

It can be seen that at $t = 0$, $Q_s = 0$ and so the measured photocurrent is entirely due to charging. At longer times the surface charge tends towards its steady state value, and the charging current falls to zero. The current due to charge transfer is proportional to Q_s, so it is zero at $t = 0$ and increases with time towards its steady state value. The recombination term also depends on Q_s, in other words the fraction of charge carriers at the surface that are transferred is determined by the ratio $k_{tr}/(k_{tr} + k_{rec})$ at all times. Analysis of the transient photocurrent response allows separation of k_{tr} and k_{rec}. A plot of $ln(j_{photo}(t) - j_{photo}(\infty))vs.\ t$ has a slope of $-(k_{tr} + k_{rec})$. The ratio $j_{photo}(\infty)/j_{photo}(0)$ is given by $k_{tr}/(k_{tr} + k_{rec})$.

Now consider the case where the incident photon flux is modulated sinusoidally:

$$I(t) = I_{dc} + I_{ac} \sin(\omega t) \tag{47a}$$

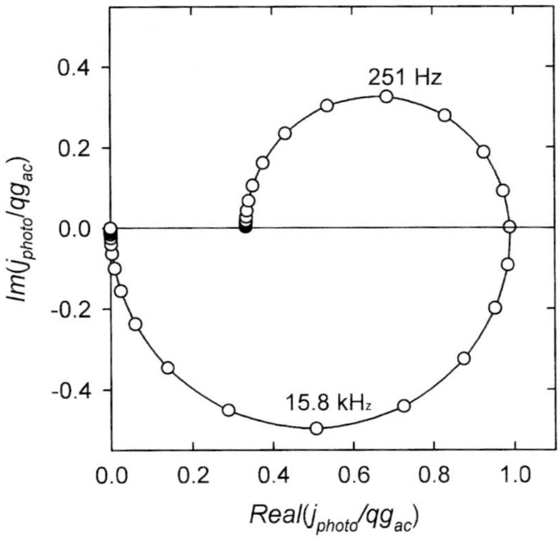

Fig. 14. IMPS plot showing the effect of τ_{cell}. Calculated for $R = 100\,\Omega$, $C_{SC} = 100\,\text{nF}$, $C_H = 10\,\mu\text{F}$. $k_{tr} = 500\,\text{s}^{-1}$, $k_{rec} = 10^3\,\text{s}^{-1}$. The upper semicircle arises from surface recombination (cf. Fig. 13) and the lower from RC attenuation (cf. Fig. 2a).

The effect of the cell time constant on the IMPS response is illustrated in Fig. 14. In this particular case, the time constant for recombination is much longer than τ_{cell} and the normalised plot crosses the real axis close to unity. If the two time constants are closer, the IMPS plot crosses the real axis at a point less than unity.

The upper limit of rate constants that can be accessed by IMPS is normally determined by τ_{cell}. Typical values of C_{sc} for a moderately doped semiconductor under depletion conditions are of the order of $10^{-7}\,\text{F cm}^{-2}$. If care is taken to minimise the sum of the solution and contact resistances below $10\ \Omega\ \text{cm}^2$, τ_{cell} will be less than 10^{-6} s. For highly doped semiconductors or for samples with poor ohmic contacts, larger *RC* time constants will restrict the range of rate constants that can be determined by IMPS.

If the series resistance is deliberately increased by adding an external resistor, then $\omega RC > 1$ in Eq. 52, and the measured response is effectively an *ac* photovoltage $U_{photo}(\omega)$ [74] given by

$$\omega U_{photo}(\omega) = \frac{qg_{ac}}{C}\left[\frac{k_{tr} + i\omega\left(\dfrac{C}{C_{sc}}\right)}{k_{tr} + k_{rec} + i\omega}\right] \tag{54}$$

This limit will only be valid if U_{photo} is sufficiently small (a few mV) that k_{rec} can be treated as constant to a first approximation]. The complex plane plot of $\omega U_{photo}(\omega)$ predicted by Eq. 54 is a semicircle which for $(RC)^{-1} \ll \omega \ll (k_{tr} + k_{rec})$ has a low frequency intercept given by

$$\omega U_{photo}(\omega)|_{\omega \to 0} = \frac{qg_{ac}}{C}\frac{k_{tr}}{k_{tr} + k_{rec}} \tag{55}$$

and a high frequency intercept

$$\omega U_{photo}(\omega)|_{\omega\to\infty} = \frac{qg_{ac}}{C_{sc}} \tag{56}$$

The maximum of the semicircle appears at $\omega_{max} = k_{rec} + k_{tr}$. It follows that frequency response analysis of the small signal photopotential response should give access not only to k_{tr}, k_{rec} but also to C_{sc} and C_H. The possibilities offered by intensity modulated photovoltage measurements remain largely unexplored, although Kamieniecki used the method to determine space charge capacitances as long ago as 1982 [91, 92]. Since the input impedances of most voltage amplifiers are at least $10^8\,\Omega$ and typical space charge capacitances are in excess of $10^{-8}\,\text{F cm}^{-2}$, the RC time constant for a typical electrochemical cell is of the order of 1 second or more under conditions where the modulated photovoltage response is measured.

4.3 Comparison of Theory with IMPS Results and Consideration of Mechanisms

The photoelectrochemical behaviour of p-InP has been studied by a number of authors [31, 35, 75, 162]. When a p-InP electrode is illuminated under depletion conditions, photogenerated electrons are driven to the surface where they can reduce protons to hydrogen. This process competes with surface recombination. Fig. 15 illustrates a set of experimental IMPS responses measured for p-InP in 1.0 mol dm^{-3} H_2SO_4 [35] at different potentials. The measurements were performed using a small *ac* modulation of the illumination intensity superimposed on a larger steady component.

It can be seen that diameters of the IMPS semicircles decrease as the potential becomes more negative. The high frequency intercept is almost constant, whereas the low frequency intercept moves away from the origin as the bend banding is increased. Although these trends correspond qualitatively to those predicted by the preceding

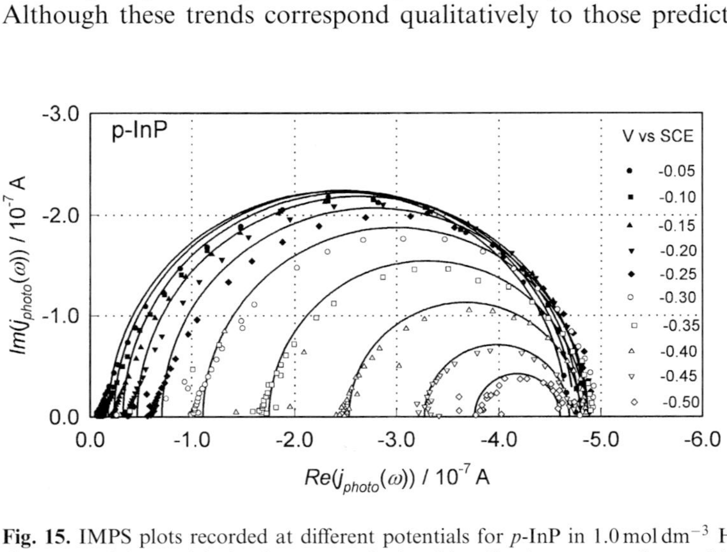

Fig. 15. IMPS plots recorded at different potentials for p-InP in 1.0 mol dm^{-3} H_2SO_4 [35]. Note that the diameters of the semicircles decrease as the band bending increases and recombination becomes less effective. In the saturation photocurrent region, the IMPS response contracts to a point.

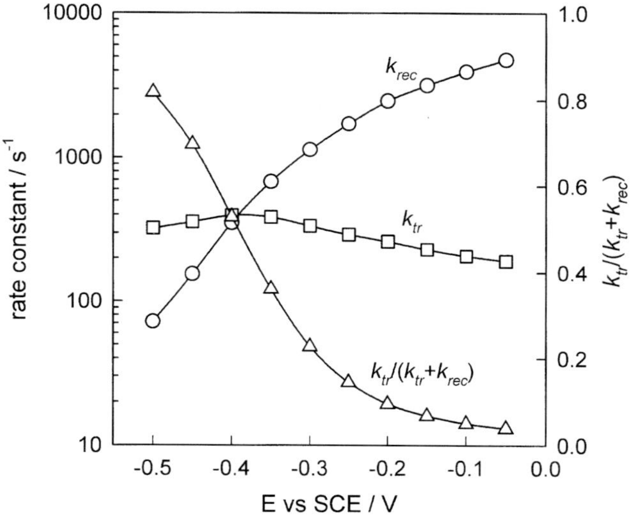

Fig. 16. Variation of k_{tr} and k_{rec} for hydrogen evolution on illuminated *p*-InP [35]. The ratio $k_{tr}/(k_{tr}+k_{rec})$ represents the fraction of photogenerated electrons transferred across the interface. Note that k_{tr} appears to depend weakly on potential. The non-ideal variation of k_{rec} with potential is interpreted as evidence for partial Fermi level pinning.

analysis, closer examination of the IMPS data shows that the experimental system does not follow the behaviour predicted by the simple one electron transfer model.

Fig. 16 shows how the derived values of k_{tr} and k_{rec} depend on electrode potential. The ratio $k_{tr}/(k_{tr}+k_{rec})$ is also shown; it represents the fraction of photogenerated electrons that react to produce hydrogen rather than recombining.

The first point to note is that k_{tr} is not independent of potential. As expected, k_{rec} decreases as the potential is made more negative and the majority carrier (hole) concentration falls. However, the plot shows that k_{rec} decreases by less than two orders of magnitude for a potential change of 0.5 V. This decrease can be compared with Eq. 34 (in the case of *p*-InP, the majority carriers are holes rather than electrons, so that n in Eq. 34 is replaced by p), which predicts that k_{rec} should decrease by more than 8 orders of magnitude if the band bending is changed by 0.5 eV. The experimental results appear to indicate that the surface hole concentration changes much more slowly with electrode potential than expected for an ideal junction, although it should be remembered that the simplified model may not provide a good approximation at high band bending (see section 3). It appears likely that not all of the change in applied potential appears across the space charge region of the semiconductor; instead some fraction must appear across the Helmholtz layer as the result of the surface charging. This phenomenon is common in semiconductor electrolyte junctions and is often referred to as Fermi level pinning.

The observed potential dependence of k_{tr} is less easily explained using the simple phenomenological approach, and at this point it is useful to consider the mechanism of the photoelectrochemical reaction. The reduction of H^+ to H_2 is a two electron process and there are two possible mechanisms:

$$h\nu \rightarrow h^+_{vb} + e^-_{cb} \qquad (57a)$$

$$H^+ + e^-_{cb} \rightarrow H^{\cdot} \qquad (57b)$$

followed either by

$$H^+ + H^{\cdot} + e^-_{cb} \rightarrow H_2 \quad \textit{mechanism I} \qquad (57c)$$

or by

$$H^{\cdot} + H^{\cdot} \rightarrow H_2 \quad \textit{mechanism II} \qquad (57d)$$

Mechanism I involves two different electron transfer steps, so that the meaning of k_{tr} needs to be clarified. Both mechanisms involve an adsorbed hydrogen intermediate, and this may have consequences for the potential distribution across the interface and hence the band bending. The formation of adsorbed intermediates is a common feature of multi-electron transfer reactions. Other examples are encountered in the photodecomposition of compound semiconductors, for example the photoanodic decomposition of *n*-CdS:

$$h\nu \rightarrow h^+_{vb} + e^-_{cb} \qquad (58a)$$

$$CdS + h^+_{vb} \rightarrow Cd^+S \qquad (58b)$$

$$Cd^+S + h^+_{vb} \rightarrow Cd^{2+} + S \qquad (58c)$$

Recombination may occur via the photogenerated intermediate

$$Cd^+S + e^-_{cb} \rightarrow CdS \qquad (58d)$$

Photoelectrochemical mechanisms of this type introduce several complications into the kinetic analysis. The first is that the steady state accumulation of the photogenerated intermediate results in charging of the surface, which reduces the band bending. The second is that photoinduced *ac* modulation of the surface intermediate concentration results in a corresponding modulation of the potential, i.e. an *ac* photovoltage is generated. This *ac* photovoltage in turn modulates the surface concentration of majority carriers and hence the rate of surface recombination. Finally the *ac* photovoltage also results in a charging current associated with charging of the space charge and Helmholtz capacitances. These effects, which lie outside the scope of this chapter, have been discussed by de Wit et al. [62] and by Peter et al. [89].

A detailed analysis of mechanisms I and II by Peter et al. [89] that takes all these factors into account has shown that the apparent or phenomenological rate constants k_{tr} and k_{rec} derived from IMPS data are functions of the rate constants for all the steps in the reaction. For both mechanisms, the analysis shows that k_{tr} is no longer potential independent because it contains a term associated with recombination *via* the intermediate. Similarly k_{rec} is determined by all three rate constants, and no longer follows a simple Boltzmann dependence on potential of the type predicted by Eq. 38.

Figs. 17a and 17b illustrate the quite complicated potential dependence of the apparent or phenomenological rate constants k_{tr} and k_{rec} predicted for mechanism I when dynamic surface charging is taken into account. More work is required to develop the analysis of experimental data to this level of sophistication.

Recently Fermín et al. [84] have used IMPS to study the photo-evolution of oxygen on n-TiO_2 and have concluded that the response is consistent with a Case II type mechanism involving $OH^{\cdot}_{ads}$ intermediates. Fig. 18 illustrates the potential dependence of k_{rec} and k_{tr} derived from the analysis of IMPS responses obtained for n-TiO_2 in 0.1 mol dm^{-3} $NaClO_4$. The same work showed that the $OH^{\cdot}_{ads}$ intermediate is also involved in the photooxidation of formic acid.

Oskam et al. [82] have used IMPS to investigate the role of surface states at the n-Si(111)/NH_4F interface. In this case, the redox reaction is simpler, and appears not to involve holes trapped at surface states. This is probably due to the presence of a surface oxide layer. However, electron transfer is evidently exceptionally slow in this case, since these authors observed a modulated photocurrent even at potentials far from the flatband potential where recombination is expected to be negligible. Accumulation of holes modifies the potential drop across the Helmholtz (and presumably also surface oxide region), leading to a capacitive charging current. This effect has also been treated by Peter et al. in more detail [89].

4.4 Relationship between IMPS and Photoelectrochemical Impedance Spectroscopy (PEIS)

IMPS uses modulation of the light intensity to produce an *ac* photocurrent that is analysed to obtain kinetic information. An alternative approach is to modulate the electrode potential while keeping the illumination intensity constant. This method has been referred to as photoelectrochemical impedance spectroscopy (PEIS), and it has been widely used to study photoelectrochemical reactions at semiconductors [30–35]. In most cases, the impedance response has been fitted using equivalent circuits since this is the usual approach used in electrochemical impedance spectroscopy. The relationship between PEIS and IMPS has been discussed by a number of authors [35, 60, 64]. Vanmaekelbergh et al. [64] have calculated both the IMPS transfer function and the photoelectrochemical impedance from first principles and shown that these methods give the same information about the mechanism and kinetics of recombination. Recombination at CdS and ZnO electrodes has been studied by both methods [62, 77]. Ponomarev and Peter [35] have shown how the equivalent circuit components used to fit impedance data are related to the physical properties of the electrode (e.g. the space charge capacitance) and to the rate constants for photoelectrochemical processes.

In a PEIS experiment, the flux of minority carriers generated by illumination is constant to a good approximation, provided that the penetration depth of the light

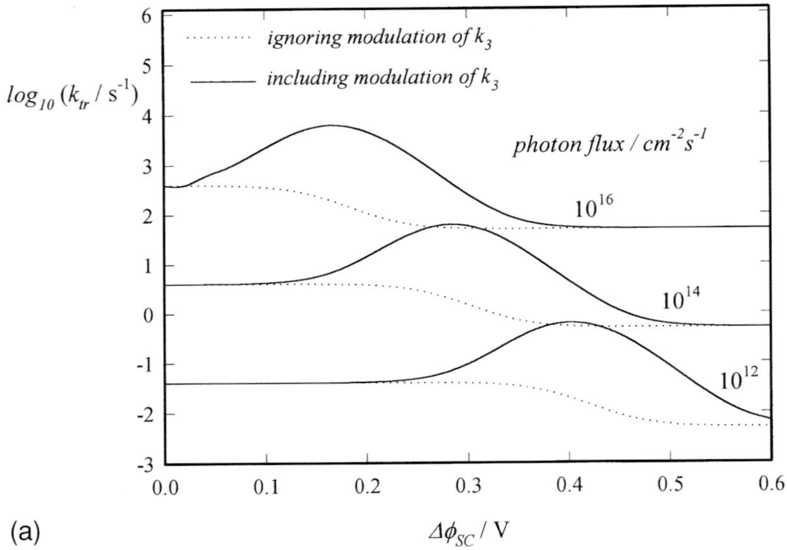

(a)

Fig. 17a. Potential dependence of the phenomenological rate constant k_{tr} derived for mechanism I (see Eq. 48) [89]. Note that for a two step mechanism, this rate constants contains terms associated with surface recombination, so that it is not the true rate constant for electron transfer. The influence of the modulation of potential due to surface charging is shown.

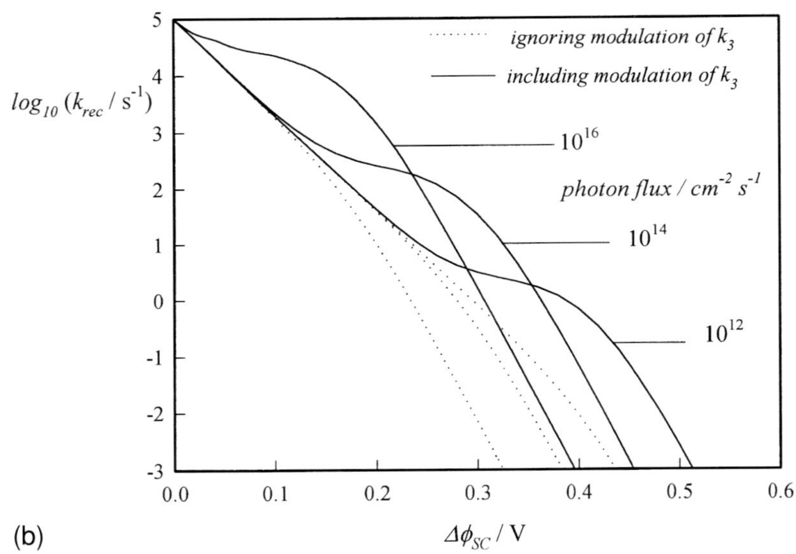

(b)

Fig. 17b. Potential dependence of the phenomenological rate constant k_{tr} for case I [89]. The influence of the dynamic modulation of surface potential by accumulated reaction intermediates is shown.

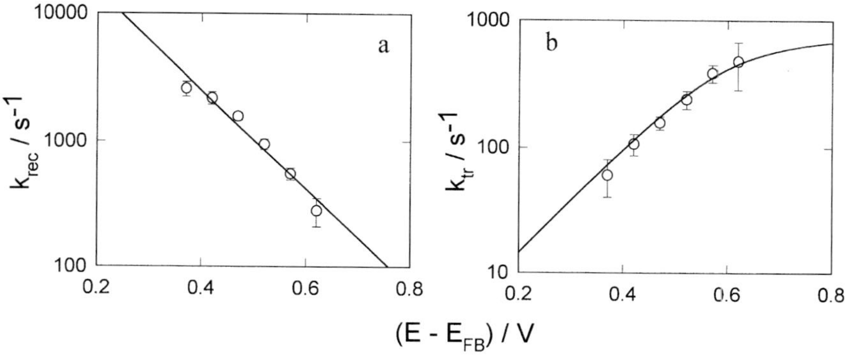

Fig. 18. Experimental variation of k_{rec} and k_{tr} with potential for photogeneration of oxygen on n-TiO_2 in 0.1 mol dm^{-3} $HClO_4$ [84]. Data derived from analysis of the IMPS response.

is sufficiently small to fulfil the condition $1/\alpha < d_{SC} + L$, where d_{SC} is the width of the space charge layer and L is the diffusion length of minority carriers (see section 2.2.1). Under these conditions, the modulation of the minority carrier flux by the applied *ac* voltage can be neglected and $g = I(0)$. The main effect of the *ac* potential perturbation is to change the density of majority carriers at the surface and consequently the value of k_{rec}. The modulation of k_{rec} about a mean (*dc*) value results in a corresponding frequency dependent *ac* photocurrent. The photoelectrochemical impedance is determined by the ratio of the perturbing *ac* voltage to the *ac* component of the total photocurrent. For the simple competition mechanism discussed in section 4.1, it can be shown [35] that the photoelectrochemical admittance ($Y_{PEIS} = 1/Z_{PEIS}$) reduces to the simple form

$$Y_{PEIS} = i\omega C_{sc} + \frac{qI(0)}{kT} \frac{k_{rec}(k_{tr} + i\omega)}{(k_{tr} + k_{rec})(k_{tr} + k_{rec} + i\omega)} \tag{59}$$

if the solution resistance is negligibly small and the space charge capacitance is much smaller than the Helmholtz capacitance. For simplicity, it is assumed that the flux of minority carriers generated by the *dc* illumination is equivalent to I_{dc} (negligible recombination in the bulk). The corresponding complex plane impedance plot exhibits two semicircles and can therefore be modelled by the series/parallel circuit shown in Fig. 19.

The R and C values for the low frequency semicircle both depend on the steady incident photon flux, $I(0)$ and are given by

$$R_{LF} = \left(\frac{kT}{q^2 I(0)}\right) \frac{k_{tr} + k_{rec}}{k_{tr}} \tag{60a}$$

and

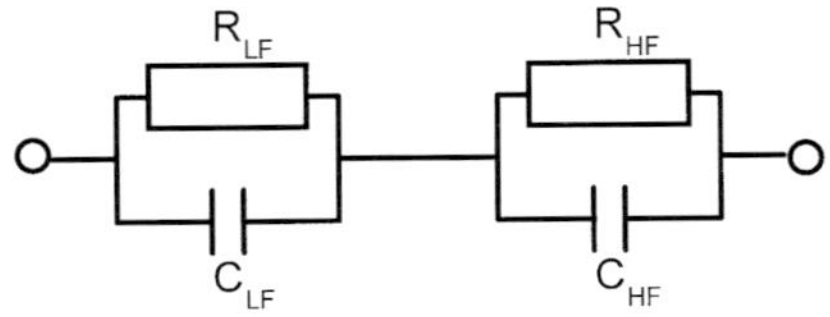

Fig. 19. Series/parallel equivalent circuit that can be used to model PEIS response (see text for definition of components).

$$C_{LF} = \left(\frac{q^2 I(0)}{kT}\right) \frac{1}{k_{tr} + k_{rec}} \tag{60b}$$

It is interesting to note that the frequency of the maximum of the low frequency semicircle, $\omega_{max}(LF)$ is equal to k_{tr} and is independent of light intensity.

The corresponding values for the high frequency semicircle are

$$R_{HF} = \left(\frac{kT}{q^2 I(0)}\right) \frac{k_{tr} + k_{rec}}{k_{rec}} \tag{61a}$$

$$C_{HF} = C_{SC} \tag{61b}$$

In this case, only the resistance component depends on light intensity. The maximum of the high frequency semicircle is therefore also dependent on light intensity:

$$\omega_{\max}(HF) = \left(\frac{q^2 I(0)}{kTC_{SC}}\right) \frac{k_{rec}}{k_{tr} + k_{rec}} \tag{62}$$

This analysis shows clearly that PEIS provides essentially the same information as IMPS (see also [64]). The physical significance of the RC components only emerges from an analysis of the kinetics, and attempts to attribute physical meaning to these elements without relating them to the kinetics and mechanism of the photoelectrochemical process will inevitably result in conceptual difficulties of interpretation.

Fig. 20 illustrates the PEIS response measured by Ponomarev and Peter [35] for p-InP in 1.0 mol dm^{-3} H_2SO_4. The two semicircles predicted theoretically are clearly evident.

The relative magnitude of the two semicircles changes with potential as the *dc* value of the band bending changes the unperturbed (*dc*) value of the recombination rate constant. The fit between theory and experiment is generally satisfactory. The flattening of the semicircles probably arises from non-uniform illumination, but it does not significantly affect the determination of the rate constants, which were found to be in excellent agreement with those obtained by IMPS and shown in Fig. 16. Further work will be necessary to extend the theory of PEIS to multiple electron transfer processes.

Schoenmakers et al. [77] have used IMPS, PEIS and potential dependent photoluminescence to study the competition between charge transfer and recombination at ZnO electrodes. Their results indicate that recombination in the bulk solid as well as via surface states is responsible for the reduction in photocurrent conversion efficiency in the photocurrent onset region.

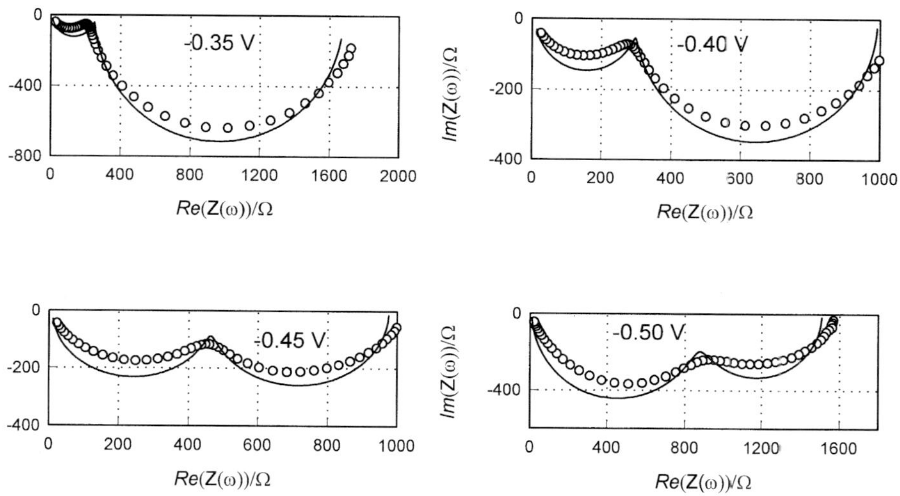

Fig. 20. Experimental PEIS plots for p-InP in 1.0 mol dm^{-3} H_2SO_4 at the potentials shown [35]. Experimental data shown as open circles. The lines show the best fit to the theory.

4.5 Light Modulated Microwave Measurements

Tributsch and co-workers [93, 94] have pioneered the application of modulated microwave reflectivity measurements to the study of the semiconductor electrolyte interface. The method is based on the fact that the microwave reflectivity of a material changes when its dielectric constant is perturbed. In the case of a semiconductor, perturbation of the density of mobile carriers (electrons and holes) by changes in potential in illumination intensity influence the conductivity and hence the imaginary component of the dielectric constant at microwave frequencies. For small perturbations, the change ΔR_m in microwave reflectivity becomes a linear function of the change in conductivity. A full discussion is given in [176].

Schlichthörl et al. [177] have used light modulated microwave reflectivity to derive the rates of interfacial electron transfer processes at the n-Si/electrolyte interface. In these measurements, the modulation frequency was constant, and the rate constants for charge transfer were derived from the potential dependent ΔR_m response. Schlichthörl et al. [73] have extended the technique considerably by introducing frequency response analysis. The technique is therefore analogous to IMPS, although, as shown below, it provides additional information.

Fig. 21 shows the experimental arrangement for IMPS and microwave measurements.

The semiconductor wafer is mounted at the end of an X-band microwave waveguide so that microwave radiation probes the reflectivity of the sample. The ohmic contact is applied as a grid of thin lines in order to minimise microwave losses. The front of the wafer is in contact with an electrolyte solution, and a modulated light source (for example a light emitting diode) illuminates the sample. The changes in

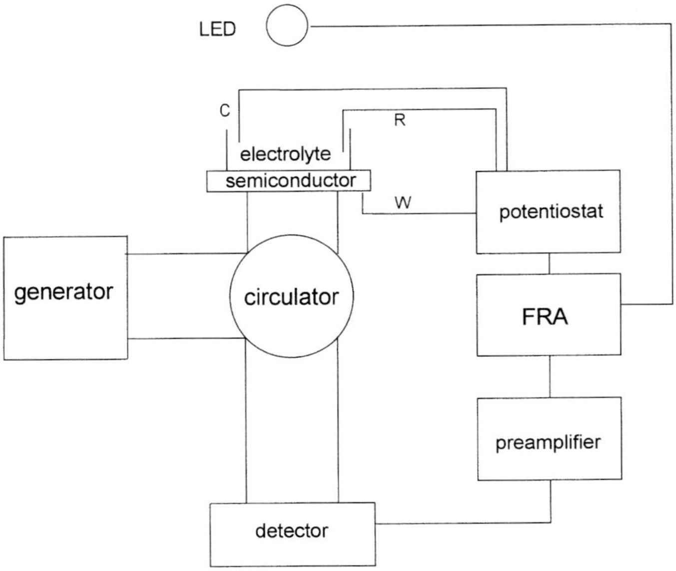

Fig. 21. Experimental set-up for light modulated microwave reflectance based on X-band microwave system [177]. The apparatus can also be used for IMPS measurements.

microwave reflectivity are small, generally a few parts per million, and an *ac* coupled low noise preamplifier is used to amplify the signal to a level where it can be detected by the frequency response analyser. The real and imaginary components of the *ac* microwave response are recorded as a function of the modulation frequency of the illumination and plotted in the complex plane.

In the case of light modulated microwave reflectivity measurements, ΔR_m is a linear function of the change in the photogenerated charge that accumulates at the semiconductor/electrolyte interface

$$\Delta R_M(t) = SQ_s(t) \tag{63}$$

where S is a sensitivity factor.

Photogenerated carriers can be transferred to the surface region (charging), be transferred across the interface or recombine, so that for modulated illumination $I_{ac}(t)$

$$qg_{ac}(t) = \frac{dQ_s(t)}{dt} + (k_{tr} + k_{rec})Q_s(t) \tag{64}$$

The photocurrent measured in the external circuit is

$$j_{photo}(t) = qg_{ac}(t) - k_{rec}Q_s(t) \tag{65}$$

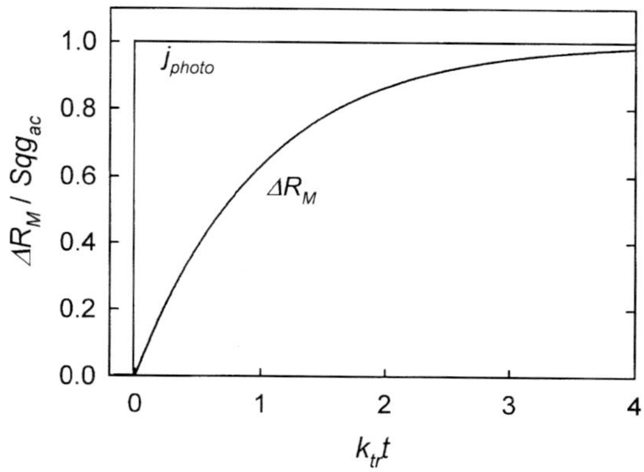

Fig. 22. Comparison of the time dependence of the photocurrent step and of the slower increase in microwavereflectivity predicted for a light step in the case where recombination is negligible. Note that no kinetic information can be derived from j_{photo}, whereas the time dependence of the microwave reflectivity provides access to the value of k_{tr} even in the potential region where recombination is negligible.

If we consider the case where the light is switched on at $t = 0$ (i.e. the initial condition is $Q_s = 0$), the time dependent microwave response is

$$\Delta R_M(t) = SQ_s(t) = Sqg\frac{(1 - e^{-(k_{tr}+k_{rec})^t})}{k_{tr} + k_{rec}} \tag{66}$$

The most interesting feature of the time dependent microwave response is the fact that it still contains information when there is no surface recombination. If $k_{rec} = 0$, Eq. 66 becomes

$$\Delta R_M(t) = Sqg\,\frac{(1 - e^{-k_{tr}t})}{k_{tr}} \tag{67}$$

This time dependence of the microwave reflectivity predicted by Eq. 67 is contrasted in Fig. 22 with the corresponding time dependent photocurrent for the case where $k_{rec} = 0$.

If $k_{rec} = 0$, the photocurrent simply follows the illumination step and contains no information about the rate of charge transfer at the interface. The comparison shows that, unlike IMPS and PEIS, light modulated microwave reflectivity measurements still provide kinetic information at high band bending where recombination is negligible and the steady state photocurrent is described by the Gärtner equation.

It can be shown that the real and imaginary components of the microwave response to sinusoidal modulation of the light intensity are [73]

$$\mathrm{Re}(\Delta R_M) = Sqg_{ac}\frac{k_{tr} + k_{rec}}{(k_{tr} + k_{rec})^2 + \omega^2} \tag{68a}$$

and

$$\mathrm{Im}(\Delta R_M) = -Sg_{ac}\frac{\omega}{(k_{tr} + k_{re})^2 + \omega^2} \tag{68b}$$

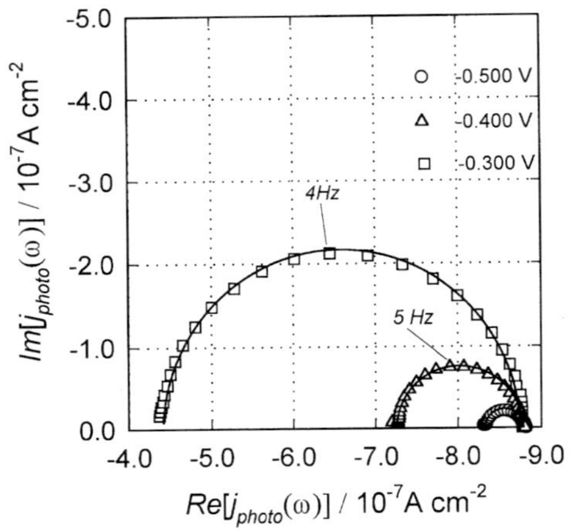

Fig. 23. Set of IMPS plots for p-Si in 1.0 mol dm^{-3} HF [177]. Note that the plots condense to a point as the band bending is increased, so that no kinetic information can be derived at potentials more negative than -0.5 V.

The microwave response is therefore a semicircle in the lower complex plane. The high frequency intercept is zero, and the low frequency intercept is

$$\mathrm{Re}(\Delta R_M)_{\omega \to 0} = \frac{Sqg_{ac}}{k_{tr} + k_{rec}} \tag{69}$$

The frequency of the minimum in the semicircle is equal to the sum of the rate constants for charge transfer and recombination: $\omega_{min} = k_{tr} + k_{rec}$. Light modulated microwave measurements therefore provide the sum of the two rate constants, but since it is possible to measure k_{tr} at high band bending where recombination is negligible, the rate constants can be separated if it is assumed that k_{tr} is independent of potential (this assumption may not be valid for multi-electron transfer reactions as noted in section 4.2).

Light modulated microwave reflectivity has been used to characterise hydrogen evolution on oxide free p-Si at low light intensities [73]. The cathodic photocurrent reaches a plateau at potentials more negative than -0.5 V, and it can be shown that this corresponds to complete collection of photogenerated electrons at the junction (i.e. no recombination). This is confirmed by the set of IMPS plots in Fig. 23 which collapse to a point on the real axis as the potential becomes more negative.

The contrast with the microwave results shown in Fig. 24 is striking. A semicircular microwave response is still observed at -1.0 V in the plateau photocurrent region, and since recombination is clearly negligible at this potential, the pseudo first order rate constant for charge transfer can be obtained directly from the frequency of the minimum of the semicircle., which in this case is 23 s^{-1}. At less negative potentials, where recombination takes place, the minimum frequency is determined by the sum of the charge transfer and recombination rate constants.

Comparison of the values of k_{tr} and k_{rec} derived from IMPS with the sum $(k_{tr} + k_{rec})$ obtained from the microwave measurements shows that they agree to better than 5% [73]. These results illustrate the power of frequency resolved microwave techniques, which should find more widespread application in future.

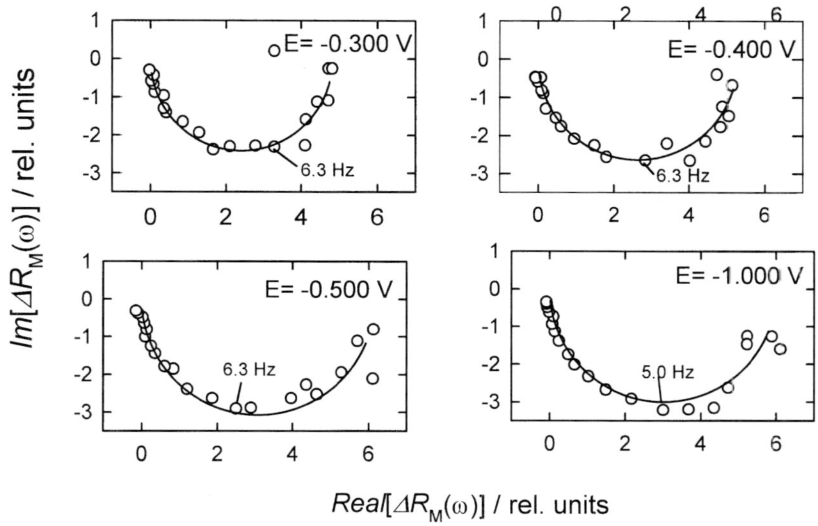

Fig. 24. Set of light modulated microwavereflectivity plots for p-Si in 1.0 mol dm^{-3} HF [177]. Contrast these plots with the IMPS plots in Fig. 23. It is clear that k_{tr} can be derived from the microwave response even at -1.0 V, where the quantum efficiency is unity and the IMPS plot has condensed to a point.

4.6 Studies of Photocurrent Multiplication by IMPS

Photocurrent multiplication processes are encountered frequently in photoelectrochemistry. Common examples include the photo-oxidation of formic acid and of secondary alcohols at n-type semiconductors [1], and the photoreduction of oxygen at p-type semiconductors [40, 41, 48]. The mechanisms are generally supposed to involve majority carrier injection by a photogenerated intermediate, and IMPS has been used to determined the rate constants for these processes. Earlier work has been reviewed previously in some detail [48]. The first example to be studied by IMPS was the photoreduction of oxygen to H_2O_2 at p-GaP [40, 41]. Subsequently, the oxidation of formic acid at n-CdS was characterised by the same method [52]. The oxidation of formic acid to CO_2 is a two step reaction which involves the following steps

$$h\nu \rightarrow h^+ + e^- \quad (70a)$$

$$HCOOH + h^+ \rightarrow COOH^{\cdot} + H^+ \quad (70b)$$

$$COOH^{\cdot} + h^+ \rightarrow CO_2 + H^+ \quad (70c)$$

or

$$COOH^{\cdot} \rightarrow CO_2 + H^+ + e^- \quad (70d)$$

Reaction 4.37d is responsible for photocurrent doubling since it results in the two electron oxidation of formic acid to CO_2 for the absorption of only one photon. The electron injection step competes with reaction 70c, hole capture, and as a result the photocurrent quantum efficiency depends on illumination intensity. At high intensities, the supply of photogenerated holes to the surface promotes reaction 70 c, and the quantum efficiency is 1. At low light intensities, electron injection becomes predominant, and the quantum efficiency rises to 2. Further examples of current doubling that have been studied by IMPS include the photo-oxidation of formic acid at TiO_2 [84] and of hydroquinone [178] at CdS.

The principle behind the application of IMPS to the study of these systems is straightforward. Electrochemical injection of majority carriers leads to a component of the photocurrent flux that lags behind the in-phase component associated with the flux of photogenerated minority carriers. The delay is associated with the first order lifetime of the injecting intermediate, and as the frequency at which the light intensity is modulated is increased, the injection component is attenuated progressively. The IMPS plot in the absence of recombination is a semicircle in the lower quadrant of the complex plane. The low frequency intercept corresponds to the *dc* quantum efficiency (2 at low light intensities). The high frequency intercept tends towards a quantum efficiency of 1. The minimum of the semicircle occurs at an angular frequency equal to the first order rate constant for majority carrier injection. Further details can be found in [41, 48, 52]. In principle the same kinetic information can be obtained from time resolved measurements, and Ponomarev et al. [179] have derived an expression for the transient photocurrent response to pulsed laser excitation in the case of photocurrent doubling. In practice, however, analysis in the frequency domain is a better diagnostic method for photocurrent multiplication reactions.

Fig. 25 illustrates potential dependent trends in the IMPS response calculated for the case where current doubling and surface recombination compete and *RC* attenuation occurs at high frequencies.

Fig. 26 illustrates the experimental IMPS response for formic acid oxidation on CdS reported by Herrasti and Peter [52]. In the photocurrent saturation region, the IMPS response exhibits a semicircle in the lower quadrant as predicted. At lower potentials, recombination becomes dominant, giving a semicircle in the upper quadrant. The rate constant for electron injection derived from $\omega_{\min}$ is $6.0 \times 10^4\ s^{-1}$. This corresponds to an activation energy of 0.47 eV, if the pre-exponential factor is taken to be $10^{13}\ s^{-1}$. Such a high value may be associated with the change in geometry from the non-linear CO_2 radical to linear CO_2. The flattening of the semicircle was modelled by a distribution of activation energies for electron injection with a standard deviation of 0.06.

It is evident from Fig. 26 that it is not possible to measure the IMPS response at sufficiently high frequencies to observe the limit where the quantum efficiency tends towards unity (i.e. where $\omega \gg k_{inj}$). The limitations arise in this case from the dynamic response of the potentionstat. In other cases, attenuation due to the *RC* time constant of the system may obscure the injection semicircle. The upper limit to the majority carrier injection rate constants that can be obtained by IMPS is around $10^5\ s^{-1}$.

IMPS has been used to investigate the involvement of electron injection in the photodissolution of *n*-Si [46, 49]. In the case of the photodissolution of *n*-Si in fluoride solutions, the quantum efficiency varies from 4 at low light intensities to 2 at

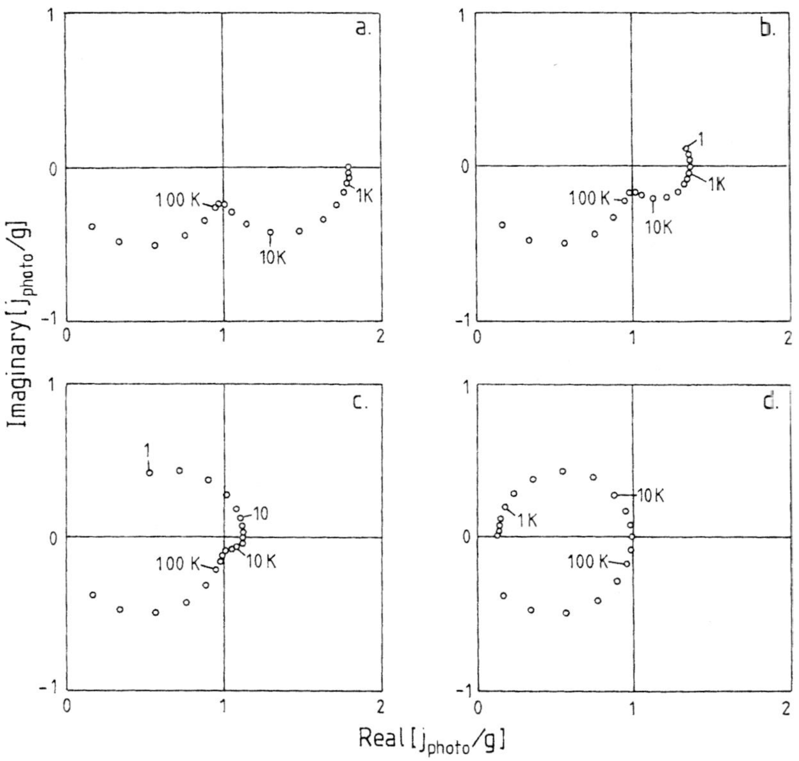

Fig. 25. Theoretical plots showing competition between recombination and current doubling [52]. Calculated for a surface state density of 5 $10^{12}\,cm^{-2}$. The surface state is located 0.3 eV below the bulk Fermi level. Donor density 1.5 $10^{16}\,cm^{-3}$. k_{inj} 5 $10^4\,s^{-1}$. $\tau_{cell} = 2\ 10^{-7}$ s. Band bending values a) 0.40 eV, b) 0.35 eV, c) 0.30 eV, d) 0.2 eV. Note the transition in the IMPS response from current doubling control at 0.4 eV to recombination control at 0.2 eV.

high light intensities. A quantum efficiency of 4 corresponds to a mechanism in which the capture of one photogenerated hole is followed by the injection of three electrons into the conduction band by dissolution intermediates. Peter *et al.* [49] have given a detailed treatment of both the steady state and periodic response expected for a mechanism involving surface bound silicon intermediates with nominal valence states between (I) and (III). The general reaction scheme shown in Fig. 27 considers competition between the capture of photogenerated holes and the injection of electrons by intermediates. Surface recombination occurring in parallel *via* surface states was also considered (a more realistic approach might be to consider recombination reactions that involve the photogenerated intermediates).

The theoretical treatment of the IMPS response predicts that if $k_a > k_b > k_c$, each electron injection step should give rise to a corresponding semicircle in the lower quadrant of the complex plane plot. The low frequency intercept corresponds to a quantum efficiency of 4 and the high frequency intercept to a quantum efficiency of unity (*RC* attenuation is neglected here). Each semicircle has a diameter corre-

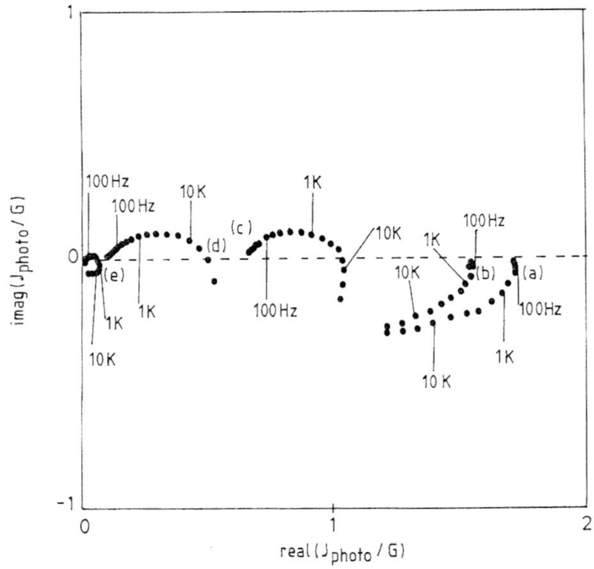

Fig. 26. Experimental IMPS for formic acid oxidation on *n*-CdS [52]. Potentials vs. RHE: a) 0.8 V, b) 0.0 V, c) −0.2 V, d) −0.4 V, d) −0.6 V. At high band bending, the response is characteristic of current doubling. The value of k_{inj} derived from ω_{min} is 6 10^4 s^{-1}.

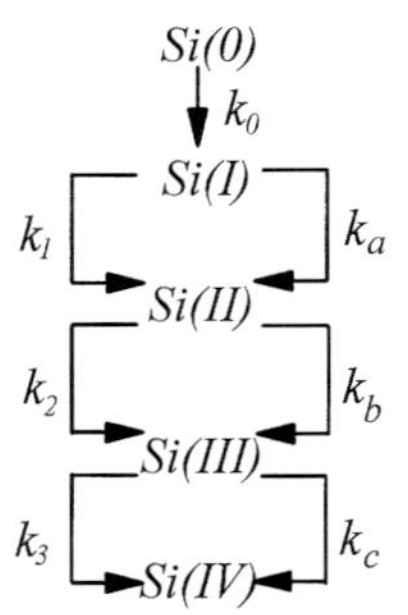

Fig. 27. Reaction scheme for photoanodic dissolution of silicon in low intensity limit illustrating the competition between hole capture steps (rate constants k_0 to k_3 and electron injection steps rate constants k_a to k_c [49]. The nominal valence states of the silicon intermediates are indicated. The final product Si(IV) is the soluble hexafluorosilicate species.

sponding to a quantum efficiency of one (Fig. 28). The rate constants for the electron injection steps are equal to the corresponding ω_{min} values. If $k_a < k_b > k_c$, two semicircles are seen (Fig. 28), one with a diameter twice that of the other and with ω_{min} values equal to k_a and k_c. If the order of the rate constants is reversed, only one semicircle is seen with $\omega_{min} = k_a$ (not shown).

IMPS measurements performed at low light intensities were analysed to obtain values of the rate constants for three electron injection steps [49]. Fig. 29 illustrates the experimental IMPS response and the fitting used to obtain the rate constants for the injection processes.

The three semicircles are flattened considerably, indicating a spread of rate constants attributed to surface heterogeneity, and this effect was modelled by assuming a plausible Gaussian distribution of activation energies with a standard deviation of the order of $k_B T$ for the electron injection steps. The rate constants obtained from the fit

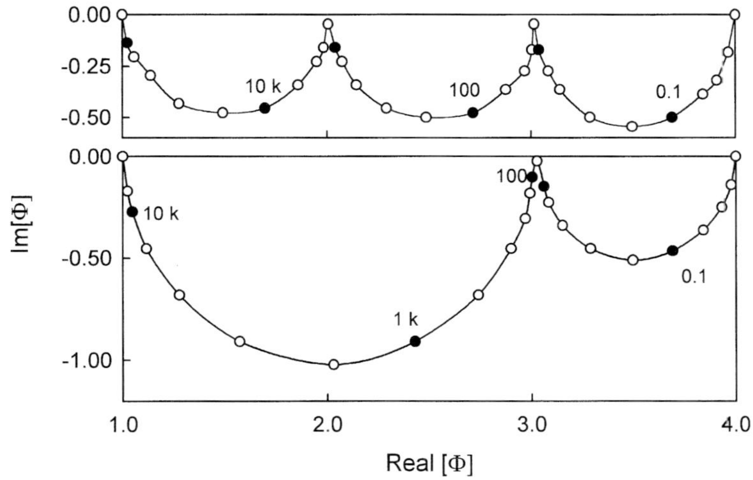

Fig. 28. Theoretical IMPS plots calculated for the scheme shown in Fig. 27. The upper plot is for the case where $k_a > k_b > k_c$. The values of these rate constants are given by the corresponding ω_{min} values. The lower figure shows the IMPS response expected for $k_a < k_b > k_c$. In this case the two higher frequency semicircle condense into one determined by k_a.

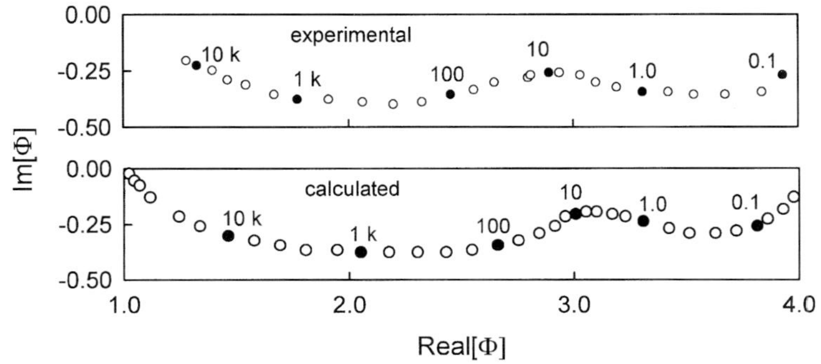

Fig. 29. The upper plot shows the experimental IMPS response measured at very low light intensities for the photodissolution of n-Si in 1.0 NH_4F at pH 4.5. The lower plot is the calculated best fit [49]. Note that the photocurrent efficiency varies from 4.0 at low frequencies to 1.0 at high frequencies as expected for the scheme shown in Fig. 27.

are $k_a = 2 \times 10^4\,s^{-1}$, $k_b = 500\,s^{-1}$ and $k_c = 0.5\,s^{-1}$. The remarkably low value of k_c explains why photocurrent quadrupling during silicon dissolution in fluoride media is only observed at very low light intensities.

The identities of the electron injecting intermediates remain unclear, but one can obtain a crude estimate of the energy levels of the injecting states relative to the conduction band by assuming that the first order rate constants for electron injection are given by

$$k_{inj} = \nu e^{(-E_a/kT)} \tag{71}$$

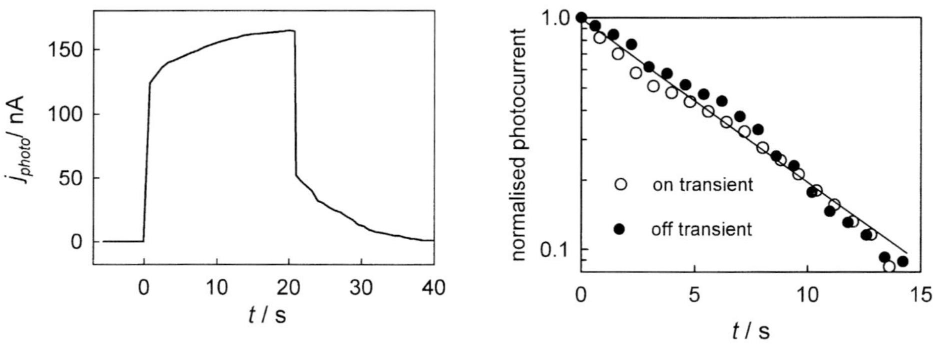

Fig. 30. Transient photocurrent response to illumination step measured at low light intensity for n-Si in 1.0 M NH_4F at pH 4.5 [49]. The slow rise and fall are attributed to the slowest step in the photocurrent multiplication scheme. The semilogarithmic plot can be used to derive the value of k_c. Compare with Fig. 29 which demonstrates the superior resolution of IMPS.

where ν is a pre-exponential factor (of the order of $10^{12}\,s^{-1}$) and the activation energy E_a is determined by the energy difference between the injecting state and the conduction band of the silicon. The activation energies estimated in this way from the measured values of k_{inj} are 0.44, 0.54 and 0.69 eV.

IMPS is certainly the most powerful method available for the study of photocurrent multiplication. It is interesting to contrast the IMPS response for the photodissolution of n-Si in ammonium fluoride with the corresponding light step measurement. Fig. 30 illustrates the photocurrent transient obtained at low light intensities [49]. The slow rising and falling parts of the transient are associated with the final and slowest electron injection step in the sequence. Analysis of the rising and falling parts of the transient gives a value of k_c in reasonable agreement with the IMPS result, but it is difficult to obtain reproducible transients at such extremely low light intensities due to fluctuations in the background current, and attempts to obtain k_a and k_b by fitting transients were unsuccessful.

Another example of photocurrent multiplication that have been studied by IMPS is the photo-anodic dissolution of n-InP in HCl [65], which is a six electron process. The IMPS analysis is an extension of the 4 electron case for silicon, and analysis of the experimental IMPS results show that in three out of the six steps, electron injection can compete with hole capture. The rate constants for the three consecutive electron injection steps were found to be $k_a > 6 \times 10^5\,s^{-1}$, $k_b = 6 \times 10^4\,s^{-1}$ and $k_c = 6 \times 10^2\,s^{-1}$.

The mechanisms of many current doubling reactions are undoubtedly more complex than the kind of scheme outlined in Eq. 64. This is illustrated by the recent work of Schoenmakers et al. [77, 88] who have reinvestigated current doubling reactions on single crystal ZnO using impedance and IMPS. The simplest mechanism of oxidation of organic species at ZnO involves capture of a valence band hole followed by injection of an electron by a radical intermediate. However it has been observed that photodissolution of ZnO takes place at the same time as current doubling, and alternative mechanisms involving formation of surface bound OH radicals by hole capture have been considered.

The upper limit to the electron injection rate constants that can be measured by IMPS is determined by τ_{cell} and potentiostat performance. If the space charge capacitance is high, it may be impossible to determine k_{inj}. This is the case, for example, for formic acid oxidation on TiO_2, where the IMPS response is dominated by the combination of the space charge and Helmholtz capacitances [84]. It would be interesting to explore the possibilities of using intensity modulated photopotential to overcome this limitation.

5 Photoelectrochemistry of Porous and Nanocrystalline Semiconductors

5.1 Introduction

Porous semiconductor electrodes consisting of interconnected structural units with dimensions in the nanometer range have attracted considerable scientific interest, and nanoporous electrodes are finding application in a range of novel devices, including dye sensitised photovoltaic cells [21], photonic crystals [181] and electrochromic windows [182]. Nanocrystalline materials can exhibit unusual optical and electrical properties that are not observed with bulk single crystal electrodes. For example, although bulk silicon has a bandgap of 1.1 eV, nanoporous silicon produced by electrochemical etching luminesces brightly in the visible region at much higher photon energies as the result of quantum confinement effects that lie outside the scope of this chapter [183]. Such effects are expected only for dimensions less than ~5 nm, and the present discussion is restricted to nanostructured networks with characteristic dimensions above this limit.

The remarkable properties of porous semiconductor electrodes can be rationalised by considering them as three-dimensional semiconductor/electrolyte interphases. The dimensions of the structural units in a nanostructured semiconductor electrode may range between 1 and 100 nm. An example of a nanostructured single crystalline GaP network consisting of large units (about 100 nm) is shown in Fig. 6. A particulate network consisting of smaller TiO_2 colloidal particles (ca. 30 nm in diameter) interconnected by slight sintering at 450 °C is illustrated in Fig. 31.

It is useful to compare the physical dimensions of nanocrystalline networks with the characteristic lengths used to describe charge distribution in the solid and electrolyte phases. The width of the depletion region for a bulk semiconductor with a doping density of 10^{17} cm^{-3} is of the order of 100 nm. By contrast, the Helmholtz electrical double layer is of molecular dimensions (the potential drop in the diffuse double layer can be neglected in concentrated electrolytes). It follows that conventional depletion and neutral regions can be formed in a porous semiconductor with a doping density of 10^{17} cm^{-3} provided the size of the structural units exceeds 100 nm, but for smaller particles this is no longer possible. If the dimensions of the pores are similar to those of the structural units, the Helmholtz electrical double layer is formed throughout the large internal surface area of the nanoporous network. As a consequence, nanoporous semiconductor electrodes behave as 'supercapacitors'

Fig. 31. SEM showing a porous nanocrystalline electrode consisting of an interconnected assembly of 30 nm colloidal TiO_2 particles. Electrodes of this type are used, for example, in dye sensitised solar cells.

under accumulation conditions. Their internal surface area can be estimated experimentally by measuring the interfacial capacitance under electron accumulation conditions, where the capacitance is determined primarily by the Helmholtz-layer, as shown for porous silicon [184].

The large surface/volume ratio characteristic of nanostructured systems can have several important consequences and applications. For example, the Grätzel dye-sensitised photovoltaic cell [21] exploits the high internal surface area to enhance light harvesting by a monolayer of sensitiser dye adsorbed on the surface of nanocrystalline TiO_2. An upper limit to the surface area/volume ratio can be estimated by assuming that the structure is built up from hexagonally close packed spheres of radius r (fractional volume occupancy 0.73). The ratio of surface area to total volume $A/V = 0.73\ 4\pi r^2/(4/3)\pi r^3 = 2.2/r$. For a $r = 15$ nm, for example, the A/V ratio is of the order of 10^6, and for a 1 cm^2 film of 10 micron thickness, this A/V ratio corresponds to an internal surface area of 1000 cm^2. In practice, lower values of internal surface area are measured experimentally because sintering introduces 'necks' between particles. A monolayer of dye adsorbed on such a high internal surface area will absorb almost all of the incident light in the wavelength region corresponding to its absorption spectrum. This efficient light harvesting contrasts with the situation at a flat electrode, where a monolayer of sensitiser dye absorbs less than 0.1% of incident radiation. The high internal surface area also has consequences for the density of interfacial electronic states. Surface state densities of 10^{12}–10^{13} cm^{-2} are common for semiconductor electrodes, and for a nanoporous electrode with an A/V ratio of 10^6, these translate into equivalent volume densities of 10^{18}–10^{19} cm^{-3}. These high densities enormously enhance light absorption by electrons in surface states. This effect is responsible for the unusually large photocurrent quantum yields for sub-bandgap light that have been measured with nanoporous GaP and SiC [127, 128, 134].

The steady state and dynamic photocurrent responses of nanostructured electrodes are clearly strongly influenced by their structure and by the interaction between the solid and solvent phases, but our present understanding of charge carrier transport is still rather limited. Since the diffusion length of photogenerated carriers, $L = (\pi D\tau)^{1/2}$, is usually larger than the physical dimensions of the structural units, holes and electrons can reach the interpenetrated electrolyte phase before being lost by bulk recombination (see section 2.1). As a result, the photocurrent quantum yield is determined entirely by competition between surface recombination and interfacial kinetics. Moreover, if, as is usually the case, one type of carrier is transferred preferentially from the solid to the solution, long-range carrier transport becomes monopolar. This situation contrasts with silicon-based diodes and other devices where electron-hole pairs may diffuse over long distances before being separated at a junction (bipolar diffusion) [158].

5.2 Carrier Generation and Transport in Nanocrystalline Electrodes

Two charge generation mechanisms can be distinguished. The first mechanism involves generation of electron-hole pairs by supra-bandgap illumination. Rapid reaction of the hole at the solid/electrolyte interface then leaves the electron to be transported to the collecting contact. The second mechanism involves photoexcitation of an adsorbed dye followed by electron injection into the conduction band of the solid. The dye is usually regenerated by reaction of its oxidised state with a supersensitiser such as I^-. In both cases, photoexcitation leads ultimately to conduction band electrons that are collected and measured as a photocurrent. The main differences in the physical description of carrier generation and collection arise from the fact that light absorption is more homogenous in the dye sensitised nanocrystalline electrodes, whereas direct band-band excitation is associated with a smaller penetration depth, at least in the case of direct bandgap semiconductors. The two photogeneration mechanisms are contrasted in Fig. 32.

Fig. 33 is a schematic illustration of a porous semiconductor electrode interpenetrated with a redox electrolyte. Two situations are shown: the dark equilibrium situation and the situation under constant illumination from the electrolyte side (comparable illustrations for a bulk semiconductor/electrolyte interface are given in Figs. 4 and 5). In the dark at equilibrium, the electron Fermi-level in the porous network, $E_{F,n}$, is equal to the Fermi-level of the redox system $E_{F,redox}(= -qU_{eq})$ and independent of the spatial co-ordinate x normal to the substrate. If an electrolyte with a sufficiently positive redox potential is chosen, $E_{F,n}$ can be located in the middle of the gap, which means that the density of electrons in the nanostructured network is

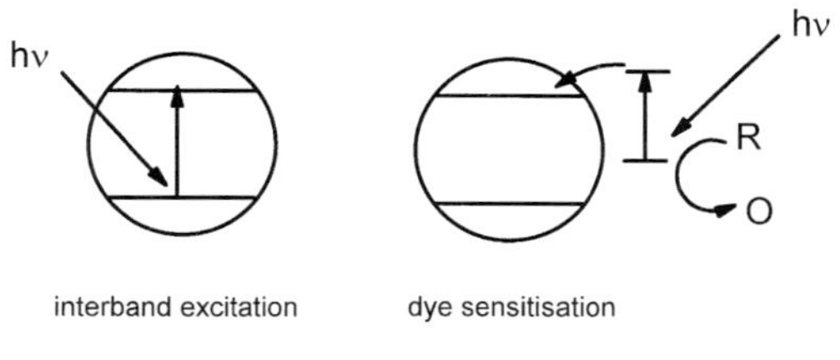

Fig. 32. Comparison of photoexcitation mechanisms in nanocrystalline electrodes. Interband excitation produces electron hole pairs in the solid. Sensitisation involves in electron injection from the photoexcited state of a dye adsorbed on the surface of the solid. The dye is regenerated by the supersensitiser R.

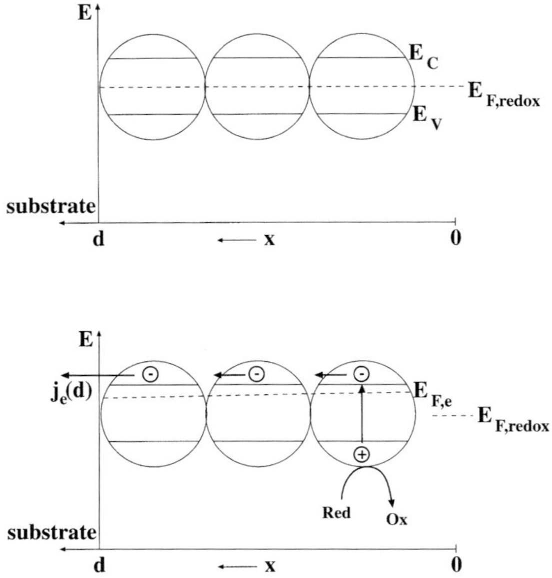

Fig. 33. Schematic representation of photoinduced current flow in a nanostructured electrode interpenetrated with a solution with a redox system Ox/Red. The band edges E_C and E_V are shown, together with the electron Fermi-level $E_{F,n}(x)$. The upper diagram illustrates the equilibrium situation in the dark when $E_{F,n}$ does not depend on x and is equal to the Fermi-level of the redox system. The lower figure shows what happens when under constant illumination from the electrolyte side. Photogenerated holes are consumed in oxidation of Red, and a gradient in $E_{F,n}(x)$ induces electron transport to the substrate. The photocurrent density is equal to $J_n(x = d)/q$.

negligibly small in the dark, i.e. the electrode is effectively insulating. Under illumination with a constant light intensity, a non-equilibrium steady state situation is established which can be described formally by the quasi-Fermi levels for electrons and holes, $E_{F,n}$ and $E_{F,p}$.

As discussed in section 2, photogenerated holes can readily reach the surface of the nanostructured network and it is assumed that they react with *Red* generating *Ox*. If the holes are consumed rapidly, $E_{F,p}$ will remain close to $E_{F,redox}$. In many systems, by contrast, it appears that electrons react much more slowly at the interface than holes. The reason for this kinetic selection of carriers is not clear at present, but it is clearly a prerequisite for the observation of substantial photocurrent efficiencies. If electrons do not react rapidly at the interface, their concentration increases substantially under illumination, so that $E_{F,n}(x)$ becomes higher than $E_{F,redox}$. Since electrons are extracted at the substrate, illumination induces an electron flux, $J_n(x)$, through the nanostructured system. Under steady state conditions the photocurrent density, j_{photo}, is equal to $qJ_n(x = d)$. The electron flux is proportional to the gradient of the electron quasi-Fermi $[dE_{F,n}(x)/dx]$, which represents a gradient of free energy, in other words a 'driving force'. Since the photopotential in a regenerative nanocrystalline solar cell is determined by $[E_{F,n}(x = d) - E_{F,redox}]$, the drop of $E_{F,n}(x)$ over the nanostructured system required to ensure unidirectional flow of charge carriers is effectively an internal voltage loss.

At this point it may be asked if the driving force for electron movement is due to a gradient in the potential energy of the electrons or to a gradient in their concentration, or to both. Unfortunately, there appears to be no clear **a priori** answer to this question. As shown below, a gradient in the concentration of free electrons induces a 'chemical' driving force of the order of $k_B T$ divided by the thickness d of the network. Photogenerated electrons can be trapped in surface states and induce a change of the potential drop over the Helmholtz-layer. If, as a result, the potential drop over the Helmholtz-layer becomes dependent on the co-ordinate x, $[dE_{F,e}(x)/dx]$ may also have an electrostatic component due to a gradient in the potential energy of the electrons. A recent study by Zaban et al. [185] has related spatially dependent dye desorption in nanoporous TiO_2 films to the local potential, and this work suggests that the potential energy of electrons may indeed vary with distance. The related question whether the photoinduced current flow can be described in terms of diffusion or migration is non-trivial: solutions of the steady state, transient and periodic photocurrent responses on the basis of the differential rate equations require appropriate boundary conditions, and these are usually specific for the type of transport mechanism.

It is interesting to compare a nanocrystalline solar cell with a solid state $p^+ - i - n^+$ solar cell. Electrons photogenerated in the intrinsic region of a $p - i - n$ cell layer travel towards the n^+ contact, while the holes travel towards the p^+ contact, and in the steady state, the flux of electrons arriving at one side is equal to the flux of holes arriving at the other side. In a cell with a nanostructured electrode, one type of carrier (usually the electron) travels through the semiconductor network. The other (the hole) reacts at the interface, so that the balancing carrier flux in the nanocrystalline cell is due to the movement of oxidised redox ions that must reach the counter electrode to be regenerated electrochemically. Under steady-state conditions, the number of electrons arriving at the substrate is equal to the number of oxidised species arriving at the counter electrode. In the case of the solid $p^+ - i - n^+$ diode, the conditions can be chosen such that the dynamics of either photogenerated electrons or photogenerated holes can be probed by transient photocurrent measurements (see e.g. [186]). In the case of porous and nanocrystalline electrodes the situation is less clear. It has generally been assumed that the transient photocurrent response is determined by the electron transport properties in the solid phase, but recent experimental evidence suggests that ion transport and electron transport may not be independent [167]. Since concentrated inert electrolytes are commonly used in nanocrystalline cells to minimise resistance losses, the transport of redox ions may be expected to occur predominantly by diffusion. However, electron transport in a porous nanocrystalline phase is likely to be strongly coupled to relaxation of the ionic atmosphere in the solution phase, so that the diffusion coefficient may be sensitive to changes in solution composition.

Electron transport in the nanoparticles may be influenced by bulk and surface scattering and trapping. If the mean free path of the electron is much smaller than the particle radius, additional surface scattering will not have much effect on the movement of electrons from particle to particle. However, if the mean free path of the electrons is larger than the particle radius, scattering at the surface becomes important. The geometry of the junction between particles is also likely to influence carrier transport; the presence of a 'neck' at the junction may inhibit movement of electrons from one particle to the other, and disorder at the junction may introduce additional scattering of electrons.

It has become clear from transient and periodic photocurrent measurements that the apparent diffusion coefficient of electrons in nanocrystalline TiO_2 is several orders of magnitude smaller than in the bulk crystalline material [80, 90, 186]. However the reasons for the remarkably slow electron transport in nanostructured porous electrodes are not clear at present. During their travel through a nanoporous system, photogenerated electrons may be trapped in localised energy levels in the forbidden gap. Such states can originate from crystal defects and from the semiconductor/electrolyte interface. As pointed out in the preceding section, the volume density of interfacial states can be very high (10^{18}–10^{19} cm^{-3}). It is therefore likely that interfacial states are more important than localised bulk states. If hole transfer is rapid, these surface states will act predominantly as electron traps. Attenuation of electron transport is mostly due to temporary localisation of electrons in the deepest unoccupied traps, i.e. those near the demarcation level ($f(E) \approx 1$ in the energy region below the demarcation level E_n; see Eq. 23).

The negative charge of electrons trapped at surface states is balanced by an ionic counter charge in the outer Helmholtz plane, and the resulting change in the potential difference across the interface shifts the energy of the band edges. Local polarisation of the solvent molecules by charge carriers may also lead to surface localisation (polaron formation), reducing surface mobility [152]. Since the surface area/volume of nanocrystalline electrodes is very large, such effects may retard electronic transport.

The novel properties of nanocrystalline electrodes are reflected in their response to pulsed or periodic perturbations of light intensity. The following sections describe in some detail the generation/collection problem, but at this point it is useful to provide some illustrations of the kind of unusual experimental behaviour that needs to be explained by any successful physical and mathematical model. The first point worth noting is that the photocurrent responses of nanocrystalline electrodes to stepped or pulsed illumination exhibit features on rather slow timescales. This is illustrated, for example, by Fig. 34, which is a set of photocurrent transients reported by Solbrand et al. [186] for band-band excitation at 308 nm of nanocrystalline TiO_2 films of differing thicknesses permeated by 0.7 mol dm^{-3} $LiClO_4$ in ethanol. The 30 ns excimer laser pulse was incident from the solution side, and since the penetration depth of the light was much smaller than the film thickness, electron-hole pairs were effectively generated in a thin sheet at the front surface of the film.

It can be seen that the photocurrent exhibits a peak on a timescale of milliseconds. The insert in the figure shows that the time at which this peak appears varies approximately with the square of the film thickness. It is immediately apparent that the timescales involved in these transients are may orders of magnitude longer than those typical for charge transport in bulk semiconductor electrodes.

The second point worth noting is that the response of nanocrystalline TiO_2 electrodes to large amplitude excursions of light intensity such as a light step is nonlinear. Furthermore, the shape of the photocurrent response is intensity dependent. The photocurrent transients for square pulse illumination of dye sensitised TiO_2 films reported by Cao et al. [80] and shown in Fig. 35 illustrate this very well.

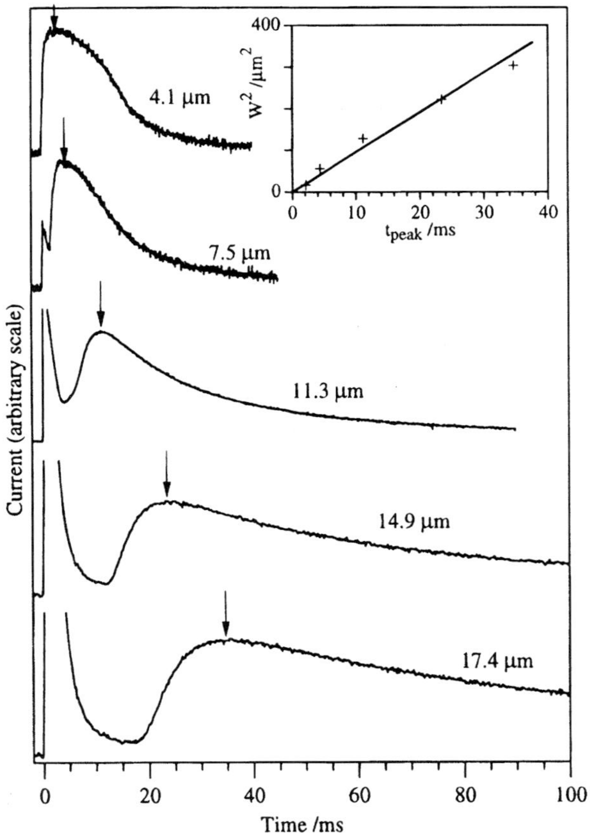

Fig. 34. Experimental photocurrent transients for pulsed excimer laser excitation of nanocrystalline TiO_2 electrodes of differing thicknesses taken from [186]. Illumination from the electrolyte side (200 mJ, 30 ns, λ 308 nm). Electrolyte 0.7 mol dm^{-3} $LiClO_4$ in ethanol. The insert shows that time t_{peak} at which the current peak occurs depends on the square of the film thickness (W), as expected for diffusion controlled electron transport.

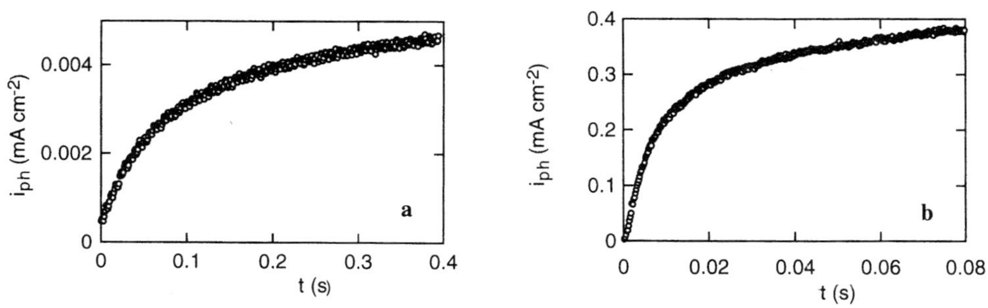

Fig. 35. Transient short circuit photocurrent response of dye sensitised TiO_2 cell to light step excitation (taken from [80]). λ 514 nm. Intensities: a) 0.05, b) 4 mW cm^2. Note that the current rises more rapidly at the higher intensity.

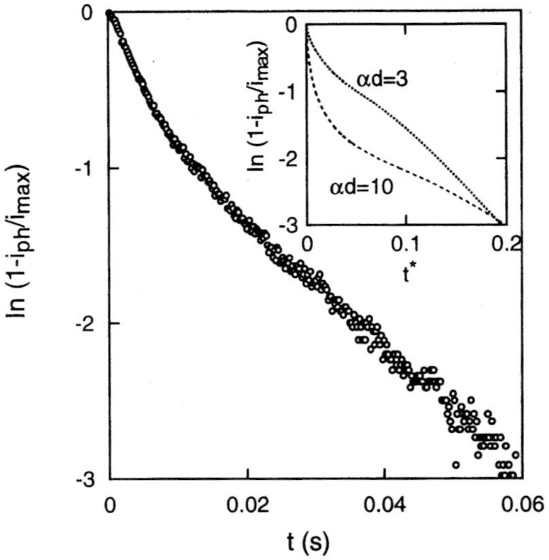

Fig. 36. Semilogarithmic plot of photocurrent transient shown in Fig. 35b plot (taken from [80]. The inset shows theoretical plots calculated from Eq. 88 (see also Fig. 40).

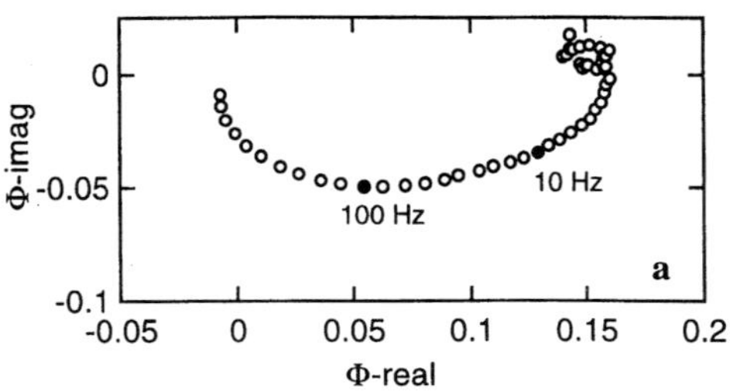

Fig. 37. IMPS response of dye sensitised TiO_2 cell (λ 514 nm, background intensity 4 mW cm^{-2}). Illumination from the substrate side (taken from [80]).

It can be seen that the risetime of the photocurrent in response to the illumination step is much faster when the light intensity is higher (b). The rise in photocurrent clearly does not correspond to a simple first order process, as is shown by the non-linear semilogarithmic plot of the photocurrent rise in Fig. 36. Marked non-linearity has also been observed in time of flight experiments of electron and hole transport in amorphous silicon [167].

The non-linear nature of the photocurrent transients suggests the use of small amplitude perturbations in order to permit linearisation of the system response. This is readily achieved in IMPS by superimposing a small modulation on a much larger dc illumination intensity. A further clear indication of the differences between bulk and nanocrystalline electrodes is given by the IMPS plot reported in the paper by Cao et al. and shown in Fig. 37. The most striking feature of the IMPS response is that it appears in a different quadrant of the complex plane than the response for a bulk semiconductor electrode in the absence of *RC* effects ($\omega < 1/\tau_{cell}$). The real component is positive, but the imaginary component is negative. This means in effect that the measured photocurrent lags behind the illumination. This behaviour corresponds to the slow rise in photocurrent in response to a step illumination function. By contrast, for $\omega < 1/\tau_{cell}$, bulk semiconductor electrodes at which surface recombination occurs exhibit an IMPS semicircle in the upper quadrant, i.e. the imaginary component is positive. In this case, the photocurrent appears to lead the illumination function. This seems odd at first sight, but in fact it is a simple consequence of the fact that recombination induces a majority carrier current that is opposite in sign to that of the photogenerated minority carrier current. This difference in sign corresponds to a phase shift of 180° that is reflected in the net IMPS response, which is the sum of the majority and minority carrier currents.

5.3 Models and Rate Equations

A quantitative interpretation of transient or periodic photocurrents in nanoporous networks requires a physical and mathematical description of the generation and collection of charge carriers. The exact treatments of the problem that have appeared in the literature are based on the assumption of either diffusion or migration as the predominant transport mechanism [78, 90]. A more general treatment that accounts for both diffusion and migration in response to a photoinduced gradient of the electrochemical potential is not yet available. Recently an attempt has been made to treat the problem within the framework of statistical mechanics [187].

In this section, rate equations are used to describe the behaviour of photogenerated electrons and holes in a nanostructured semiconductor electrode. It is assumed that holes are trapped in interfacial states or transferred close to their point of generation to a reducing species in the solution permeating the network. Photogenerated electrons diffuse (or migrate) through the network and are collected in the substrate. The rate equation describing the hole concentration is

$$\frac{dp(x,t)}{dt} = \alpha I(x,t) - k_V p(x,t)[Red] - p(x,t)\beta_p \int_{E_{VB}}^{E_{CB}} X(E) f(E,x,t)\, dt \tag{72}$$

The first term describes the photogeneration of holes for illumination via the electrolyte, the second term represents reaction of holes with the reduced species and the third term describes irreversible trapping of holes by a distribution of trap states. The trapped holes may be consumed in the oxidation of the reducing agent or by recombination with electrons (see section 5.2).

Photogenerated electrons move through the network towards the substrate due to a gradient of the electrochemical potential $dE_{F,n}(x,t)/q\,dx$, where $E_{F,n}(x,t)$ is given by

$$E_{F,n}(x,t) - E_F^{ref} = -q[\phi(x,t) - \phi^{ref}] + k_B T\, ln\left(\frac{n(x,t)}{n^{ref}}\right) \tag{73}$$

$E_{F,n}(x,t)$ is given with respect to a reference level E_F^{ref}; ϕ^{ref} and n^{ref} are reference values of electric potential and electron concentration. During their transport through the nanoporous network, electrons can be trapped and released from localised states or they may recombine. Generation, transport and recombination of electrons are described by the expression

$$\frac{dn(x,t)}{dt} = \alpha I(x,t) - \frac{dJ_n(x,t)}{dx} - \int_{E_{VB}}^{E_{CB}} X(E)\frac{df(E,x,t)}{dt}\,dE - J_{R,n}(x,t) \tag{74}$$

where $J_n(x,t)$, the electron flux in the direction of increasing x, is proportional to the gradient of the electron quasi Fermi level:

$$J_n(x,t) = \mu_n n(x,t)\left[\frac{-dE_{F,n}(x,t)}{qdx}\right] \tag{75}$$

The term $-\int X(E)[df(E,x,t)/dt]\,dE$ in Eq. 74 accounts for trapping in and thermal release from electronic levels distributed in energy in the band gap. Only those states are taken into account from which trapped electrons are released thermally without loss by recombination; these are states of energy equal to or above the demarcation energy E_n (see section 2.3.3). The term $J_{R,n}(x,t)$ accounts for irreversible loss of photogenerated electrons by recombination. 'Recombination' here is defined to include not only electron hole recombination mediated by bandgap states below E_n, but also electron transfer to the oxidised species of the redox system in solution [188]. The redox species plays a role analogous to a surface trapping state in the sense that it captures successively a hole and an electron. In the case of dye sensitised nanocrystalline electrodes, $J_{R,n}$ may also contain a contribution due to back reaction of electrons with the oxidised state of the dye that is generated by electron injection [189, 190].

5.4 Steady State Solutions: The Diffusion Controlled Case

Solution of Eq. 74 for the steady state gives the concentration profiles of photogenerated carriers in the nanostructured electrode. A priori separation of migration and diffusion is difficult, and most analytical models have been based on the assumption that diffusion is predominant. Therefore in order to simplify the analysis, the boundary conditions are chosen to be appropriate for diffusion controlled transport. Initially it is assumed that recombination is absent. With $dn(x,t)/dt$ and $df(E,x,t)/dt$ equal to zero, Eq. (74) simplifies to

$$\alpha I(x) - \frac{dJ_n(x)}{dx} = 0 \tag{76}$$

It follows that $dJ_n(x)/dx$ is equal to the generation profile $\alpha I(x) = \alpha I(0)\,exp(-\alpha x)$. Since the boundary condition $J_n(0) = 0$ [i.e. $[dn(x)/dx] = 0$ for $x = 0$] must hold, the flux of electrons at x is obtained as

$$J_n(x) = I(0)[1 - e^{-\alpha x}] \tag{77}$$

The steady state photocurrent density, j_{photo}, is determined by the flux of electrons arriving at and collected by the substrate $(x = d)$

$$j_{photo} = qJ_n(x = d) = qI(0)[1 - e^{-\alpha d}] \tag{78}$$

Since it has been assumed that the driving force responsible for the electron flux in the direction towards the substrate is the gradient in the chemical potential of the electrons (i.e. diffusion controlled transport), it follows from Eq. 73 that the gradient of the quasi Fermi level of electrons is given by

$$\frac{dE_{F,n}(x)}{dx} = \frac{k_BT}{n(x)}\frac{dn(x)}{dx} \tag{79}$$

Substitution of Eq. 79 in Eq. 75 gives Fick's first law of diffusion in the form

$$J_n(x) = \mu_n n(x)\left[-\frac{dE_{F,n}(x)}{qdx}\right] = -\mu_n\frac{k_BT}{q}\frac{dn(x)}{dx} = -D\frac{dn(x)}{dx} \tag{80}$$

where $D = \mu_n(k_BT/q)$ is the diffusion coefficient of electrons in the nanoporous electrode. The concentration gradient $[dn(x)/dx]$ $(= -J_n(x)/D)$ is shown in Fig. 38.

The concentration profile of free electrons can be calculated from Eqs. 78 and 80

$$n(x) = n(d) + \left[\frac{I(0)}{\alpha D}\right]\{\alpha(d - x) - (e^{-\alpha x} - e^{-\alpha d})\} \tag{81}$$

It should be noted that this equation describes the concentration profile of free electrons diffusing towards the substrate for the case where the electrode is illuminated from the electrolyte side (i.e. $x = 0$). The concentration profile according to Eq. 81 is also plotted in Fig. 38.

The gradient of free energy, i.e. the driving force for electron flow, due to the concentration gradient follows from Eq. (79), (80) and (81)

$$\frac{dE_{F,n}}{dx} = \frac{k_BT}{n(x)}\frac{dn(x)}{dx} = \frac{-k_BT\dfrac{I(0)}{D}[1 - e^{-\alpha x}]}{n(d) + \dfrac{I(0)}{\alpha D}\{\alpha(d - x) - (e^{-\alpha x} - e^{-\alpha d})\}} \tag{82}$$

It is interesting to note that if $n(d)$ and D are sufficiently small, the driving force is independent of the incident light intensity $I(0)$. In addition, the driving force does not depend strongly on the spatial co-ordinate x in the region $1/\alpha < x < d$. It follows from Eq. 82 that if $n(d)$ can be neglected, the average driving force associated with the concentration of the free electrons is of the order of k_BT divided by the thickness d of the nanostructured film.

The steady state solution of Eq. 74 is still valid if it is assumed that electrons are trapped/detrapped in localised states. However, electrons trapped in interfacial states induce a counter charge in the interfacial double layer and hence a change in the potential drop over the Helmholtz-layer and the energy of the conduction band edge at the surface. Trapping could therefore affect transport of photogenerated electrons by introducing a non-negligible field-component into the driving force.

Photogenerated electrons in a nanocrystalline electrode may react with redox species before they are collected at the back contact. Södergren et al. [189] have considered generation, diffusion and back reaction of electrons with I_3^- ions in the

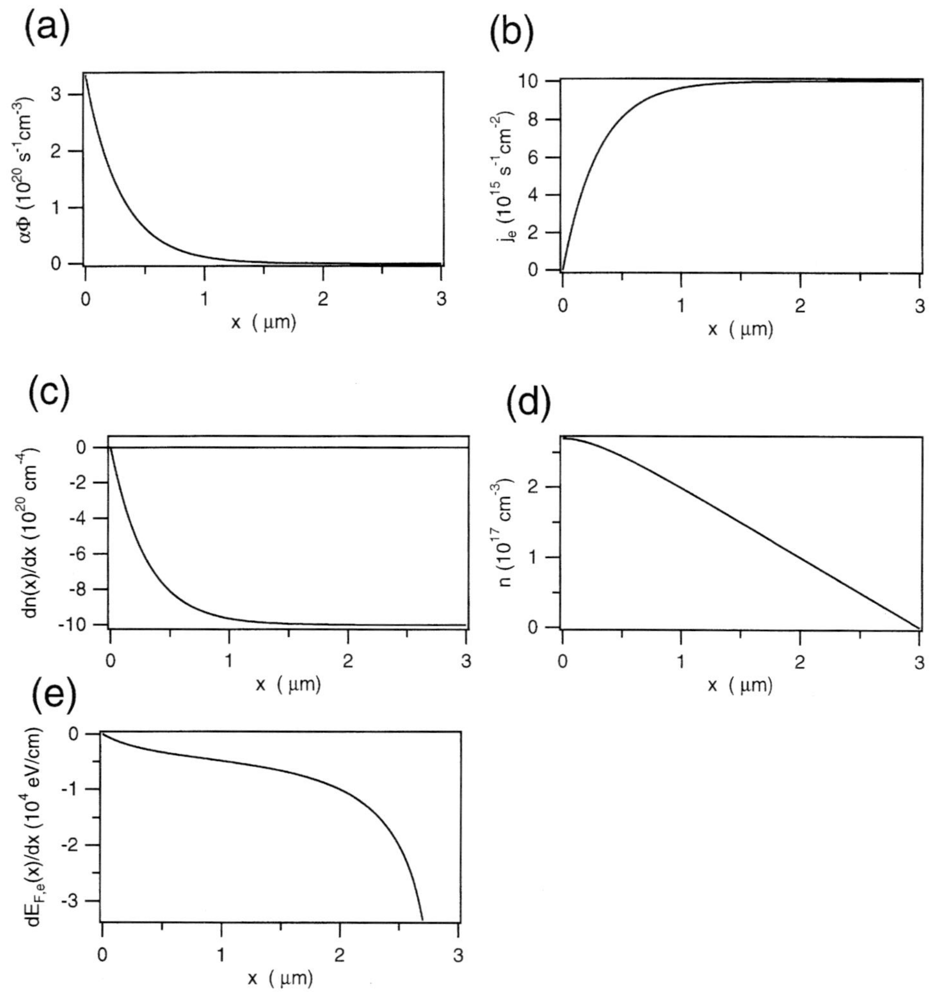

Fig. 38. Steady state solutions for boundary conditions typical for diffusion controlled transport of photogenerated electrons through the substrate. The film thickness is 3 microns, and illumination is from the electrolyte side. The absorption depth $(1/\alpha)$ is 1/10 of the thickness d. a) the generation profile $\alpha I(x) = dJ_n(x)/dx$, b) the electron flux $J_n(x)$, c) concentration gradient $dn(x)/dx$, d) the profile of excess free carrier concentration $n(x)$, e) the driving force for electron diffusion in eV cm^{-1}.

dye sensitised nanocrystalline TiO_2 cell. The time dependence of the electron density, $n(x, t)$, is given by

$$\frac{\partial n(x,t)}{\partial t} = \eta \alpha I_0 e^{-\alpha x} + D\frac{\partial^2 n(x,t)}{\partial x^2} - \frac{n(x,t) - n_0}{\tau} \tag{83}$$

where D is the diffusion coefficient of electrons, n_0 is the electron density in the dark, τ is the pseudo-first order lifetime of electrons determined by back reaction with triiodide ions in the electrolyte, α is the effective absorption (cm^{-1}) coefficient of the dye sensitised TiO_2 (determined by the absorption cross section and surface coverage of the dye), I_0 is the incident photon flux corrected for reflection losses and η is the net efficiency of electron injection. Eq. 83 is also appropriate for bandgap excitation of nanocrystalline materials if one type of photogenerated carriers reacts very rapidly, leaving the other type as excess carriers to be collected at the back contact. Solution of Eq. 83 presents a number of difficulties, even if it is assumed that electron transport is by diffusion. It is not possible to assume **a priori** that electron diffusion in TiO_2 is independent of the total electron density since the effective diffusion coefficient of electrons depends on trapping/detrapping, so that D is a function of n and consequently also of x. Similarly the effective electron lifetime τ depends on trapping/ detrapping so that it too may depend on n and x. Finally the net efficiency, η, for electron injection is likely to be affected by the total electron density as a consequence of back reaction of electrons with the oxidised state of the sensitiser dye. Ideally, the dye should be regenerated from its oxidised form by capture of an electron from iodide, but under strong electron accumulation, electrons can instead be captured from the TiO_2, so that η becomes dependent on the total electron density [188].

Södergren *et al.* obtained the steady state solution of Eq. 83 by assuming that D and τ are constant and $\eta = 1$. Under these conditions, the photocurrent due to excess carriers is independent of voltage, and the steady state photocurrent is given by

$$j_{photo}\Big|_{t\to\infty} = \frac{qI_0\alpha L\left(-sinh\left(\frac{d}{L}\right)\right) - \alpha L\,exp(-\alpha d) + \alpha L\,cosh\left(\frac{d}{L}\right)}{(\alpha^2L^2 - 1)cosh\left(\frac{d}{L}\right)} \tag{84}$$

where d is the film thickness and $L = (D\tau)^{1/2}$ is the electron diffusion length. It should be noted that the photocurrent predicted by this model is independent of voltage. This is physically implausible, of course. Neglect of the back reaction of electrons with the oxidised dye is no longer reasonable under conditions of strong electron accumulation, in other words η is expected to decrease. Also the assumption that there is no barrier to electron extraction at the substrate is questionable. If such a barrier exists, the boundary condition at $x = d$ must be modified to take into account the (possibly voltage dependent) electron exchange kinetics [90].

5.5 Transient Photocurrent Response of Porous and Nanocrystalline Electrodes

One of the most remarkable features of the transient photocurrent response of nanocrystalline TiO_2 electrodes is the very slow rise and fall times observed at low light intensities; in some cases the risetime can extend to seconds [80, 191]. Interestingly, the risetimes became faster with increasing light intensity. The influence of electron trapping and release on the transient photocurrent response of dye sensitised nanocrystalline cells was discussed in an early paper by Schwarzburg and Willig [191], who considered a two level trap model to explain the experimental observation that the photocurrent risetime in response to a square light pulses is non-linear and intensity dependent. By contrast, Solbrand *et al.* [186] have discussed time resolved photocurrent transients in nanocrystalline TiO_2 in terms of an electron diffusion model that ignores trapping effects (the current transients that they observed are illustrated in Fig. 34). In this work, a relatively intense 200 mJ 30 ns excimer laser pulse (308 nm) was used to generate electron-hole pairs in a thin layer on the electrolyte side of nanocrystalline TiO_2 films of different thicknesses. It was assumed that the photogenerated holes react rapidly leaving excess electrons in the nanocrystalline particles. Subsequent transport of electrons towards the back contact is determined by the gradient of the electron quasi-Fermi level in the film. Solbrand et al. propose that initially electrostatic repulsion between electrons will force some of the electrons towards the rear contact giving rise to an exponentially decaying current, $j_{photo}(t)$ that can be approximated by an RC equivalent circuit.

$$j_{RC}(t) = \frac{Q'}{\tau_{RC}} e^{-(t/\tau_{cell})} \tag{85}$$

where Q′ is the charge density induced at the back contact and τ_{cell} is the RC time constant of the cell. It is assumed that the transport of electrons after the RC charging pulse occurs by diffusion, since electron-electron repulsion terms become negligible due to effective screening by the electrolyte. Electron trapping and back reaction of electrons with species in the electrolyte is neglected, and electron-hole recombination is presumed to be important only at times much shorter that τ_{RC}. The initial concentration distribution is approximated by a delta function since the penetration depth of the illumination is small compared with the film thickness. Solution of Eq. 83 for $\tau = \infty$ and D independent of n and x gives $n(x, t)$:

$$n(x, t) = \frac{\Delta n}{2\sqrt{\pi D t}} e^{-(x-d)^2/4Dt} \tag{86}$$

where Δn is the number of photogenerated electrons, D is the electron diffusion coefficient, x is the distance from the substrate and d is the film thickness. The diffusion controlled current at the back contact is found to be

$$j_{diff} = \frac{qd\Delta n}{2\sqrt{\pi D} t^{3/2}} e^{-(d^2/4Dt)} \tag{87}$$

The total photocurrent is described by the sum of Eq. 85 and 87.

Fig. 39 illustrates the theoretical electron density profiles and photocurrent transients calculated from Eq. 85 and 87. If the RC time constant is small, differentiation of Eq. 87 with respect to t shows that the photocurrent transients exhibit a maximum at a time $t_{peak} = d^2/6D$. In the limit of small perturbations, this time is expected to be inversely related to the characteristic frequency in the IMPS response.

The experimental photocurrent transients obtained by Solbrand *et al.* for TiO_2 films of different thickness are shown in Fig. 34. Similar transients have been reported by Hoyer and Weller [192] for nanocrystalline ZnO films. The inset in

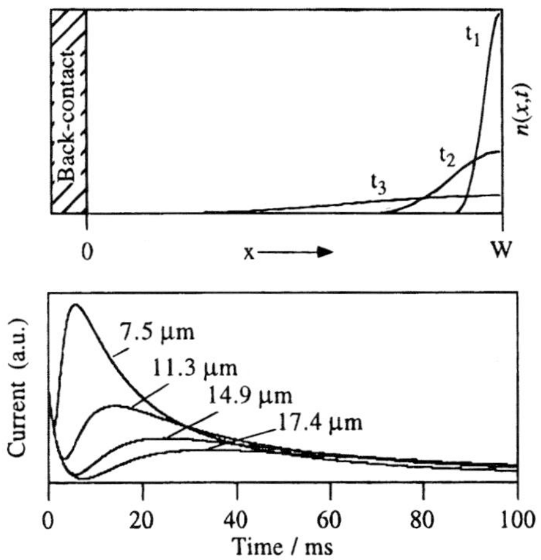

Fig. 39. Illustration of electron density profiles at different times $t_1 < t_2 < t_3$ following pulsed excitation from the electrolyte side. W is the film thickness. The lower part of the figure shows transients calculated from Eq. 86 and 87 for different film thicknesses (taken from [186])

Fig. 34 shows that t_{peak} varies approximately with d^2 as predicted (the authors use W rather than d to denote the film thickness), and the slope of the plot gives a value of $1.5 \times 10^{-5}\,cm^{-2}\,s^{-1}$ for the diffusion coefficient of electrons. This value is orders of magnitude below that measured for the electron diffusion coefficient in bulk anatase and rutile. Solbrand *et al.* observed that t_{peak} and hence D appear to depend on electrolyte conductivity, and this together with the magnitude of D suggests that diffusing electrons interact strongly with the solvent and ions in the electrolyte that permeates the film.

Cao *et al.* [80] have also discussed time and frequency dependent solutions of the generation collection equation for the case where the cell is illuminated from the substrate side. Like Solbrand et al., they assume that electron transport occurs predominantly by diffusion at relatively low light intensities where the electric field arising from reconfiguration of electrons is small. The expression derived for the response to a light step shows that the rise of the photocurrent is multi-exponential.

$$
\begin{aligned}
j_{photo}(t) &= qD\frac{\partial n(x,t)}{dt} \\
&= j_{photo}\Big|_{t\to\infty} - \sum_{k=0}^{\infty} C_k \frac{\pi}{2d}(2k+1)\,exp\left[-\left(\frac{1}{\tau}+\frac{D\pi^2(2k+1)^2}{4d^2}\right)t\right] \qquad (88)
\end{aligned}
$$

where the first term is the steady state photocurrent defined by Eq. 84. Figure 40 illustrates normalised transients predicted by Eq. 88 for different values of the product αd, where α is the absorption coefficient and d is the film thickness (the normalised time t^* in the figure is defined by $t^* = Dt/d^2$). These are also shown in semi-logarithmic form as an insert in Fig. 36.

The theoretical transients can be compared with the experimental photocurrent transients shown in Fig. 36. It is clear that the general features of the experimental transient are reproduced in the theoretical plots.

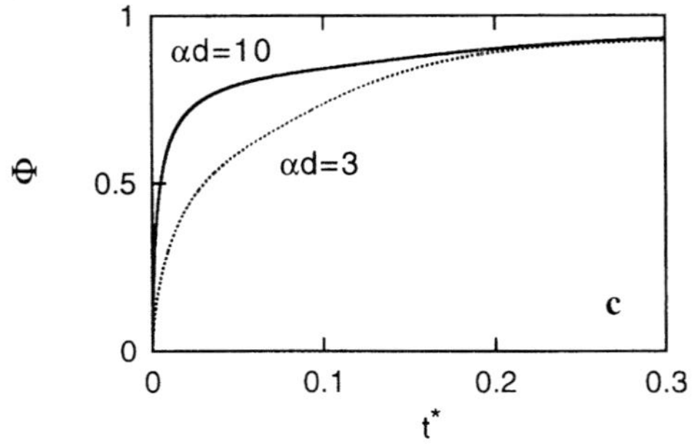

Fig. 40. Theoretical dimensionless photocurrent transients calculated from Eq. 88 for two different values of α d, where α is the absorption coefficient and d is the film thickness. The dimensionless time t^* is defined by $t^* = Dt/d^2$ (taken from [80]).

Cao *et al.* and other authors [191] have observed that the photocurrent risetime decreases as the light intensity is increased. The risetime appears to follow a power law of the form $t_{1/2} \propto I_0^n$ with $n = 0.5$–0.6. This effect may arise from intensity dependent occupancy of electron traps. Cao et al. therefore assumed as a first approximation that the diffusion coefficient of electrons is a linear function of electron concentration. Numerical solution of Eq. 88 then yields transients that depend on light intensity. In principle this approach allows the incorporation of any arbitrary dependence of D on n., but a more satisfactory approach would be to relate the diffusion coefficient directly to the trap occupancy or to separate diffusion and trapping. Cao et al. estimate an upper limit of $D_n = 10^{-7}\,\mathrm{cm^2\,s^{-1}}$ for the diffusion of electrons in the dark, whereas the D_n value for intensities corresponding to solar illumination levels are two orders of magnitude higher.

The dynamics of charge transfer and back reaction in dye sensitised nanocrystalline systems such as SnO_2 have also been studied extensively by transient optical and microwave absorption spectroscopies [193–199].

5.6 IMPS Response of Porous and Nanocrystalline Electrodes

Interpretation of photocurrents generated by large amplitude perturbations is complicated by the fact that it is not possible to treat D and τ as constant if the electron density is changing by orders of magnitude during the perturbation. Meaningful analytical non steady state solutions of Eq. 74 or 83 can, however, be obtained for small amplitude perturbation conditions where D and τ can be treated as constants to a first approximation. In practice, small amplitude means that the amplitude of the light pulse or modulation are small compared to the *dc* illumination level. It is common practice to use small modulation amplitudes for IMPS measurements in order to ensure that linearisation of the response is a valid approximation. By contrast, light pulse or light step measurements invariably involve perturbation from the dark state. Even in the case of IMPS, the variation of excess carrier density with distance poses a problem, since it may cause D and τ to vary across the film. Nevertheless, it is clear that IMPS is a better than time resolved methods for the deconvolution of the factors affecting the photocurrent response.

IMPS measures the photocurrent density $j_{photo}(\omega)$ due to the movement of electrons photogenerated by a harmonically varying light intensity $I_{ac} = I_m\, exp(-i\omega t)$ superimposed on a background light intensity $I(0)$. The relationship between the ac photocurrent response and the flow of carriers in the semiconductor is discussed in section 3.2. Here, the low frequency range $0 < \omega < 1/\tau_{cell}$ is considered. It follows from Eq. 32 that the flux of photogenerated electrons through the nanoporous network, $J_n(x, \omega)$ is equivalent to $j_{photo}(\omega)$ in the external circuit.

$$j_{photo(\omega)} = \frac{q}{d}\int_0^d J_n(x, \omega)\, dx \tag{89}$$

From Eq. 75, it follows that the electron flux at any point in the film is given by

$$J_n(x, \omega) = \mu_n n(x, \omega)\left[\frac{-dE_{F,n}(x)}{qdx}\right] + \mu_n n(x)\frac{-dE_{F,n}(x, \omega)}{qdx} \tag{90}$$

As shown above, the 'chemical' part of the driving force does not depend on the light intensity if the substrate is a perfect sink for photogenerated electrons. In this case, the second term on the right hand side of Eq. 90 can be neglected to a first approximation. Also since the driving force depends only weakly on the spatial coordinate x, $[dJ_n(x, \omega)/dx]$ can be approximated by

$$\frac{dJ_n(x, \omega)}{dx} = \mu_n \frac{d}{dx}\left\{n(x, \omega)\left[\frac{-dE_{F,n}(x)}{qdx}\right]\right\} \cong \mu_n \frac{dn(x, \omega)}{dx}\left[\frac{-dE_{F,n}}{qdx}\right] \tag{91}$$

For the IMPS case, the rate equation for the ac component of the electron concentration can be derived from Eq. 74:

$$\frac{dn(x, \omega)}{dt} = i\omega n(x, \omega) = \alpha I(x, \omega) - \frac{dJ_n(x, \omega)}{dx}$$
$$- \int_{E_V}^{E_C} X(E) i\omega f(E, x, \omega)\, dE - J_{n,R}(x, \omega) \tag{92}$$

It has not yet proved possible to obtain general solutions of Eq. 92 without making **a priori** assumptions about whether electrons move primarily by diffusion or migration.

The experimental results shown in Fig. 37 indicate that the transport of electrons from the front of the film to the substrate is associated with a measurable transit time, $\tau(d)$, that depends on the film thickness, d. A simplified zero order treatment of the problem is considered first to illustrate the gross features of the IMPS response and to show that in the absence of recombination and trapping, the electron transit time should in principle be accessible from IMPS measurements. It is assumed that the electrode is illuminated from the solution side and that the penetration depth of the light is much smaller than the film thickness. The driving force for electron flow is taken to be constant throughout the nanoporous film, and it is assumed that the

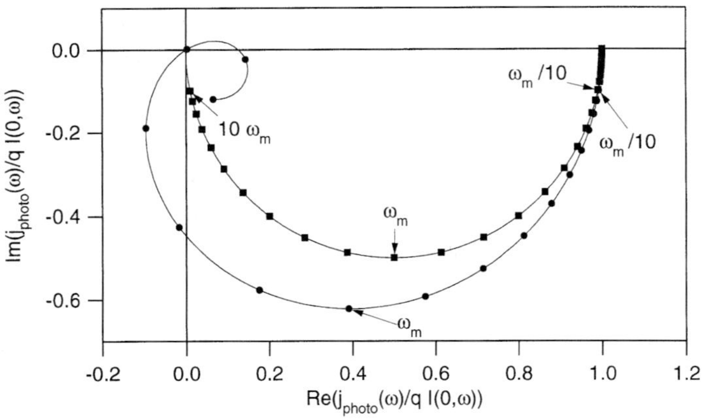

Fig. 41. The simple semicircular IMPS response is predicted by Eq. 93 for the case where photogenerated carriers diffuse or migrate through a nanocrystalline network without being trapped. The second IMPS plot is predicted by a more exact treatment of field driven electron transport [78]. Note that in this case, the plot crosses the imaginary axis at high frequencies and spirals into the origin.

harmonic component of the electron concentration decreases linearly in magnitude with decreasing x towards the substrate such that the derivative $dn(x,\omega)/dx$ can be approximated by $-n(x,\omega)/(d-x)$. The spatial derivative of the electron flux is then linear in $n(x,\omega)$, and the normalised IMPS response can be obtained from Eq. 92 as

$$\frac{j_{photo}(\omega)}{qI(0,\omega)} = \frac{1}{1+i\omega\tau_d} \tag{93}$$

Fig. 41 shows that the complex plane IMPS plot of Eq. 93 is a semicircle with a positive real and a negative imaginary part. The low and high frequency intercepts are 1 and 0 respectively, and the inverse of the characteristic frequency ω_{min} gives the transit time τ_d.

Vanmaekelbergh et al. [78] have given a more exact treatment of the transport problem for boundary conditions pertaining to field-driven electron transport (migration) and illumination through the electrolyte. As shown in Fig. 41, the calculated IMPS response in this case is a distorted semicircle characterised by a change in the real component from positive to negative values in the frequency range above $1/\tau_d$. It can be seen that the IMPS signature spirals into the origin at high frequencies. This reflects the fact that the transit time is effectively a propagation delay for the signal.

Transport of photogenerated electrons through nanostructured networks has been studied with porous single crystalline GaP networks [78] (see Fig. 42). A more detailed SEM micrograph is shown in section 2.2. Single crystalline networks of GaP with strongly interconnected structural units in the 100 nm range can be prepared on a n-GaP substrate by anodic etching under conditions of dielectric breakdown. The thickness of the porous film can be controlled coulometrically, and films of thickness between a few μm and 200 μm have been investigated.

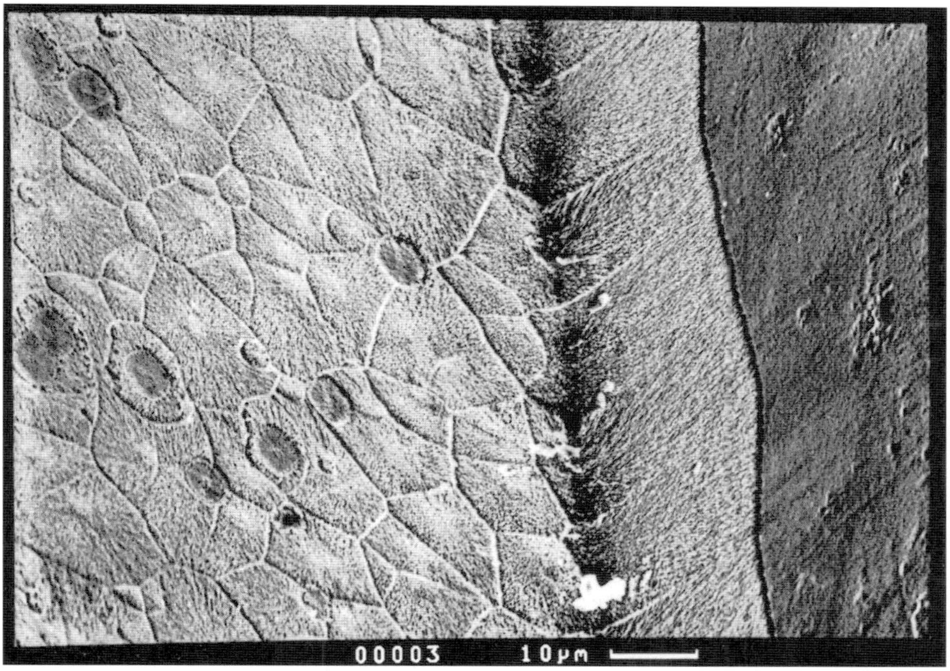

Fig. 42. SEM micrograph (top side view) of a nanostructured porous GaP film formed by anodic etching of crystalline *n*-type substrate (dark area). (see also Fig. 4).

IMPS measurements of electronic transport in a GaP networks have been studied by using a modulated UV laser [78]. The light was incident from the electrolyte side, and in all cases the penetration depth of the light ($1/\alpha \approx 100\,\text{nm}$) was much smaller than the thickness of the porous film d. Photogenerated holes are driven to the semiconductor surface and consumed in photoanodic oxidation of the sample. Photogenerated electrons are driven to the interior of the GaP filaments by an electric field perpendicular to the surface. Transport of photogenerated electrons through a fully depleted GaP network can be measured with IMPS in the frequency domain well below $(1/\tau_{cell})$ [78]. A typical IMPS plot is shown in Fig. 43.

The experimental IMPS plot has the shape predicted for the case where charge carriers are generated at the electrolyte side and diffuse (or migrate) through the network to the substrate. The inverse of the characteristic frequency ω_{min} (measured with a given background light intensity) is plotted vs. d^2 in Fig. 44.

It can be seen that $(1/\omega_{min})$ increases linearly with d^2 as expected for diffusive transport of electrons. $(1/\omega_{min})$ corresponds to the average transit time of photogenerated electrons through a film of thickness d. Similar results have been reported by Solbrand et al. for electron transport through a particulate TiO_2 electrode [186]. If electronic transport is not strongly attenuated by trapping/detrapping effects, the diffusion controlled transit time is given by

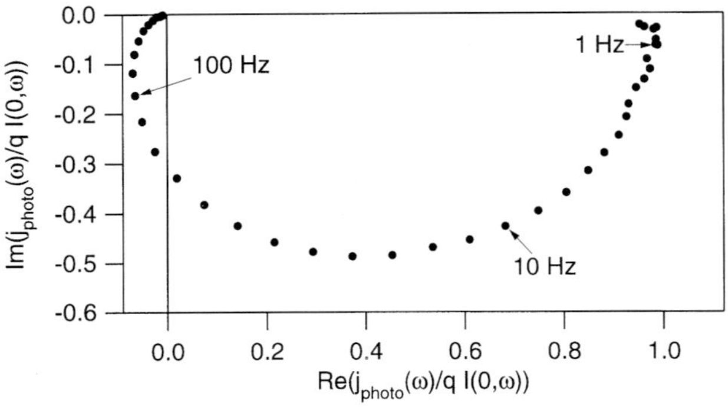

Fig. 43. Complex plane IMPS plot for 14 micron thick nanoporous GaP layer on n-GaP under depletion conditions (potential 2.5 V vs. SCE) in acid electrolyte (pH = 1.0). Illumination from the electrolyte side (λ 350 nm). The steady state photocurrent efficiency is unity. The transit time $\tau(d)$ derived from ω_{min} is 5 10^{-3} s.

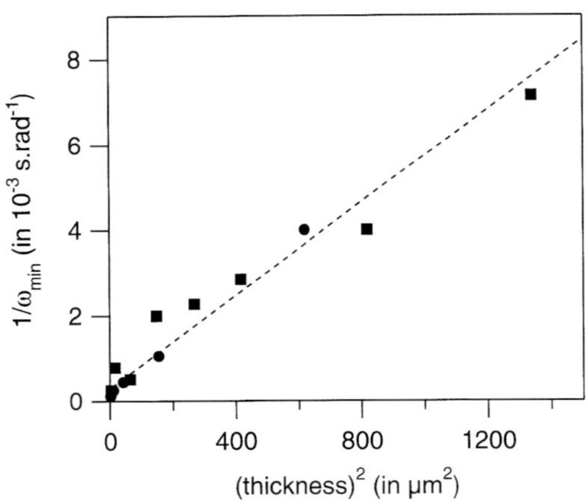

Fig. 44. Plot of $1/\omega_{min}$ vs. the square of the thickness, d, of the porous GaP film. The IMPS response was measured using modulated illumination from the electrolyte side (λ 35 nm) and a constant background illumination level.

$$\tau_d = \frac{d}{\mu_n \frac{dE_{F,n}}{qdx}} = \frac{d^2}{\mu_n \frac{k_B T}{q}} = \frac{d^2}{D_n} \tag{94}$$

From Fig. 44, a value of 1.6 $10^{-3}\,\mathrm{cm^2\,s^{-1}}$ is found for D_n. At present, it is not clear if this value pertains to the true diffusion constant of free electrons in the nanoporous network, since recent investigations show that the characteristic frequency also depends on the background light intensity [200]. In any case, this value is about three orders of magnitude lower than the diffusion constant ($2\,\mathrm{cm^2\,s^{-1}}$) for electrons in bulk GaP.

Bare and sensitised nanoporous TiO_2 electrodes with thicknesses between 0.1 and 20 μm have been studied extensively by several research groups, both by measuring photocurrent transients and IMPS responses. Although the excitation mechanisms differ (band-band excitation as opposed to dye sensitised electron injection), the

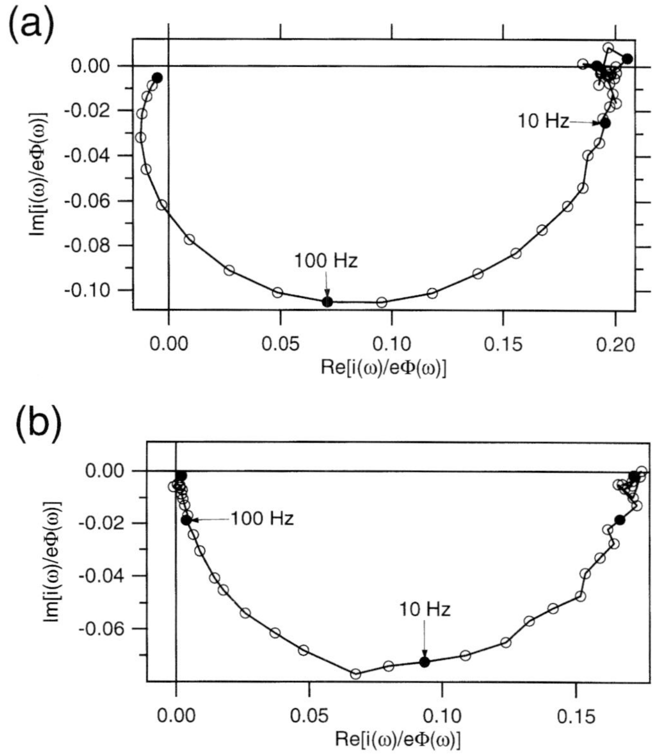

Fig. 45. IMPS response measured for a 0.6 μm thick particulate TiO_2 electrode (lightly sintered 30 nm particles; Degussa P25) interopenetrated with an aqueous electrolyte (0.2 mol dm^{-3} citrate) to act as a hole scavenger. Illumination from the electrolyte side at two different background light intensities (λ 35 nm). (a) 0.83 10^{15} cm^2 s^{-1}, (b) 2.8 10^{13} cm^{-2} s^{-1} [78].

results are remarkably similar. Examples of the IMPS plots observed by Cao et al. [80] for dye sensitised nanocrystalline TiO_2 electrodes are shown in Fig. 37. Similar results have been reported by Dlocik et al. [90]. Cao et al. and Dlocik et al. both found that the characteristic frequency of the IMPS plot varied with light intensity according to the empirical relationship $\omega_{min} \propto I^n$ with $n < 1$.

de Jongh and Vanmaekelbergh [79, 143] have studied bare particulate TiO_2 electrodes interpenetrated with aqueous electrolyte. Electron-hole pairs were generated by 350 nm light from an argon laser with an absorption depth of the light of about 0.3 μm. The holes were partly captured by a reducing agent in the electrolyte and partly lost by recombination. Electron transport was investigated with IMPS with illumination from the electrolyte and substrate sides, and typical IMPS plots are shown in Fig. 45. It is interesting to note that at high frequencies, the plot crosses the imaginary axis before converging on the origin. This behaviour is similar to, though less extreme, than with nanoporous GaP networks. It was found that the $\omega_{\min}$ varied with intensity according to the relationship $\omega_{min} \propto I^n$ with $n = 0.5$, which is similar to the results reported for dye sensitised electrodes [80, 90]. ω_{min} was found to be independent of temperature in the range 0–50 °C and to decrease with increasing thickness of the electrode when the light was incident from the electrolyte side.

Cao *et al.* [80] give numerical solutions of Eq. 42 that are appropriate for IMPS, but although the calculated complex plane plots results resemble their experimental data for dye sensitised. TiO_2 electrodes, the numerical solutions do not lead readily to identification of diagnostic features in the IMPS response, and more convenient analytical solutions can be obtained. Dlocik *et al.* [90] have taken a general approach to the generation collection problem which considers voltage dependent rate constants for electron extraction/injection at the TiO_2/substrate interface. Here attention is restricted to the diffusion controlled limit obtained when the rate constant for electron extraction is sufficiently large. Illumination from the substrate and from the solution sides has been considered. The thin film of nanocrystalline semiconductor (thickness d) is presumed to be uniformly sensitised with dye so that the absorption coefficient α is independent of distance. Light absorption by I_3^- in the pores is not considered, and light scattering by the nanocrystalline phase is treated semi-empirically by defining an effective absorption coefficient. The electrolyte phase is assumed to penetrate throughout the porous film so that injected electrons are screened by counter-ions and solvent dipoles in the electrolyte. Coulombic interactions between injected electrons are assumed to be negligible. The lifetime of excess electrons in the nanocrystalline solid is presumed to be determined by back reactions with the oxidised component of the supersensitiser couple (I^-/I_3^-). Regeneration of the dye from its oxidised state by reaction with I^- is assumed to be fast compared with the time scale of the modulation. Depletion of I^- and accumulation of I_3^- in the pores of the film are neglected. The effects of electron trapping are considered and related to the effective diffusion coefficient of electrons. The possibility of electron recombination with I_3^- *via* trap states is also included.

As pointed out earlier, it has been observed that the diffusion coefficient of electrons in dye sensitised nanocrystalline TiO_2 cells increases with light intensity, probably as the result of light-dependent trap filling. The use of small relative modulation amplitudes in IMPS measurements has the advantage that the diffusion coefficient of electrons can be considered to be determined primarily by the *dc* illumination intensity.

For illumination from the electrolyte side, the *ac* photocurrent conversion efficiency $\Phi(\omega) = j_{photo}/qI_o$ is given by [90]

$$\Phi(\omega) = \frac{\alpha}{\alpha+\gamma} \cdot \frac{e^{(\gamma-\alpha)d} - e^{-(\alpha+\gamma)d} + 2\alpha \dfrac{e^{(\gamma-\alpha)d} - 1}{\gamma - \alpha}}{e^{\gamma d} + e^{-\gamma d}} \tag{95a}$$

For illumination from the substrate side,

$$\Phi(\omega) = \frac{\alpha}{\alpha+\gamma} \cdot \frac{e^{\gamma d} - e^{-\gamma d} + 2\alpha \dfrac{e^{-\alpha d} - e^{-\gamma d}}{\gamma - \alpha}}{e^{\gamma d} + e^{-\gamma d}} \tag{95b}$$

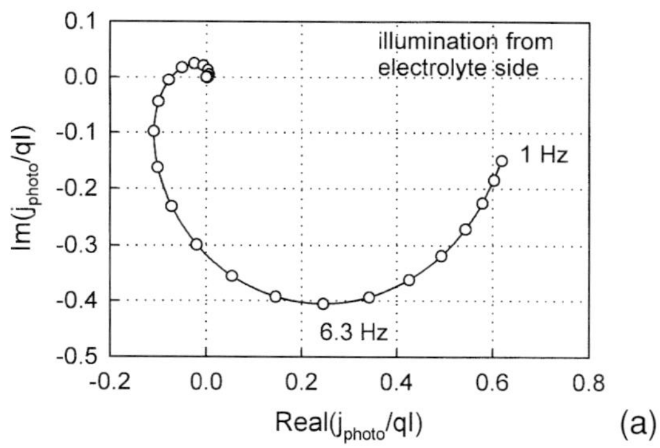

Fig. 46a. Diffusion controlled IMPS response calculated for illumination from the electrolyte side for small penetration depth. $\alpha = 10^5\,cm^{-1}$, $d = 10$ microns, $D = 10^{-5}\,cm^2\,s^{-1}$, $\tau = 0.1$ s. The value of ω_{min} is equal to the inverse of the transit time for electrons to reach the substrate. Note how the response spirals into the origin at high frequencies (compare Fig. 41).

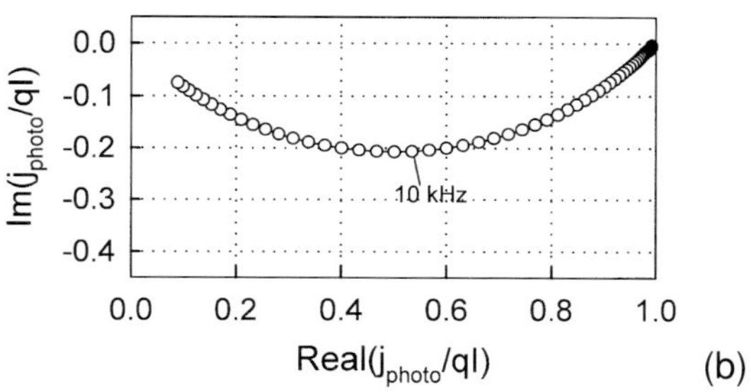

Fig. 46b. IMPS response calculated for the same values Fig. 45, except that illumination is from the substrate side. In this case electrons have to diffuse only a very short distance, and this is reflected in the much higher value of ω_{min}.

Here α is the effective absorption coefficient, d is the film thickness and γ is given by

$$\gamma = \sqrt{\left(\frac{1}{D\tau} + \frac{i\omega}{D}\right)} \tag{96}$$

The corresponding *dc* solutions obtained by setting $\omega = 0$ in Eq. 95 are identical with those derived by Södergren *et al.* [189].

The influences on the IMPS response of the illumination geometry, absorption coefficient, and electron recombination lifetime predicted for the diffusion controlled case are illustrated in Fig. 46 and Fig. 47. Two cases are considered corresponding to interband excitation and to dye sensitisation. In the first the penetration depth of the light is chosen to be much smaller than the film thickness. The calculated IMPS responses for illumination from the electrolyte and from the substrate side are contrasted in Fig. 46a and 46b respectively. The first thing to note is that the imaginary component of the IMPS response is negative under most conditions. This contrasts

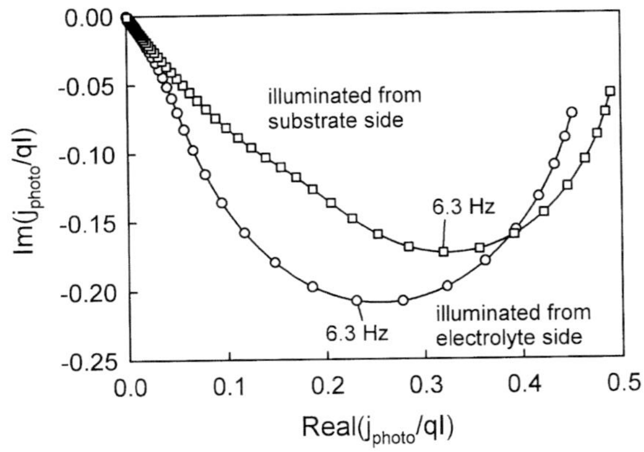

Fig. 47. IMPS responses calculated for the case where the penetration depth of the exciting light is comparable with the film thickness as is usually the case for dye sensitised cells. $\alpha = 10^3\,\text{cm}^{-1}$, $D = 10^{-5}\,\text{cm}^2\,\text{s}^{-1}$, $d = 10$ microns, $\tau = 0.1$ s. Direction of illumination as shown. Note that the ω_{min} values for the two directions of illumination are similar (compare Fig. 46).

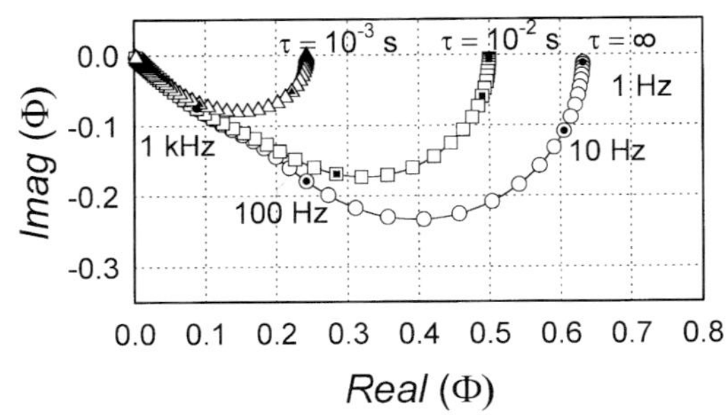

Fig. 48. Calculated IMPS responses showing the influence of the electron recombination lifetime τ. $d = 12$ microns, $D = 10^{-4}\,\text{cm}^2\,\text{s}^{-1}$, $\alpha = 10^3\,\text{cm}^{-1}$. Illumination from the substrate side ($\lambda = 514$ nm). Note that the characteristic frequency tends towards $1/\tau$ as τ becomes smaller.

with the IMPS response for a minority carrier device, which has a positive imaginary component arising from the flux of majority carriers involved in recombination. Illumination from the electrolyte side gives a characteristic spiral in the high frequency response that is associated with the transit time required for carriers to move form the front face to the rear contact. This behaviour is similar to that predicted by Vanmaekelbergh et al. [78]. By contrast, illumination through the substrate generates electrons very close to the contact and the transit time is very small. This is reflected in the high value of the characteristic frequency. The flattening of the semicircle is characteristic of a diffusion controlled process.

The difference between the two illumination geometries is much less marked when the penetration depth of the light is comparable with the film thickness, as is usually the case for dye sensitised cells. As Fig. 47 shows, the characteristic frequency is quite similar in both cases since carriers are generated throughout the whole of the film.

The effect of the electron recombination lifetime is illustrated in Fig. 48. If the

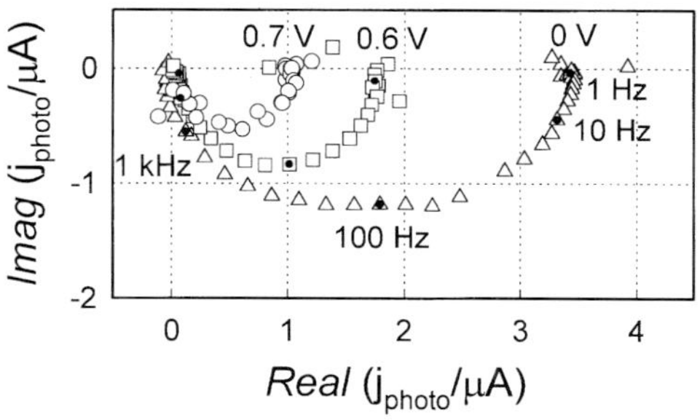

Fig. 49. Experimental IMPS responses for dye sensitised cells (illumination from the substrate side, $\lambda = 514\,\text{nm}$) measured at different voltages [90]. Dc current 6.14 mA. Note that the modulated photocurrent is 3 orders of magnitude lower than the dc photocurrent.

lifetime is sufficiently long, all photo-injected electrons are collected, and the *dc* value of Φ is determined by the fraction of incident light that is absorbed, and the characteristic frequency is determined by the mean electron transit time. For a finite electron lifetime, the *dc* photocurrent conversion efficiency depends on the ratio between the electron diffusion length $L = (D\tau)^{1/2}$ and the film thickness, and in the limit that $L < d$, the characteristic frequency is determined by the electron lifetime.

The IMPS response of nanocrystalline electrodes is evidently influenced by τ_{cell}. The capacitance of nanocrystalline TiO_2 electrodes is strongly dependent on potential, and under accumulation conditions capacitance values in the mF cm^{-2} range are common. In fact, the recent work of Dloczik et al. [90] shows that the IMPS response of dye sensitised TiO_2 cells is modified by the τ_{cell} even under short circuit conditions. (Vanmaekelbergh et al. [78] have discussed the effect of τ_{cell} for porous GaP, where the space charge capacitance only becomes important close to flatband). *RC* attenuation alters the shape of the IMPS response, and in the limit $\tau_{cell} > \tau_d$, the IMPS plot becomes semicircular and is dominated by τ_{cell}. The potential dependence of the IMPS response reported by Dloczik et al. is illustrated in Fig. 49.

It has become common practice to fit IMPS data in the complex plane, but this approach is rather unsatisfactory when the effects of *RC* attenuation are convoluted with those due to carrier transport and recombination. A more sensitive method is to use Bode plots. The magnitude and phase angle plots provide an excellent diagnostic analysis. An example from the work of Dloczik et al. [90] is shown in Fig. 50.

As shown in section 4, if τ_{cell} is made large enough, the intensity modulated photovoltage (IMVS) response is obtained. Schlichthörl et al. [188] have recently studied band edge movement and recombination kinetics in dye sensitised TiO_2 solar cells by this method in which a small intensity modulation is superimposed on a much larger steady background illumination in order to permit linearisation of the ac photovoltage response. The observed open circuit IMVS response is a semicircle with negative real and positive imaginary components. The relaxation of the open circuit photovoltage can be attributed to back reaction of free and trapped electrons with I_3^- ions. Theoretical considerations predict that two semicircles could be observed, but experimentally only one is seen. The maximum of the semicircle, which can be associated with the electron recombination lifetime τ, appears at a frequency ranging

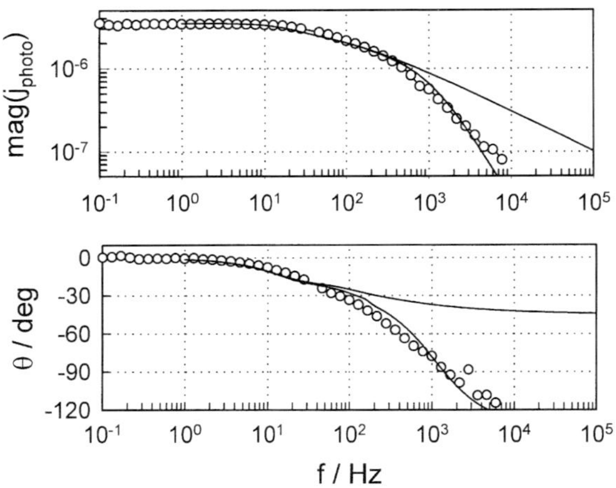

Fig. 50. Bode plot showing experimental data and best and fit to the short circuit IMPS response of a dye sensitised cell [90]. Dc photocurrent 6.3 mA. The upper solid line in each plot shows the fitted response in the absence of RC attenuation. Note the limiting high frequency phase shift is 45°, which is characteristic of diffusion control. The lower line illustrates the improved fit obtained by including the influence of τ_{cell}. $R = 10\,\Omega$, $C = 5\ 10^{-5}$ F.

from a few Hz at low intensities (10^{-3} sun) to a more than 10 Hz at 1 sun. This variation is attributed to a change in the average energy of occupied trapping levels involved in the electron transfer. However, it seems more probable that the intensity dependence is related in some way to the variation of the trap occupancy as discussed below.

One of the key questions that remains unanswered is the role of temporary localisation of electrons in traps in determining the rate of electron transport in porous and nanocrystalline electrodes. Dlocik et al. [90] have taken electron trapping and detrapping into account by defining effective values for the electron diffusion coefficient and lifetime as

$$D_{eff} = \frac{D}{1 + \dfrac{k_{trap} k_{det\,rap}}{\left(k_{det\,rap} + \dfrac{1}{\tau_{trap}}\right)^2}} \tag{97}$$

and

$$\frac{1}{\tau_{eff}} = \frac{1}{\tau_{cb}} + \frac{k_{trap}}{1 + k_{detrap}\tau_{trap}} \tag{98}$$

where k_{trap} and k_{detrap} are the first order rate constants for trapping and detrapping of electrons, and τ_{cb} and τ_{trap} are the lifetimes of electrons in the conduction band and in traps respectively. It should be noted that both D_{eff} and τ_{eff} are expected to depend on the illumination level as well as on the voltage, since they depend on trap occupancy. If carrier trapping and release are fast compared to the modulation frequency, the IMPS response retains the shape predicted for the trap free case, but the frequencies are scaled by the reduction in transit time defined by Eq. 98. If carrier relaxation is slow, it is predicted to alter the shape of the IMPS response, but this has not been observed experimentally. In fact, de Jongh and Vanmakelbergh [143] have reported that IMPS plots for particulate TiO_2 films are nearly semicircular at low background intensities, and change progressively to a distorted semicircle that crosses the imaginary axis at higher light intensities. These authors suggest that the influence of deep trapping on the IMPS response diminishes with increasing background light intensity due to an increase in the electron demarcation level.

A more detailed consideration of trapping in energy levels distributed in energy has been given by de Jongh et al. [79, 143]. They concluded that the lifetime for trapping of a free electron in the deepest available localised state (at the demarcation level E_n) is given by

$$\tau_{trap}(E_n) = \frac{1}{k_B T \beta_n [X(E_n)]} \tag{99}$$

If the electron life time for deep trapping $\tau_{trap}(E_n)$ is considerably larger than the transit time τ_d of free electrons, deep electron trapping can be neglected, and the characteristic frequency of the IMPS response will be determined by the transit time τ_d. If, however, $\tau_{trap}(E_n)$ is smaller than τ_d, trapping/detrapping will influence the IMPS response over a large frequency range. In this case, ω_{min} is given by

$$\omega_{min} = 2\left(\frac{\tau_{trap}(E_n)}{\tau_d}\right) B_n N_C e^{(E_n - E_C)/k_B T} \tag{100}$$

It follows that $(1/\omega_{min})$ is equal to the average number of deep trapping events experienced by photogenerated electrons during their travel through the network times the residence time in a deep trap at E_n. The thermal release rate, $\beta_n N_c\, exp$ $exp[(E_n - E_C)/k_B T]$, which is equal to $\beta_n \langle n(x) \rangle$, where $\langle n(x) \rangle$ is the spatially averaged concentration of excess electrons, increases with the concentration of photogenerated electrons and hence with the background light intensity. This means that if trapping is important, ω_{min} should increase with the background light intensity. In the study of de Jongh and Vanmaekelbergh, $\langle n(x) \rangle$ was considered to be proportional to the square root of the illumination intensity due to strong electron-hole pair recombination in order to explain the dependence of ω_{min} on light intensity. However, electron-hole recombination does not occur as in the case of the dye sensitised cell, so that $\langle n(x) \rangle$ and hence ω_{min} should be linearly proportional to the light intensity. However, a sub-linear dependence of the type $\omega_{min} \propto I^n$ with n between 0.5 [90] and 0.7 [80] has been observed experimentally. Present models clearly fail to provide an adequate explanation of the intensity dependence of the electron transport and recombination kinetics in nanoporous systems, which remain a challenging area for investigation by transient and periodic methods.

The survey in this section, which describes progress in a rapidly developing field, demonstrates that many problems remain to be solved. It is clear, however, that time and frequency resolved photocurrent measurements hold the key to understanding the behaviour of these novel electrode/electrolyte systems.

6 Acknowledgments

The authors acknowledge support from the British Council and NWO (Anglo Dutch Joint Scientific Research Programme JRP401).

7 References

1. For a general introduction see S. R. Morrison, 'Electrochemistry at Semiconductor and Oxidized Metal electrodes', Plenum, New York (1980).
2. See also Yu. V. Pleskov and Yu. Ya Gurevich, 'Semiconductor Photoelectrochemistry', Consultants Bureau, New York (1986).
3. 'Photoelectrochemistry: fundamental processes and measurement techniques'. Proc. Symp. on Photoelectrochemical Processes and Techniques for Photoelectrochemical Solar Cells. Ed. W. L. Wallace. The Electrochemical Society, Pennington NJ (1982).
4. 'Photoelectrochemistry'. Proc. Symp. Photoelectrochemistry. Editors K. Rajeshwar, L. M. Peter, A. Fujishima, D. Meissner and M. Tomkiewich. The Electrochemical Society Pennington, N.J. (1997).
5. F. Williams and A. J. Nozik, Nature 312, 21 (1984).
6. A. J. Nozik, K. Honda, P. Salvador, R. Tenne and Z. W. Tian, New Journal of Chemistry 11, 205 (1997).
7. A. Heller, Acc. Chem. Res. 14, 154 (1981).
8. A. Heller, Solar Energy 29, 153 (1982).
9. G. Hodes, S. J. Fonash, A. Heller and B. Miller, Adv. Electrochem. Electrochem. Eng. 13, 113 (1984).
10. H. Gerischer, Pure Appl. Chem. 52, 2649 (1980).
11. A. J. Bard, J. Phys. Chem. 86, 172 (1982).
12. A. J. Bard, J. Electroanal. Chem. 168, 5 (1984).
13. Y. V. Pleskov, Prog. Surf. Sci 15, 401 (1984).
14. R. Memming, Topics in Current Chemistry 153, 79 (1988).
15. N. S. Lewis, Nature 345, 293 (1990).
16. Y. V. Pleskov. 'Solar Energy Conversion – a photoelectrochemical approach'. Springer. Berlin 1990.
17. H. Gerischer, Faraday Disc. Chem. Soc. 70, 137 (1980).
18. A. Heller, Science 223, 1141 (1984).
19. A. J. Nozik, J. Less Common Metals 103, 1 (1984).
20. M. K. Nazeeruddin, P. Liska, J. Moser, N. Vlachopoulos and M. Gräzel, Helv. Chim. Acta 73, 1788 (1990).
21. B. O'Regan and M. Grätzel, Nature (London), 353, 737 (1991).
22. M. Grätzel, Coord. Chem. Rev. 111, 167 (1991).
23. M. K. Nazeeruddin, A. Kay, I. Rodicio, R. Huphrey-Baker, E. Muller, P. Liska, N. Vlachopoulos and M. Grätzel, J. Am. Chem. Soc. 115, 6382 (1993).

24. M. Grätzel and K. Kalayasundaram, Current Science 66, 706 (1994).
25. Kay and M. Grätzel, Solar Energy Mat. and Solar Cells 44, 99 (1996).
26. T. Gerfin, M. Grätzel and L. Walder, Prog. Inorg. Chem. 44, 345 (1997).
27. D. D. MacDonald 'Transient Techniques in Electrochemistry', Plenum, New York 1977.
28. C. Gabrielli 'Identification of electrochemical processes by frequency response analysis' Technical report No 004/83. Solartron Instruments, UK 1980.
29. J. R. Macdonald (Editor). 'Impedance Spectroscopy emphasising solid materials and systems'. Wiley, New York 1987.
30. P. Allongue and H. Cachet, J. Electroanal. Chem. 119, 371 (1981).
31. J. Schefold and H. M. Kühne, J. Electroanal. Chem. 300, 211 (1991).
32. F. Cardon and D. Vanmakelbergh, Electrochim. Acta 37, 837 (1992).
33. J. Schefold, J. Electrochem. Soc. 142, 850 (1995).
34. W. P. Gomes and D. Vanmaekelbergh, Electrochim. Acta 41, 967 (1996).
35. E. A. Ponomarev and L. M. Peter, J. Electroanal. Chem. 397, 45 (1995).
36. W. J. Albery and P. N. Bartlett, J. Electrochem. Soc. 129, 2254 (1982).
37. J. Li and L. M. Peter, J. Electroanal. Chem. 193, 27 (1985).
38. J. Li and L. M. Peter, J. Electroanal. Chem. 199, 1 (1986).
39. L. M. Peter in 'Trends in Interfacial Electrochemistry', p. 523. Editor F. A. Silva. Reidel, Dordrecht (1986).
40. R. Peat and L. M. Peter, Electrochim. Acta 31, 731 (1986).
41. R. Peat and L. M. Peter, J. Electroanal. Chem. 209, 307 (1986).
42. R. Peat and L. M. Peter, Ber. Bunsenges. Phys. Chem. 91, 382 (1987).
43. R. Peat and L. M. Peter, J. Electroanal. Chem. 228, 351 (1987).
44. R. Peat and L. M. Peter, Appl. Phys. Lett. 51, 328 (1987).
45. L. M. Peter in 'Photocatalysis and the Environment'. p. 275. Editor M. Schiavello, Kluwer, Dordrecht (1987).
46. H-J. Lewerenz, J. Stumper and L. M. Peter, Phys. Rev. Lett. 61, 1989 (1988).
47. L. M. Peter, J. Li, R. Peat, H. J. Lewerenz and J. Stumper, Electrochim. Acta 35, 1657 (1990).
48. L. M. Peter, Chem. Rev. 90, 753 (1990).
49. L. M. Peter, A. N. Borazio, H. J. Lewerenz and J. Stumper, J. Electroanal. Chem. 290, 229 (1990).
50. L. M. Peter, Croat. Chem. Acta 63, 401 (1990).
51. B. P. Minks, N. A. M. Verhaegh, J. J. Kelly, J. Stumper and L. M. Peter, J. Electroanal. Chem. 303, 277 (1991).
52. P. Herrasti and L. M. Peter, J. Electroanal. Chem. 305, 241 (1991).
53. H. K. Song and D. D Macdonald, J. Electrochem. Soc. 138, 1408 (1991).
54. Z. A. Rotenberg and O. A. Semenikhin, J. Electroanal. Chem. 316, 165 (1991).
55. O. A. Semenikhin, Z. A. Rotenberg and A. G. Pschenichnikov, Soviet Electrochemistry 27, 317 (1991).
56. O. A. Semenikhin, Z. A. Rotenberg and G. L. Teplitskaya, Soviet Electrochemistry 27, 190 (1991).
57. J. J. Kelly, B. P. Minks, N. A. M. Verhaegh, J. Stumper and L. M. Peter, Electrochim. Acta 37, 877 (1992).
58. R. S. Hutton and L. M. Peter, J. Electroanal. Chem. 332, 315 (1992).
59. P. Herrasti and L. M. Peter, J. Electroanal. Chem. 334, 133 (1992).
60. J. Schefold, J. Electroanal. Chem. 341, 111 (1992).
61. P. C. Searson, D. D. MacDonald and L. M. Peter, J. Electrochem. Soc. 139, 2538 (1992).
62. A. R. de Wit, D. Vanmaekelbergh and J. J. Kelly, J. Electrochem. Soc. 139, 2508 (1992).
63. Z. A. Rotenberg, Soviet Electrochemistry 28, 1317 (1992).
64. D. Vanmaekelbergh, A. R. de Wit and F. Cardon, J. Appl. Phys. 73, 5049 (1993).
65. B. H. Erné, D. Vanmaekelbergh and I. E. Vermeir, Electrochim. Acta 38, 2559 (1993).

66. I. E. Vermeir, W. P. Gomes, B. H. Erné and D. Vanmaekelbergh, Electrochim. Acta 38, 2659 (1993).
67. A. M. Chaparro, P. Salvador, A. Tabernero, R. Navarro, B. Coll and V. Caselles, Surf Sci. 295, 457 (1993).
68. P. C. Searson and D. D. Macdonald, Electrochim. Acta 38, 1913 (1993).
69. A. Goossens and D. D Macdonald, J. Electroanal. Chem. 352, 65 (1993).
70. A. Goossens and D. D. Macdonald, Electrochim. Acta 38, 1965 (1993).
71. A. D. Modestov, G. D. Zhou, H. H. Ge and B. H Loo, J. Electroanal. Chem. 375, 293 (1994).
72. R. S. Hutton and D. E. Williams, Electrochim. Acta 39, 701 (1994).
73. G. Schlichthörl, E. A. Ponomarev and L. M. Peter, J. Electrochem. Soc. 142, 3062 (1995).
74. E. A. Ponomarev and L. M. Peter, J. Electroanal. Chem. 396, 209 (1995).
75. J. Schefold, J. Electroanal. Chem. 394, 35 (1995).
76. A. M. Chaparro, P. Salvador and L. M. Peter, J. Phys. Chem. 99, 6677 (1995).
77. G. H. Schoenmakers, D. Vanmaekelbergh and J. J. Kelly, J. Phys. Chem. 100, 3215 (1996).
78. D. Vanmaekelbergh, F. Iranzo Marín and J. van de Lagemaat, Ber. Bunsenges. Phys. Chem. 100, 616 (1996).
79. P. E. de Jongh and D. Vanmaekelbergh, Phys. Rev. Lett 77, 3427 (1996).
80. F. Cao, G. Oskam, G. J. Meyer and P. C. Searson, J. Phys. Chem 100, 17021 (1996).
81. A. Goossens, Surf. Sci. 365, 662 (1996).
82. G. Oskam, J. C. Schmidt and P. C. Searson, J. Electrochem. Soc. 142, 2538 (1996).
83. E. A. Meulenkamp and A. R. deWit, Electrochim. Acta 41, 109 (1996).
84. D. J. Fermín, E. A. Ponomarev and L. M. Peter, ref. 4, p 62.
85. E. A. Ponomarev, A. Katty and C. Levy-Clement, ref. 4, p 269.
86. D. Vanmaekelbergh, P. E. de Jongh and J. van de Lagemaat, ref. 4, p 72.
87. P. E de Jongh and D. Vanmaekelbergh, J. Phys. Chem. B 1010, 2716 (1997).
88. G. H. Schoenmakers, D. Vanmaekelbergh and J. J. Kelly, J. Chem. Soc. Faraday Trans. 93, 1127 (1997).
89. L. M. Peter, E. A. Ponomarev and D. J. Fermín, J. Electroanal. Chem. 427, 79 (1997).
90. L. Dloczik, O. Ileperuma, I. Lauermann, L. M. Peter, E. A. Ponomarev, G. Redmond, N. J. Shaw and I. Uhlendorf, J. Phys. Chem. B 101, 10281 (1997).
91. E. Kamieniecki, J. Vac. Sci. Technol., 20, 811 (1982).
92. E. Kamieniecki, J. Appl. Phys. 54, 6481 (1983).
93. G. Schlichthörl and H. Tributsch, Electrochim. Acta 37, 919 (1992).
94. H. Tributsch, G. Schlichthörl and L. Elstner, Electrochim. Acta 38, 141 (1993).
95. H. J. Lewerenz and G. Schlichthörl, J. Appl. Phys. 75, 3544 (1994).
96. F. Wunsch, Y. Nakato, M. Kunst and H. Tributsch, J. Chem. Soc. Faraday Trans. 92, 4053 (1996).
97. C. Gabrielli and M. Keddam, Electrochim. Acta 41, 957 (1996).
98. C. Gabrielli, M. Keddam, H. Perrot and R. Torresi, J. Electroanal. Chem. 378, 85 (1994).
99. M. Kalaji and L. M. Peter, J. Chem. Soc. Faraday Trans. 87, 853–860 (1991).
100. J. J. Kim, T. Ameiya, K. Hasimoto and A. Fujishima, J. Electroanal. Chem. 435, 31 (1997).
101. C. Gabrielli and B. Tribollet, J. Electrochem. Soc. 141, 1147 (1994).
102. Y. Tachibana, J. E. Moser, M. Grätzel, M. Klug and D. R. Durrant, J. Phys. Chem. 100, 20056 (1996).
103. T. Hannappel, B. Burfeindt, W. Storck and F. Willig, J. Phys. Chem. B, 101, 6799 (1997).
104. W. Schmickler 'Interfacial Electrochemistry' Chapters 8 and 19. Oxford University Press, Oxford (1996).

105. R. J. Dwayne Miller, G. L. McLendon, A. J. Nozik, W. Schmickler and F. Willig. 'Surface Electron Transfer Processes'. Chapter 4. VCH, Weinheim (1995).
106. A. M. Kuznetsov. 'Charge Transfer in Physics, Chemistry and Biology'. Chapter 8. Gordon and Breach, Amsterdam (1995).
107. K. R. Kneten and R. L. McCreey, Anal. Chem. 64, 2518 (1992).
108. G. A. Shreve and N. S. Lewis, J. Electrochem. Soc. 142, 112 (1995).
109. K. E. Pomykal, A. M. Fajardo and N. S. Lewis, J. Phys. Chem. 99, 8302 (1995).
110. K. E. Pomykal, A. M. Fajardo and N. S. Lewis, J. Phys. Chem. 100, 3652 (1996).
111. K. E. Pomykal and N. S. Lewis, J. Phys. Chem. 101, 2476 (1997).
112. I. Uhlendorf, R. Reinekekoch and R. Memming, J. Phys. Chem. 100, 4930 (1996).
113. R. Memming, Solid State Ionics 94, 131 (1997).
114. A. Meier, S. S. Kocha, M. C. Hanna, A. J. Nozik, K. Siemoneit, R. Reinekekoch and R. Memming, J. Phys. Chem. 101, 7038 (1997).
115. see for instance J. L. Pankove 'Optical processes in semiconductors'. Prentice Hall, Englewood Cliffs, N.J. (1971).
116. D. E. Aspnes and A. A. Studna, Phys. Rev. B 27, 985 (1983).
117. D. Vanmaekelbergh, B. H. Erné, C. W. Cheung and R. W. Tjekstra, Electrochim. Acta 40, 686 (1995).
118. G. Chiarotti, G. Del Signore and S. Nannarone, Phys. Rev. Lett. 21, 1170 (1968).
119. J. Lagowski, Surf. Sci. 299.300, 92 (1994).
120. A. O. Evwaraye, S. R. Smith and W. C. Mitchel. J. Appl. Phys. 77, 4477 (1995).
121. N. G. Semaltianos, G. Karczewski, B. Hu. T. Wojtowicz and J. K. Furdyna, Phys. Rev. B 51, 17499 (1995).
122. H. Beckmann and R. Memming, J. Electrochem. Soc. 116, 368 (1969).
123. J-N. Chazalviel, M. Stefenel and T. B. Truong, Surface Science 134, 865 (1983).
124. G. Nogami, S. Okhubo, L. Avalle and K. Hongo, J. Electrochem. Soc. 143, 3600 (1997).
125. J. Li, R. Peat, and L. M. Peter, J. Electroanal. Chem. 165, 41 (1984).
126. D. Vanmaekelbergh and L. van Pieterson, Phys. Rev. Lett. 80, 821 (1998).
127. B. H. Erné, D. Vanmaekelbergh and J. J. Kelly, Adv. Mater. 7, 739 (1995).
128. F. Iranzo Marín, M. A. Hamstra, and D. Vanmaekelbergh, J. Electrochem. Soc. 143, 1137 (1997).
129. see also ref. 104, Chapter 5.
130. W. W. Gärtner, Phys. Rev. 116, 84 (1959).
131. D. Vanmaekelbergh, W. Rigole, W. P. Gomes and F. Cardon, J. Chem. Soc. Faraday Trans. 179, 2813 (1983).
132. D. Vanmaekelbergh, B. H. Erné, C. W. Cheung and R. W. Tjerkstra, Electrochim. Acta 40, 686 (1995).
133. G. H. Schoenmakers, D. Vanmaekelbergh and J. J. Kelly, J. Phys. Chem. 100, 3215 (1996).
134. J. van de Lagemaat, D. Vanmaekelbergh and J. J. Kelly, J. Appl Phys. 83, 6089 (1998).
135. D. Vanmaekelbergh and J. J. Kelly, Electrochim. Acta 43, 2773 (1998).
136. B. H. Erné, D. Vanmaekelbergh and J. J. Kelly, J. Electrochem. Soc. 143, 305 (1996).
137. J. van de Lagemaat, M. Plakman, D. Vanmaekelbergh and J. J. Kelly, Appl. Phys. Lett. 69, 2801 (1996).
138. A. O. Konstantinov, C. I. Harris, and E. Janzén, Appl. Phys. Lett. 65, 2699 (1994).
139. G. Oskam, A. Natarajan, P. C. Searson and F. M. Ross, Appl. Surf. Sci. 119, 160 (1997).
140. A. Anedda, A. Serpi, V. A. Karavanskii, I. M. Tiginyanu, and V. M. Ichizli, Appl. Phys. Lett. 67, 3316 (1995).
141. I. M. Tiginyanu, G. Irmer, J. Monecke, A. Vogt, and H. L. Hartnagel, Semicond. Sci. Technol. 12, 491 (1997).
142. G. Hodes, I. D. Howell, and L. M. Peter, J. Electrochem. Soc. 139, 3136 (1992).
143. P. E. de Jongh and D. Vanmaekelbergh, J. Phys. Chem. B, 101, 2716 (1997).

144. H. Rensmo, K. Keis, H. Lindström, S. Södergren, A. Solbrand, A. Hagfeldt, S.-E. Lindquist, L. N. Wang, and M. Muhammed, J. Phys. Chem. B, 101, 2598 (1997).
145. F. Willig, Ber. Bunsenges. Phys. Chem. 92, 1312 (1988).
146. P. V. Kamat and M. A. Fox, J. Phys. Chem. 87, 59 (1983).
147. R. A. Marcus, Angewandte Chemie Int. Ed. Engl. 32. 1111 (1993).
148. H. Gerischer, J. Phys. Chem. 95, 1356 (1991).
149. D. Vanmaekelbergh, Electrochim. Acta 42, 1121 (1997).
150. A. M. Fajardo and N. S. Lewis, Science 274, 969 (1996).
151. W. P. Gomes and H. H. Goossens in Advances in Electrochemical Science and Engineering, Vol. 3. Edited by H. Gerischer and W. Tobias, VCH Publishers, Inc. (1994).
152. D. Vanmaekelbergh and W. P. Gomes, J. Phys. Chem., 94, 1571 (1990).
153. W. Shockley and W. T. Read, Phys. Rev. 87, 835 (1952).
154. R. N. Hall, Phys. Rev. 87, 387 (1952).
155. A. Rose, Prog. In Semicond. 2, 211 (1956).
156. R. H. Bube and F. Cardon, J. Appl. Phys. 35, 2712 (1964).
157. J. G. Simmons and G. W. Taylor, Phys. Rev. B 4, 502 (1971).
158. C. N. Kenyon, M. X. Tang, O. Krüger, and N. S. Lewis, J. Phys. Chem. B, 101, 2850 (1997).
159. F. Willig, K. Bitterling, K.-P. Charlé, and F. Decker, Ber. Bunsen-Ges. Phys. Chem. 88, 374 (1987).
160. K. Bitterling and F. Willig, J. Electroanal. Chem. 204, 211, (1986).
161. P. Perone, J. H. Richardson, S. B. Deutscher, J. Rosenthal and J. N. Ziemer, J. Electrochem. Soc. 127, 2580 (1980).
162. R. L. Cook, P. F. Dempsey and A. F. Sammells, J. Electrochem. Soc. 133, 2287 (1986).
163. R. G. Kepler, Phys. Rev. 119, 1226 (1960).
164. S. M. Sze 'Semiconductor devices. Physics and Technology.' J. Wiley and Sons, New York (1985).
165. Bousse, S. Mostarshed, D. Hafeman, M. Sartore, M. Adami, and C. Nicolini, J. Appl. Phys. 75, 4000 (1994).
166. O. G. Koshelev and V. A. Morozova, Solid-State Electronics 39, 1379 (1996).
167. R. A. Street, 'Hydrogenated Amorphous Silicon'. Cambridge Solid State Science Series Editors R. W. Cahn, E. A. Davis, and I. M. Ward. (1991).
168. D. Vanmaekelbergh, J. van de Lagemaat, and R. E. I. Schropp, Solar Energy Materials and Solar Cells 41/42, 537 (1996).
169. D. Han, D. C. Melcher, E. A. Schiff, and M. Silver, Phys. Rev. B 48, 8658 (1993).
170. H. M. Branz and E. A. Schiff, Phys. Rev. B 48, 8667 (1993).
171. R. Brüggeman, C. Main, J. Berkin and S. Reynolds, Philos. Mag. B 62, 29 (1990).
172. F. Seynhaeve, R. P. Barclay, G. J. Adriaenssens, and J. M. Marshall, Phys. Rev. B, 39, 10196, (1989).
173. C. Longeaud and J. P. Kleider, Phys. Rev. B 48, 8715 (1993).
174. L. M. Peter in 'Electrochemistry', Specialist Periodical Report. P. 66. Editor D. Pletcher. Royal Society of Chemistry. London (1984).
175. B. H. Erné, M. Stchakovsky, F. Ozanam and J-N Chazalviel, J. Electrochem. Soc. (submitted).
176. M. Kunst and G. Beck, J. Appl. Phys. 63, 1093 (1988).
177. G. Schlichthörl, Ph.D. Thesis, Freie Universität Berlin (Fachbereich Chemie) (1992).
178. W. J. Albery, N. L. Dias and C. P. Wilde, J. Electrochem. Soc. 134, 601 (1987).
179. E. A. Ponomarev and S. D. Babenko, J. Electroanal. Chem. 371, 27 (1994).
180. E. S. Kooj and D. Vanmaekelbergh, J. Electrochem. Soc. 144, 1296 (1997) and references therein.
181. J. D. Joannopoulos, P. R. Villeneuve and S. Fan, Nature 386, 143 (1997) and references therein.

182. C. G. Granquist, 'Handbook of Inorganic Electrochromic Materials', Elsevier, Amsterdam (1995).
183. L. T. Canham, Appl. Phys. Lett. 57, 1046 (1990).
184. L. M. Peter, D. J. Riley and R. I. Wielgosz. Appl. Phys. Lett. 66, 2355 (1995).
185. A. Zaban, A. Meier and B. A. Gregg, J. Phys. Chem. 101, 7985 (1997).
186. A. Solbrand, H. Linström, H. Rensmo, A. Hagfeldt, S-E. Lindquist and S. Södergren, J. Phys. Chem. B 101, 2514 (1997).
187. M. O. Cáceres, H. Matsuda, T. Odagaki, D. P. Prato and W. Lambert, Phys. Rev. B 56, 5897 (1997).
188. S. Y. Huang, G. Schlichthörl, A. J. Nozik, M. Grätzel and A. J. Frank, J. Phys. Chem. B 101, 2576 (1997).
189. S. Södergren, A. Hagfeldt, J. Olsson and S.-E. Lindquist, J. Phys. Chem. 98, 5552 (1994).
190. S. G. Yan and J. T. Hupp, J. Phys. Chem. 100, 6867 (1996).
191. Schwarzburg and F. Willig, J. Phys. Chem. B 101, 2451 (1997).
192. P. Hoyer and H. Weller, J. Phys. Chem. 99, 14096 (1995).
193. R. W. Fessenden and P. V. Kamat, J. Phys. Chem. 99, 12902 (1995).
194. P. V. Kamat, J. Bedja, H. Hotchandani and L. K. Patterson, J. Phys. Chem. 100, 4900 (1996).
195. D. Liu, R. W. Fessenden, G. L. Hug and P. V. Kamat, J. Phys. Chem. 101, 2583 (1997).
196. P. V. Kamat, Prog. Inorg. Chem. 44, 273 (1997).
197. P. V. Kamat, Studies in Surface Science and Catalysis 103, 237 (1997).
198. S. T. Martin, H. Herrmann and M. R. Hoffman, J. Chem. Soc. Faraday Trans. 90, 3323 (1994).
199. B. O'Regan, J. Moser, M. Anderson and M. Grätzel, J. Phys. Chem. 94, 8720.
200. J. van de Lagemaat (personal communication).

Chemical Deposition of Chalcogenide Thin Films from Solution

Daniel Lincot[1], Michel Froment[2] and Hubert Cachet[2]

[1] Laboratoire d'Electrochimie et de Chimie Analytique, Unité Associée au CNRS, Ecole Nationale Supérieure de Chimie de Paris, 11 rue Pierre et Marie Curie, 75231 Paris cedex 05, France
[2] Laboratoire de Physique des Liquides et Electrochimie, Unité Propre du CNRS n° 15, Université Pierre et Marie Curie, 4 place Jussieu, 75252 Paris cedex 05, France.

Contents

List of Symbols

a	activity
A	surface area
G	Gibbs free energy
J	nucleation rate
k	Boltzmann's constant
K	equilibrium constant
K_s	solubility constant
[L]	ligand concentration
n	number of ions or molecules
P	molecule in solution or in a condensed state
R	Boltzmann's constant
S	supersaturation ratio
T	absolute temperature
v	volume of a molecule
μ	chemical potential
γ	surface or interface tension

Abbreviations

AFM	atomic force microscopy
ATU	allylthiourea
CBD	chemical bath deposition
CSD	chemical solution deposition
CVD	chemical vapor deposition
CIS	copper indium diselenide
CIGS	copper indium gallium diselenide
DEN	diethylenetriamine
DMSU	dimethylselenourea
ECALE	electrochemical atomic layer epitaxy
ED	electrodeposition
EDTA	ethylenediaminetetraacetate
ELD	electroless deposition
EN	ethylenediamine
EOCPC	electrochemical open-circuit potential change
ETU	ethylthiourea
FWHM	full width at half maximun
HEMT	high electron mobility transitor
HREM	high resolution electron microscopy
LED	light emitting diode
MBE	molecular beam epitaxy
MIS	metal insulator semiconductor

MISFET	MIS field effect transitor
MSM	metal semiconductor metal
NTA	nitrilotriacetate
OMCVD	organometallic chemical vapor deposition
QCM	quartz crystal microbalance
RHEED	reflection high energy electron diffraction
SAD	selective area diffraction
SEM	scanning electron microscopy
SES	selenosulfite
SILAR	successive ionic layer adsorption and reaction
STA	silicotungstic acid
STEM	scanning transmission electron microscopy
STM	scanning tunneling microscopy
SU	selenourea
TA	thioacetamide
TC	thiocarbazide
TEA	triethanolamine
TEM	transmission electron microscopy
TRIEN	triethylenetriamine
TS	thiosulfate
TSC	thiosemicarbazide
TU	thiourea
XPS	X-ray photoelectron spectroscopy
XRD	X-ray diffraction

1 Introduction

Chemical deposition from solutions of chalcogenide compound thin films finds its origin more than one century ago when for instance Emerson-Reynolds, in 1884 [1], reported that thiourea is easily desulfurated by certain metallic oxides and hydroxides (silver, mercury, lead) to form the corresponding sulfide. When the reaction proceeds in basic solutions, a specular and strongly adherent layer is formed on the sides of the vessel, in parallel with the precipitation in the volume of the solution, according to the global reaction:

$$ML_n{}^{2+} + SC(NH_2)_2 + 2\,OH^- \rightarrow MS + 2\,H_2O + nL + CN_2H_2 \qquad (1)$$

where L represent a ligand for the metallic cation.

Reaction (1) is the analogue in solution of CVD (Chemical Vapor Deposition) reactions in the vapor phase. The method in solution is usually called CBD for Chemical Bath Deposition or CSD for Chemical Solution Deposition. As compared to CVD which is widely used in the semiconductor industry, the development of CBD has remained limited.

CBD belongs to the same class of deposition processes from solutions as "Electroless deposition" (ELD) widely used for metal deposition [2] in the sense that it is a chemical process which does not involve electron exchange with a conducting substrate as in the electrodeposition technique. It can also be used on insulating substrates. However in CBD there is no change in the oxidation state of the elements in solution and in the film. They are deposited together as in precipitation reactions, whereas ELD is an oxydo-reduction process, the presence of the reductor element in the film (as P or B) being a side reaction.

CBD was used first in the 1940's for the fabrication of infra-red photoconducting detectors based on PbS [3]. n-p junctions were also formed by introducing the appropriate dopants in the bath [4].

Replacing thiourea by its selenium analogues (selenourea, dimethylselenourea) allowed the fabrication of selenide compounds as first shown for PbSe films [5, 6]. At the same period CBD was initiated for other important sulfides as CdS [7], Ag_2S [8] and selenides as CdSe [9] and ZnSe [10]. In 1982, as reported in the review by Chopra et al. [11], about 20 different compounds were synthesized by CBD. From this date other materials have been prepared and the method is presently receiving increased attention. The main example concerns CdS which enters in the fabrication of thin film solar cells based on copper indium diselenide [12] or cadmium telluride [13, 14]. CBD CdS displays superior qualities as compared to classical vapor phase techniques both for the fabrication of high efficiency cells and for the extension to large area processing (m^2 level). It emerges also in integrated microlectronics as interface passivation layers in III–V MISFET devices [15].

Presently, interest for chemical deposition from solutions is increasing as an alternative or complementary technique to more established ones for thin film deposition, especially for chalcogenide compounds. In addition to the characterization of materials, more attention is currently paid to basic aspects of the deposition and reaction mechanisms. As is also the case for "Electroless deposition" [2], CBD is often considered as a "black magic". The aim of this paper is to present a state-of-the-art of this area and to give clues for a more comprehensive approach of chemical bath deposition, as has been done for a long time in CVD.

The following section will give an overview of the literature, after which we will consider successively basic aspects of CBD, deposition mechanisms, structure of CBD films and epitaxial growth, and finish with an overview of the main contemporary applications of this technique.

2 Overview of Chemical Deposition of Chalcogenides

Table 1 provides a list of chalcogenide compounds that have been prepared up to now by CBD. Sulfide compounds are by far the most studied category but an increasing number of selenide compounds have been also synthesized recently. In contrast, very few works have been reported for telluride compounds. Beside binaries, significant works have been devoted to ternaries. In Table 2 are listed the chalcogenide precursors used for most of these deposition studies.

Table 1. List of chalcogenide compounds deposited chemically with selected references. The chalcogenide precursors utilized are indicated. TU: thiourea, TA: Thioacetamide, ATU: Allylthiourea, TS: thiosulfate ions $S_2O_3{}^{2-}$. SU: selenourea, SES: selenosulfate ions $SeSO_3{}^{2-}$.

Binaries	
As_2S_3 [16] TS	Bi_2Se_3 [71, 72], SES [71], SU [72]
Ag_2S [8, 16–19] TS [16], TU [8, 19], TA [17]	CdSe [9, 41, 73–81] SES [41, 73–78, 81], SU [79]
Bi_2S_3 [18, 20–24] TU [20, 21], TA [22–24], TS [18]	CoSe [82] SES
CdS [7, 16, 25–45] TU, ATU [30], TA [36, 38, 39, 42], TS [16], H_2S [40]	Cu_xSe [83–84] SES
CoS [46]	NiSe [54] SES
Cu_xS [47–50], TU [47, 48], TS [49, 50]	PbSe [73, 85, 86] SES, [5, 6] SU
In_2S_3 [51, 52] TA	Sb_2Se_3 [87]
MnS [53] TA	$SnSe_x$ [89] SES
NiS TA [54]	TlSe [88]
PbS [3, 4, 18, 55–60]	ZnSe [90–93] SU [90, 93], SES [91], DSU [92]
PdS_2 [16, 61], TS [16]	CdTe [94]
Sb_2S_3 [16, 18, 62] TA [62], TS [16]	PbTe [95]
SnS_x [16, 63–65] TA [63, 65], S [64], TS [16]	
TlS [66], TU	
ZnS [67–70], TU [67–70], TA [68]	
Ternaries	
Cd_xPb_{1-x} S [96, 97] TU	$CuInSe_2$ [110, 111]
$Cd_xZn_{1-x}S$ [98–101] TU	CdS_xSe_{1-x} [112, 113] TU and SES
$CuBiS_2$ [102] TU	ZnS_xSe_{1-x} [114] TU and SES
Cu_3BiS_3 [103] TU and after annealing	InS_xOH_{1-x} [115] TA
$CuInS_2$ [104] TU	SnS_xOH_{1-x} [115] TA
$Hg_xPb_{1-x}S$ [105, 106] TU	ZnS_xOH_{1-x} [117, 118] TU
Cd_xZn_{1-x} Se [107–109] SES	

Table 2. Chalcogenide Precursors. TU: thiourea, TA: thioacetamide, TC: thiocarbazide, TSC: thiosemicarbazide, ETU: ethylthiourea, ATU: allylthiourea, SU: selenourea, DSU: N,N dimethyl selenourea, TS: thiosulfate, SES: selenonilfate

TU: $SC(NH_2)_2$	SU: $SeC(NH_2)_2$
TA: $SCNH_2CH_3$	DSU: $(CH_3NH)_2CSe$
ETU: $CH_3CH_2NHCSNH_2$	SES: $SeSO_3{}^{2-}$
ATU: $CH_2CHCH_2NHCSNH_2$	
TSC: $SC(NH_2)NHNH_2$	
TC: $SC(NH(NH)_2)_2$	
TS: $S_2O_3{}^{2-}$	

2.1 Sulfide Binaries

Pioneering work has been done on the deposition of PbS [3, 4]. Only slight modifications of the initial formulations have been made up to now, which are mainly based on the use of thiourea and lead salt (tartrate, acetate or nitrate) in basic solutions at pH from 10 to 12. Following Davis and Nohr [55], Isshiki et al. [59] have studied in detail the growth under conditions that allow epitaxial growth at room temperature. The growth is strongly temperature activated with an activation energy of 65 $kJ.mol^{-1}$ which denotes a kinetic limitation. In recent works, triethanolamine has been introduced in the bath [57, 58]. The use of thiosulfate as the sulfur source in acidic conditions (pH 4–6) has also been reported [18].

CdS is the most studied material. Tens of papers have been published during the last years and a wide range of deposition conditions has been explored. Ammonia baths are most commonly used since ammonia is an efficient complexing agent for cadmium ions and allows (when combined with ammonium ions) to fix the pH in basic conditions suitable for the reaction [25, 26, 30, 32, 41, 43]. Other complexing agents have been used as TEA [27], EN [42], NTA [41]. The sulfur source is mainly thiourea, but thioacetamide [36, 39, 38, 75] and allylthiourea [30] can be used, TA being more reactive and ATU less reactive than TU. The use of thiosulfate ions in acidic conditions has been also reported [16]. It can be noted that deposition using directly dissolved hydrogen sulfide is also possible in a restricted range of pH (1–2) and deposition conditions [40]. For processes involving thiourea, kinetic studies show that the growth is, as for PbS, under kinetic control with activation energies in the range of 85–100 $kJ.mol^{-1}$ [34, 31]. In contradistinction for the case of H_2S precusor the growth is under diffusion control [40]. Epitaxial growth of CdS has been also demonstrated (see Section 6).

ZnS films have been prepared [68, 69, 70]. When using formulations based on ammonia thiourea baths similar to those used for CdS, as compared to CdS the growth rate is much lower [69, 70]. The addition of hydrazine has been found to accelerate the deposition allowing also the growth of close to stoichiometric films

[69]. Thioacetamide, which is well established for the precipitation of ZnS in solutions, can be also used [68] in which case the films have been deposited from acidic solutions. The addition of urea has a beneficial effect on the adherence [68]. Some attempts have been made to deposit ZnS by using thiosulfate-based solutions [16]. As compared to CdS and PbS it appears that the deposition of ZnS films is not yet optimized and in addition presents some differences in the growth mechanism. This is illustrated by the lower activation energy values (≈ 20 kJ.mol^{-1}) which has been determined in the ammonia-thiourea-hydrazine process, which is more likely characteristic of a diffusion limited growth [69]. The deposition of indium sulfide has been also reported in acidic solutions using TA [52], along with a detailed study of the influence of the deposition conditions on the structural and optical properties of the films.

Bismuth sulfide (Bi_2S_3) films have been deposited by using either TU, TA or thiosulfate processes with various complexing agents (TEA, NTA, EDTA) [16, 20–24]. The deposition is generally carried out in basic solutions (pH $\approx$ 10). In the case of thiosulfate, acidic media (pH $= 2 - 3$) can be used [16]. A tendency of the films to peel off have been noted by several authors [18, 24]. This phenomenon is generally due to a lack of adhesion to the substrates and can be suppressed by treating the substrate surface before the growth. Huang et al. report an original method with a treatment of glass by functionalized silane groups [24].

Interesting cases of deposition concern compounds where the oxidation state of the metal can take different values, such as copper or tin. Several reports concern the formation of Cu_xS with x between 1 and 2 [18, 47–50]. Varkey et al. [48] show that cuprous sulfide (Cu_2S) is formed in TU solutions, using a Cu(I) precursor (CuCl). However it can be obtained also when starting from a cupric salt (Cu(II)) due to the reducing properties of thiourea (and other sulfur precursors) as shown by Nair et al. [47]. Grozdanov et al. [18, 49, 50] report a direct relation between the TS concentration and the final compound composition. Low concentrations tend to form cuprous sulfide and high ones tend to form cupric sulfide (CuS). It is possible that in the first case the reduction of cupric ions preceeds the deposition step. Similar effects may occur in the tin sulfide system. Engelken et al. [64] report the formation of Sn_xS with some variations in x using S dissolved in propionic acid as the sulfur precusor; in this study the solution was non-aqueous which is rather rare in CBD studies.

Non-metallic sulfides have been deposited successfully, such as Sb_2S_3 [16, 18, 61] and As_2S_3 [16]. Chloride or oxides can be used as precursors for Sb and As, TA [62] or TS [16, 18] for sulfur.

2.2 Selenide Binaries

Many binary selenide compounds may be formed by CBD, lead and cadmium selenides being the most studied compounds. Contrary to thiourea for sulfides, the use of selenourea and its analogues is limited since these precursors are much more reactive and unstable toward oxidation. Selenosulfate ions which can be prepared easily are the most used precursors. Hodes et al. have studied in depth the deposition of PbSe with this precursor [86, 87] and different complexing agents. An interesting aspect of these studies is the possibility of growing nanocrystalline films, with grain size down to the nm level, by adjusting the deposition temperature (from -10 °C to 60 °C). Such films display quantum size effects, initially evidenced for CdSe [76], with an important increase of the band gap value from the infrared range to the visible range. Similar effects are reported for sulfides, such as In_2S_3 [52].

Like CdS, CdSe deposition has been widely studied. The solutions used are very similar to those used for CdS and the reaction proceeds in basic solutions at pH 10–11. CBD CdSe layers have been deposited for photoelectrochemical cells by Boudreau and Rauh [74]. The use of selenosulfate is well established by an in depth study by Hodes et al. of the deposition mechanism [41]. As for PbSe, quantum size effects have been demonstrated for nanocrystalline films [76]. Epitaxial films have been also recently deposited on InP single crystals (see Section 6).

Copper selenides films have been deposited and, interestingly, when starting from a copper (II) precursor, the composition of the final compound (from CuSe to Cu_2Se) appears to depend on the initial concentrations of the selenide precursor concentration as also observed for Cu_xS [18].

As shown in Table 1, other metallic (ZnSe, CoSe, Bi_2Se_3, NiSe) and non metallic (Sb_2Se_3) selenide binaries have been deposited successfully.

2.3 Telluride Binaries

Works on chemical deposition of tellurides are very limited, despite in principle the expectation that deposition should be similar to the sulfide or selenide systems. As indicated by Bode [3], telluride precusors (such as tellurourea) are very unstable which renders the deposition very difficult to achieve. Deposition has been reported with using a different process [94, 95]. Telluride precursor is introduced as dissolved TeO_2 (TeO_3^{2-} in basic solutions) in the solution containing complexed metallic ions. Upon the addition of hydrazine, the reduction to Te (-II) can be slowly achieved, leading to the formation of the metallic telluride as for PbTe [95]. More recently CdTe films have been prepared this way [94]. This process can be considered as an extension of the electroless process used for the deposition of metals [2]. It is probable that other routes similar to the selenosulfite route for selenides are possible for tellurides too, but have not yet been investigated in depth.

2.4 Ternaries

As compared to binaries, ternary compounds offer extended possibilities to engineer precise properties (in particular the band gap for semiconductors) by modifying the composition of the film. This technologically significant feature explains why the synthesis of these compounds is attracting great interest.

The deposition of $Pb_{1-x}Hg_xS$ alloys has been studied in depth by Chopra et al. [105, 106]. The bath composition was varied by changing the ratio of lead to mercuric ions concentrations. Single phase films were obtained up to $x = 0.35$. Epitaxial growth was even obtained on Si and Ge with $x = 0.33$ at deposition temperatures as low as 15 °C [106].

$Cd_{1-x}Zn_xS$ ternaries are intensively studied for their potential use in solar cells. The presence of zinc allows one to increase the band gap and to change the electron affinity of the layers, which should be beneficial for solar cell performance. Using acetate salts Padam et al. [98] reported the fabrication of the alloy over the whole composition, as verified by XRD. Others indicate the possibility of obtaining the alloys with other conditions, using iodide metallic salts for instance [100]. Recently the use of triethanolamine as a ligand has been reported [101]. Nevertheless it appears that some difficulties arise from the competition between the formation of the ternary compound and that of the binaries [99–101]. Yamagushi et al. [100] report that these phenomena depend on the sequence of preparation of the bath. Dona et al. [99] also encountered the problem of phase segregation and found that $Cd_{0.3}Zn_{0.7}S$ tend to form along with the binaries whatever the solution composition. They explain this on the basis of structural arguments based on the distinct growth habit of CdS and ZnS (hexagonal vs cubic).

In $Cd_xPb_{1-x}S$ formation [97], below 15% of Cd, a single phase ternary film is formed but higher concentrations lead to the additional formation of CdS. For other ternaries like $Sb_{2-x}Bi_xS_3$ single phase films are formed over the entire composition domain [18].

Other ternary sulfides with a definite composition have been reported, such as $CuInS_2$ [104] which belongs to the important family of chalcopyrite semiconductors.

Another approach to the formation of ternaries by CBD is the sequential process where two layers are formed in different baths, the single phase film being obtained after a post thermal annealing treatment. The formation of Cu_3BiS_3 [103] was accomplished in this fashion.

Fewer reports are available for the formation of selenide ternaries. $Cd_xZn_{1-x}Se$ deposition has been reported by Pandey et al. [107, 108]. Sharma et al. also demonstrated the formation of the alloy over the entire composition range when using selenosulfate precursor [109]. Attemps have been made to grow $CuInSe_2$ [109].

Ternaries can also be formed by substituting the chalcogenide element, leading to other series of materials. CdS_xSe_{1-x} has been deposited in single phase films for x higher than 0.5, using TU and SES precursors together in various ratios [113]. In the case of ZnS_xSe_{1-x} alloys, only films with $x = 0.84$ have been synthesized [114]. Recently, the formation of ternaries involving the substitution of chacolgenide elements by hydroxyl groups have been reported, such as Zn(OH, S) [117, 118], In(OH, S) [115] and Sn(OH, S) [116]. These hydroxy compounds are formed when the sulfide formation is in competition with the hydroxide formation. The films appear to present very good properties for applications such as interface buffer layers in thin film solar cells.

From this literature survey it appears that despite the appearance of a relatively large number of papers concerning chemical bath deposition, most are focused on a few materials, especially sulfide binaries such as PbS, CdS, which are used for applications. If we exclude these materials, in fact the work on other compounds is often limited to a few papers, or appears in screening studies. There is a trend towards the elaboration of ternary compounds by CBD, but the competition between the formation of the binaries represent a major difficulty. It seems that the composition of the solution, especially the nature of the complexing agents used plays a role. A general statement is that many works are restricted to characterization aspects (properties of the films). The chemical aspects are often recipe oriented. The consequence is that basic chemical aspects related to the growth mechanism, especially the interplay between deposition conditions and properties are often not considered. This is an obvious limitation to the development of the CBD method on comprehensive grounds.

3 Basic Aspects of Chemical Deposition of Chalcogenide Compounds

3.1 Nucleation Processes From Solutions

Nucleation mechanisms in solutions belong to two categories:

(i) Homogeneous nucleation which corresponds to the case where the new phase is formed in a solution containing initially no foreign phases.
(ii) Heterogeneous nucleation which is due to the presence of a foreign interface on which the growth proceeds selectively.

Thin films deposition falls mainly in the second category. Classical treatments based on thermodynamic analyse give a good description of the processes [119–121].

3.1.1 Homogeneous Nucleation

The change in free energy accompanying a phase transition between n molecules in a solution and in a cluster:

$$nP \rightarrow P_n \tag{2}$$

is composed of two terms, one related to the change of the chemical potential in the solution and in the solid, and the second to the variation of the surface energy:

$$\Delta G = n(\mu_s - \mu_l) + \gamma A \tag{3}$$

where γ is the surface tension between the solid and the solution, and A is the surface area. At equilibrium the chemical potential in the solid equals that in the solution ($\mu_{l,e} = \mu_s$) and equation (3) becomes:

$$\Delta G = -nkT\,\mathrm{Log}(S) + \gamma A \quad \text{with } S = a(P)/a(P)_e \tag{4}$$

where S is the supersaturation ratio, a is the activity corresponds to the concentration in diluted solutions and e subscript denotes the equilibrium state between the solution and the solid.

$$S = (a(A)^m a(B)^n / K_S)^{1/(m+n)} \tag{5}$$

where K_S is the solubility product of the compound (equal to $a(P)_e$ in equation (4)). By assuming that the cluster is a sphere, Equation (3) transforms to:

$$\Delta G = -nkT\,Log(S) + (36\pi v^2)^{1/3}\gamma n^{2/3} \tag{6}$$

The critical nucleus is associated to the value of n for which $d(\Delta G)/dn = 0$, leading to:

$$n^* = 32\pi v^2\gamma^3/3(kT\,Log\,S)^3 \tag{7}$$

The corresponding free energy change is considered as the activation energy for the homogeneous nucleation process:

$$\Delta G^* = 16\pi v^2\gamma^3/3(kT\,Log\,S)^2 = n^*kT\,Log\,S/2 \tag{8}$$

The nucleation rate is thus given by the expression:

$$J = J_0\exp(-\Delta G^*/kT) = J_0\exp(-16\pi v^2\gamma^3/3(kT)^3(Log\,S)^2) \tag{9}$$

The value of J_0 is the collision frequency between the precursors typically 10^{30} cm^{-3} s^{-1} [18]. When a chemical reaction between the precursors in solution is needed the value of J_0 can be replaced by $J_0^* = J_0\exp(-\Delta G_R/kT)$ where ΔG_R is the activation energy of the chemical reaction [122].

From Eq. (9) it may be seen that the nucleation rate is very sensitive to the supersaturation ratio and the surface tension. Figure 1 represents calculated variation of the nucleation rate with respect to the supersaturation ratio. Due to the exponential term the nucleation rate increases steeply from a supersaturation beyond a threshold value. The calculations have been made for a surface tension of 10^{-5} $J.cm^{-2}$ and a molecular volume of the solid of 4. 10^{-23} cm^3. With these values, typical of solid compounds, the critical supersaturation S* is about 12. Below this value the nucleation rate becomes rapidly very low and the solution can remains in a supersaturated situation with negligible homogeneous nucleation. The nucleation rate decreases markedly when the surface tension increases, for instance with a factor of 3 on γ (i.e. $\gamma = 3.10^{-5}$ $J.cm^{-2}$) leads to $S^* = 10^5$. The surface tension being related to the adsorption of species on the surface, an action on this parameter (and thus S*) is possible through the composition of the solution (pH, nature of ions, additives) [120].

The previous treatment can be extended to the case where a compound as A_m3_n is formed instead of an elemental material. The supersaturation ratio is taken as:

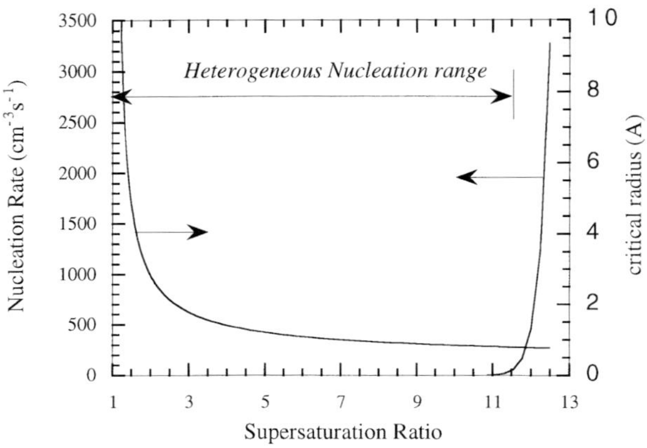

Fig. 1. Calculated homogeneous nucleation rate and critical nuclei radius as a function of supersaturation for $\gamma = 10^{-5}$ J.cm^{-2} and v = 4.10^{-23} cm^3.

3.1.2 Heterogeneous Nucleation

When nucleation occurs on a foreign surface a modification of the surface energy term takes place in equation (4). The surface tension at the interface between the substrate and the nucleus has to be considered. Assuming for example a flat rectangular nucleus as shown in figure 2, equation (4) is replaced by:

$$\Delta G_h = -nkT\,\mathrm{Log}\,S + (\gamma_d - \gamma_s + \gamma_{sd})L^2 + 4\gamma_d hL \tag{10}$$

The presence of the substrate reduces the surface energy needed for the nucleation process through the values of γ_s and γ_{sd}. In particular when the substrate is the same material, $\gamma_d = \gamma_s$ and $\gamma_{sd} = 0$ and the second term is zero. This means that the nucleation energy barrier tends to be suppressed in presence of a foreign substrate which matches with the nucleus material. The last term on the right introduces an unbalanced surface energy term, explaining in some case the existence of a residual activation energy for heterogeneous nucleation on the same material. It is reduced if the nucleation takes place on edges (figure 2) or even better on kink sites.

As a consequence, contrary to homogeneous nucleation, heterogeneous nuclea-

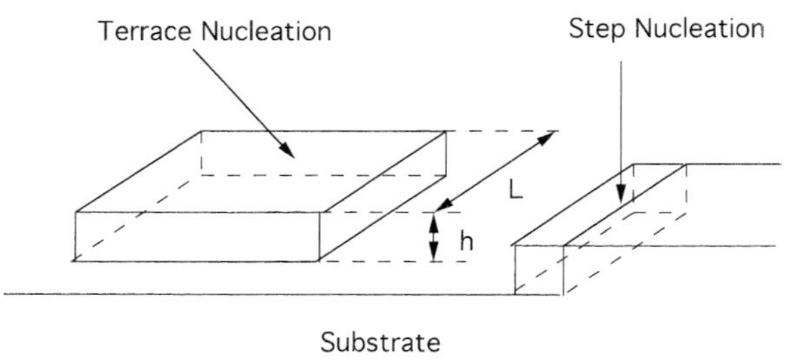

Fig. 2. Schematic representation of heterogeneous nucleation on a terrace and on a step.

tion will theoretically take place once the supersaturation ratio exceeds one. Thus, as illustrated in figure 1, a window exists for supersaturation values for $1 \leq S < S^*$ where the growth will proceed heterogeneously without nucleation in the solution. This is clearly demonstrated by experiments of seeded precipitation [119]: below the critical supersaturation ratio, the number of particles in the solution remains the same as the number of seeds introduced and only the size increases whereas beyond it the number increases rapidly due the homogeneous nucleation. These aspects are extremely important to understand and control chemical bath deposition which appears as a particular case of heterogeneous nucleation and growth processes.

In the case of heterogeneous growth from the vapor phase the growth mode of the deposit has been related to the interface tension between the two materials γ_{sd} [123, 124]. If this tension is very low there is no nucleation barrier ($\Delta G < 0$) and the growth takes place via a 2D layer by layer mode. If ΔG is positive a tridimentionnal (3D) growth mode (island growth) is expected. In a narrow range for ΔG values close to 0 the growth is first 2D and then 3D.

3.2 Solubility Diagrams

The foregoing relations show that supersaturation plays a key role both on nucleation as well as on growth steps for a given material. Solution chemistry provides a useful tool for predicting the occurence of a deposition reaction by using the approach of conditional solubility diagrams.

The deposition of a compound AB, for example, is related to the chemical equilibrium:

$$A^{2+} + B^{2-} \leftrightarrow AB \tag{11}$$

At equilibrium, the concentrations in solutions are fixed by the relation:

$$[A^{2+}][B^{2-}] = K_s \tag{12}$$

The left hand term is the ionic product in solution and K_s is the solubility product related to the free energy change associated to the reaction (11) by the equation:

$$K_s = \exp(\Delta G_0/RT) \tag{13}$$

The more stable is the compound, the more negative is the free energy of formation and the smaller the solubility product.

In table 3 are reported solubility products of chalcogenide compounds [125–128]. The values of the solubility products in real systems can vary of a few units since they depend on various parameters related to the solution (i.e. the ionic strength) and the structural state of the deposited material (amorphous, microcrystalline, well crystallized). Most chalcogenides are very insoluble compounds, and extremely low concentrations of free ions will lead to high supersaturation and thus to homogeneous precipitation.

Table 3. Solubility products of chalcogenide compounds calculated from [125, 126]. Values in parentheses are taken from [127]. $pK_s = -\log_{10} K_s$

Compound	pK_S	Compound	pK_S	Compound	pK_S
Ag_2S	49.9 (49.2)	Ag_2Se	54	Ag_2Te	73
Bi_2S_3	117.7 (97)	Bi_2Se_3	130.6	Bi_2Te_3	168.4
CdS	28.2 (26.1)	CdSe	34.2	CdTe	42
CoS	(24.7)				
CuS	34.6 (35.2)	CuSe	39.6	CuTe	52.4
Cu_2S	48.2	Cu_2Se	52.6	Cu_2Te	64
FeS	15.4 (17.2)				
FeS_2	42.5				
GaS	21	GaSe	20	GaTe	30
Ga_2S_3	82	Ga_2Se_3	87.2	Ga_2Te_3	113
HgS	50.6 (52.4)	HgSe	58.4	HgTe	71
In_2S_3	76	In_2Se_3	94	In_2Te_3	119
MnS	11.8 (12.6)	MnSe	8.5	MnTe	16
NiS	20.3 (25.7)	NiSe	25.7	NiTe	37.7
PbS	27.9 (26.6)	PbSe	35.9	PbTe	46.5
SnS	29 (25)	SnSe	33.3	SnTe	44.7
SnS_2	56.4				
ZnS	(24.3)	ZnSe	26.4	ZnTe	33.1

However free ions generally form complex species with ligands present in solution. In aqueous solutions the ligands are the protons for the chalcogenide ions and the hydroxide ions for the metallic cations. Additional ligands can be introduced, especially for the complexation of the metallic cations [127]. Table 4 shows a matrix of selected complexing agents and metal cations. As a consequence additional chemical equilibria have to be considered:

$$X + nL \leftrightarrow XL_n \tag{14}$$

where L is a ligand (OH^-, H^+ or a foreign species) and X denotes A^{2+} or B^{2-} ions. In addition, we have to consider the water dissociation equilibrium: $[H^+][OH^-] = K_w$ ($= 10^{-14}$ at 25 °C).

Table 4. Decimal logarithm of overall complexation coefficients for selected cations and ligands. The values are calculated for a free ligand concentration of 0.1 M, excepted for OH^- which was 10^{-3} M (pH = 11). EN = ethylenediamine, $NH_2CH_2CH_2NH_2$; DEN = diethylenetriamine, $(NH_2CH_2CH_2)_2NH$; TRIEN = triethylenetriamine $(NH_2CH_2CH_2NHCH_2)_2$. TEA = triethanolamine, $(CH_2CH_2OH)_3N$; NTA = Nitrilotriacetate, $N(CH_2COO)_3^{3-}$: EDTA = ethylenediaminetetraacetate, $(OOCCH_2)_2NCH_2CH_2N(CH_2COO)_2^{4-}$; TU = thiourea, $SC(NH_2)_2$. (–) indicates weak complexing effect.

	OH^- pH11	NH_3	EN	DEN	TRIEN	TEA	F^-	I^-	NTA	EDTA	TU
Ag^+	2	5.4	5.7	5.1	6.7	1.9	–	12.8	–	6.3	10.1
Bi^{3+}	9.4	–	–	–	–	–	–	12.8	–	21.8	5.9
Cd^{2+}	1.9	3.4	9.1	11.8	11.9	–	–	2.4	12.4	15.5	0.8
Co^{2+}	2.2	–	–	12.1	10	–	–	–	12.4	14.7	–
Cu^{2+}	3	8.6	17.6	19.3	19.4	3.4	0.17	–	14.3	17.8	–
In^{3+}	4	–	–	–	–	–	5.9	0.96	22.6	24	–
Mn^{2+}	0.5	0.26	3.05	4.8	3.9	–	–	–	6.4	13	–
Ni^{2+}	1.6	4.1	15.5	16.9	17.4	1.7		—	13.8	17.6	–
Pb^{2+}	4.6	–	–	–	9.4	–	–	1.1	10.8	17	–
Zn^{2+}	5.4	5.1	9.1	12.5	11.1	1.04	0.17	–	9.5	15.5	–

The equilibrium concentrations are related to the stability constants $\beta^{n}_{X,1}$ by:

$$[XL_n]/[X][L]^n = \beta^{n}_{X,1} \tag{15}$$

As a consequence the total concentration of dissolved X, $[X]_T$ under all forms is given by:

$$[X]_T = [X]\left(1 + \sum_{L,n} \beta^{n}_{X,1} L^n\right) = [X]\alpha_X \tag{16}$$

where α_X is the overall complexation coefficient of X species, and [X] the concentration of non complexed ions. In table 4 are indicated the values of $\log \alpha_X$ calculated for different metal ligand pairs at pH = 11. Replacing the concentrations of ions in equation (4) gives the apparent solubility product:

$$[A]_T[B]_T = K_s\alpha_A\alpha_B = K'_s \tag{19}$$

The complexation serves to increase the solubility. K'_S is called the conditional solubility of the compound.

In figure 3a are shown for example the variations with pH of the overall complexation coefficient of sulfide and cadmium ions. For sulfide ions (or the other

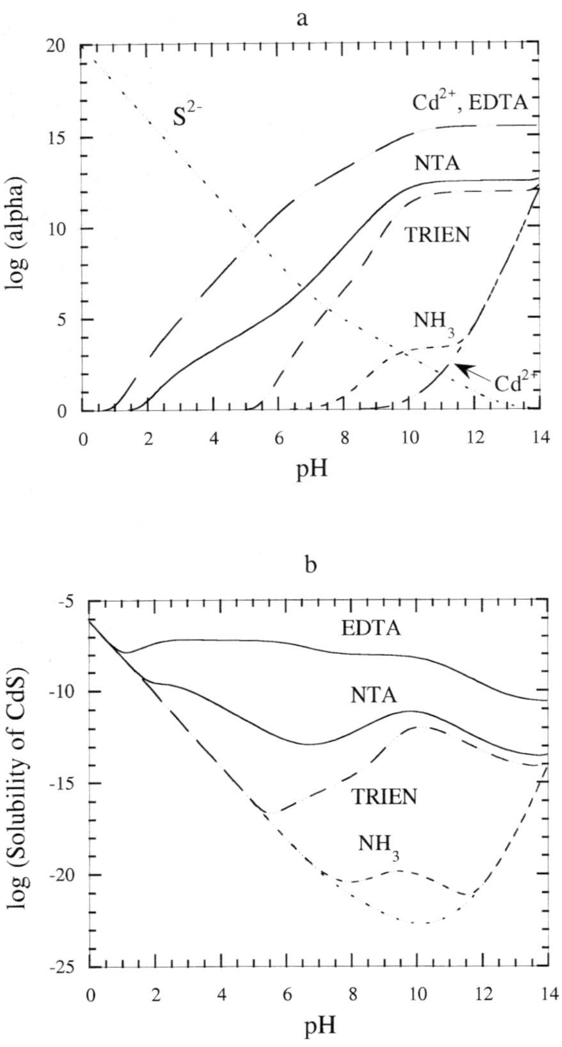

Fig. 3. a) Decimal logarithm of the overall complexation coefficient as a function of pH of S^{2-} and Cd^{2+} in presence of different ligands (at a free concentration of 0.1 M). b) Solubility of CdS obtained from the data of figure 3a and taking a solubility product of 10^{-26}.

chalcogenide ions) it increases towards low pH due to the complexation with protons (HS^-, H_2S). The reverse variation is observed for metallic cations. The effect of the addition of different ligands at a concentration of 0.1 M is shown. Large differences arise with the strength of the complexes, ammonia is a low strength ligand while EDTA is one of the highest available for cadmium ions.

The variation with pH of the complexation is due to acido-basic properties of the ligands, generally only the basic forms are active. In the case for ammonia, there is only one equilibrium:

$$NH_3 + H^+ \leftrightarrow NH_4{}^+ \quad K_A = 10^{-9.2} \quad \text{at } 25\ ^\circ\text{C} \tag{20}$$

The complexation is only effective for pH higher than the pK_A value of 9.2.

From relation (19) the solubility diagram can be established. Figure 3b shows that of CdS as a function of pH. It gives the maximum value $[A]_T[B]_T$ acceptable without precipitation. Without added ligands, the solubility is minimal at around pH 10 but increases markedly towards low and high pH values due to the complexation effects.

For chemical bath deposition, the precursor concentrations in solution ($[A]_T$, $[B]_T$) are conveniently situated in concentration windows between 10^{-4} to 10^{-2} M (i.e. 10^{-8} to 10^{-4} for K_s'). To obtain specifically the heterogeneous growth we have seen in the previous section that the supersaturation must remain sufficiently low. This allows to define deposition concentration windows from the solubility diagrams. As shown in figure 3b, in the case of CdS without added ligands, only one window appears in acidic situation. Heterogeneous deposition in acidic conditions has been indeed realized for CdS and PbS [40].

In the presence of complexing agent the solubility can be increased by many orders of magnitude in the low solubility range. In the case of EDTA, the values are close to the CBD window defined previously in a wide pH range.

3.3 Chalcogenide Precursors in CBD

Efficient complex formation for metal cations is mainly achieved in basic solutions while chalcogenide ions are mainly in the uncomplexed state. For a better control of the deposition it is also convenient to limit the concentration of free chalcogenide ions. The most common chalcogenide precursors are reported in Table 2. They are mainly soluble organic molecules belonging to the carbamide groups based on thiourea and selenourea. Non molecular inorganic precursors can also be used as thiosulfate and selenosulfate ions. Selenosulfate is increasingly used instead of selenourea [41, 129]. Selenosulfate is simply made by reacting sulfite in excess with elemental selenium (detailed procedure in [41]):

$$Se + SO_3^{2-} \leftrightarrow SeSO_3^{2-} \tag{21}$$

3.3.1 Hydrolysis Reactions

Contrary to the metal precursors, chalcogenide precursors undergo complex chemical reactions. Shaw and Walker have studied the hydrolysis of thiourea at temperatures between 90 and 130 °C [130]. They show that the rate increases both in acidic and basic solutions with the formation of H_2S and S^{2-} resp. In the range pH 2-8, the rate is pH independent and the hydrolysis products are thiocyanate and ammonium ions. Marcotrigiano et al. [131, 132] carried out a detailed study in basic solution and give the following scheme for the occuring reactions:

$$SC(NH_2)_2 + 2\,OH^- \rightarrow S^{2-} + CN_2H_2 + 2\,H_2O \tag{22}$$

$$CN_2H_2 + H_2O \rightarrow OC(NH_2)_2 \tag{23}$$

$$OC(NH_2)_2 + OH^- \rightarrow CNO^- + NH_3 + H_2O \quad (24)$$

$$CNO^- + OH^- + H_2O \rightarrow CO_3{}^{2-} + NH_3 \quad (25)$$

The first reaction leads to the formation of sulfide ions and cyanamide. Cyanamide can be further decomposed to urea, and then to carbonate anions and ammonia, which are the final reaction products. The associated rate constants have been estimated. In basic solutions the overall rate of hydrolysis has the following form:

$$v = k[TU][OH^-] \quad (26)$$

with $k = 10^{-4}$ mol^{-1}.L. s^{-1} at 100 °C. The process is strongly temperature dependent with an activation energy of 76 kJ.mol^{-1}. Similar reactions take place with the other precursors (thioacetamide, ethylthiourea). For thioacetamide in basic solutions first acetamide and then acetate anions and ammonia are formed together with released sulfide ions. The reaction rate is higher than with TU. Acid hydrolysis, which leads to the formation of hydrogen sulfide, is also much faster than with thiourea, explaining the wide use of TA for sulfide precipitation in acidic conditions [133]. Contrary to TA, ethylthiourea reacts much slower than TU but the activation energy is similar [134].

This shows that the reactivity of the precusor depends on the chemical groups bonded to carbon or nitrogen. Attempts have been made to predict the reactivity between different precursors by quantum chemistry calculations, as for instance the negative electronic charge on the sulfur atom [135, 136]. According to this particular criterion, the reactivity should decrease along the series TA > TC > TU > TSC > ATU, in agreement with some observations.

Selenide precursors based on selenourea display similar hydrolysis reactions and effects of substitution; N,N dimethylselenourea is for instance more stable than selenourea [5]. However, the reactivity of selenoureas, especially in contact with oxygen, is much larger than their sulfur analogues, making the use of these precursors more difficult. Tellurium compounds are even more reactive. To overcome this problem, reducing agents, like sulfite ions, have to be added to the bath [5].

When selenosulfate ions are used as precursors, selenide ions are produced by the hydrolysis reaction [129]:

$$SeSO_3{}^{2-} + 2\,OH^- \rightarrow Se^{2-} + SO_4{}^{2-} + H_2O \quad (27)$$

Similar reaction can be proposed for the release of sulfide ions from thiosulfate ions ($S_2O_3{}^{2-}$) [16, 18]:

$$S_2O_3{}^{2-} + 2\,OH^- \rightarrow S^{2-} + SO_4{}^{2-} + H_2O \quad (28)$$

All these reactions present a certain degree of reversibility. For instance the reverse reaction (22) is used to produce thiourea [137, 138], with a maximum rate at $pH = 8.5$, and selenoureas [139]. Equilibrium constants have been given. Values of K around 10^{-13} are given for thiourea [140] and around 10^{-14} for selenourea [141]. These values are also strongly temperature dependent, with activation energies in the range 50 to 100 $kJ.mol^{-1}$. The high values of the activation energies both for the reaction rates and the equilibrium constants are characteristic of these chalcogenide precursors. This is due to reaction mechanisms involving the existence of activated intermediate species, contrary to standard complexation-decomplexation reactions which are very labile.

3.3.2 Solubility Diagrams

The complexity of the chemistry of the chalgogenide precursors makes difficult the analysis of sulfide compound formation. Kitaev and coworkers [9,10] have used the equilibrium approach for various sulfides and selenides. For instance, in the case of sulfide formation with thiourea, reaction (22) is considered as reversible. Then it is possible to establish solubility diagrams. Assuming as an example that the total released sulfide ion concentration equals the cyanamide concentration gives:

$$[S^{2-}] = K^{0.5}[TU]^{0.5}[OH^-]\alpha_S^{-0.5} \tag{29}$$

Solubility diagrams, of cadmium sulfide for example, can be calculated from:

$$[Cd]_T[TU]^{0.5} = K_S K^{-0.5}[OH^-]^{-1}\alpha_{Cd}\alpha_S^{0.5} \tag{30}$$

with α_{Cd} and α_S corresponding to the functions defined previously. Figure 4 shows the solubility diagram calculated with $K = 10^{-15}$ in presence of ammonia 1.5 M. The

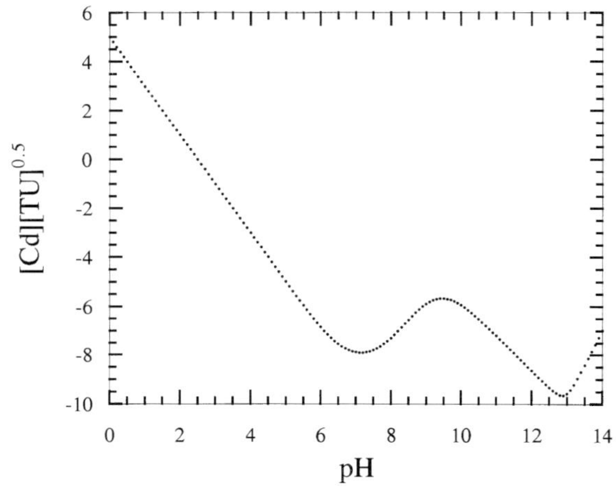

Fig. 4. Solubility of CdS in presence of thiourea and ammonia (1.5 M) as a complexing agent. The value of the dissociation constant of thiourea is taken as 10^{-15} and $K_S = 10^{-26.5}$.

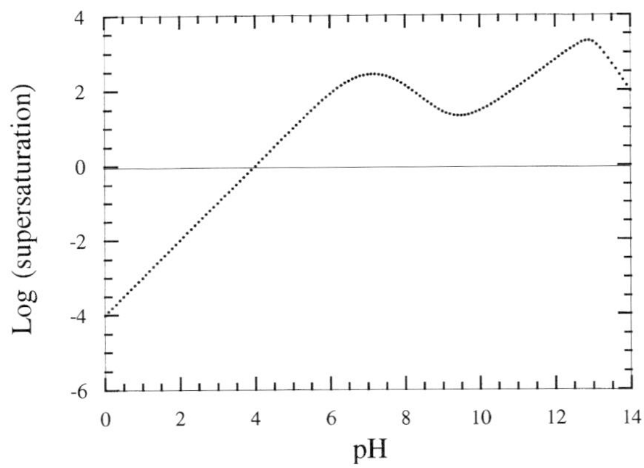

Fig. 5. Logarithm of supersaturation for CdS, in presence of total concentrations of thiourea and cadmium of 0.01 M and a concentration of ammonia of 1.5 M.

allowed concentrations of precursors are much higher than without thiourea. In these conditions it is possible to define solution composition suitable for deposition in a wide range of pH, between 6 and 14. The deposition conditions can be also expressed in terms of supersaturation diagrams. Figure 5 shows log S as a function of pH for [TU] = [Cd] = 0.01 M. At pH 11.5 the supersaturation S is about 10^3. Playing with temperature allows to control further the supersaturation through the modification of the hydrolysis constant. In standard CBD conditions the solutions are generally stable at room temperature for hours but react within a few minutes at 90 °C with homogeneous precipitation. In an intermediate range heterogeneous growth can dominate.

These figures, which are typical of CBD bath formulation, show that the conditions are compatible with a heterogeneous growth process. The diagrams can be completed by considering also acid-base properties of cyanamide CN_2H_2, which gives CN_2^{2-} ions in basic solutions ($pK_{a,1} = 10.28$ and $pK_{a,2} = 11.82$ [142]). Acid-base properties of the chalcogenide precursor may be considered too. However thiourea can be considered as stable over the whole usual pH range.

Similar treatments can be used for the other precursors of this family. It appears that with the use of these precursors a good control of the supersaturation can be achieved over the whole pH range.

3.3.3 Competitive Precipitation

In many cases the desired compound is not the only one which can be formed in solution. In that case competitive precipitation of several compounds may take place [128]. In aqueous solutions, metal hydroxides or oxides are the most common ones. If we consider again the case of CdS, the solubility product of $Cd(OH)_2$ ($K_s = 10^{-17.5}$) corresponds to the equilibrium:

Temperature effects on solution chemistry have to be considered if not working at 25°C.

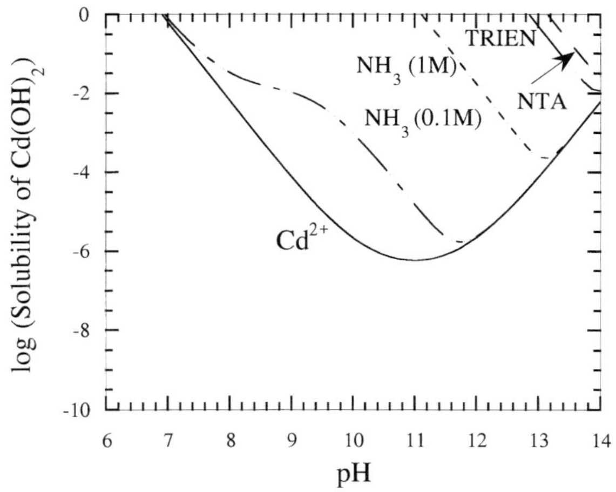

Fig. 6. Solubility of cadmium hydroxyde as a function of pH, without and with different ligands of figure 3 (concentrations of 0.1 M if unspecified).

$$Cd^{2+} + 2\,OH^- \leftrightarrow Cd(OH)_2 \tag{31}$$

Figure 6 shows the conditional solubility of cadmium hydroxide as a function of pH. The hydroxide ion concentration is in that case fixed by the pH.

The line corresponds to the maximum value of the total concentration of cadmium ions in solution without precipitation of the hydroxide. If the experimental value is higher, then cadmium hydroxide clusters will be formed in the solution which can be converted into sulfide clusters (chapter 4). This must be taken into account in fixing the metal concentration in the CBD bath. It is possible to modify the solubility limit by changing the strength of the ligand or its concentration as also shown in figure 6.

Competitive precipitation may also involve the formation of compounds with reaction products. In the case of thiourea, the basic form of cyanamide and carbonate ions can react with metallic ions to form solid compounds (i.e. $CdCN_2$ and $CdCO_3$ with $K_s = 10^{-14,2}$ and 10^{-14} respectively). The formation of these compounds at the surface and their inclusion in the deposit is enhanced by the higher concentration of the reaction species at the surface during the deposition reaction.

The existence of competitive precipitation phenomena have a direct influence on the film composition which is presently a matter of increasing attention [44]. Cyanamide and hydroxide groups have been found as major impurities in CdS films at levels which can reach 20% [140]. Carbonate groups have been recently evidenced too [44]. The concentrations depend on the solution composition, for instance cadmium sulfide films prepared at lower pH are close to stoichiometry whereas those prepared at higher pH have a ratio [Cd]/[S] about 1.1 due to the presence of hydroxide species ($Cd(OH)_2$) [45]. The formation of other hydroxy-sulfide films (of Zn [105, 106], In [116], Sn [116]) can be also related to competitive precipitation phenomenon. As a consequence, the analysis of the solubility diagrams of all solid compounds in a given CBD system is a useful tool to improve the composition control of the films. This approach can be generalized to analyse the formation of ternary compounds as $Cd_xZn_{1-x}S$ or CdS_xSe_{1-x}. Composition is also affected by the deposition rate [45] and the structure of the film. In microcrystalline films a high grain surface to volume ratio favors the insertion of adsorbed foreign species.

4 Studies of Deposition Mechanism

4.1 Literature Survey of Growth Kinetics Studies

CBD is a process which can take place over a period of time that can range from a few minutes to over several hours. The thickness of the films is typically in the tens of nm to the micron range. Kinetic studies are classically based on *ex situ* film thickness determinations, using a mechanical profilometer [11], optical absorbance [42], or film composition [35]. Such a strategy for studying deposition kinetics was largely employed for a number of binary and ternary chalcogenides ([11] and references therein and more recent references about CdS [25, 31, 35, 42], ZnS [69], CdSe [77], ZnSe [92, 93], CuSe [92], Sb_2S_3 [62]).

However *in situ* approaches are better adapted for studying in details the growth of CBD films. One possibility is to use electrical measurements during the growth when the films are deposited on conducting substrates. High frequency capacitance measurements have been used for in situ thickness determinations in conditions where the film is fully depleted [30, 143]. Open circuit voltage measurements have been used by Dona and Herrero [31]. It was assumed that the EOCPC signal is proportional to the square of the film thickness. The most significant progresses were achieved by the introduction in the field of the Quartz Crystal Microbalance (QCM) technique allowing *in situ* mass measurements [30, 34]. This was made possible by the recent developments of lab-made or commercially available oscillators working in liquid solutions [144–146]. Figure 7a shows an example of a growth curve obtained *in situ* by QCM and *ex situ* by profilometry showing different stages of the growth process: an induction range, a linear growth range defining a mean growth rate r_m, and a third range where growth saturation is generally observed. Figure 7b illustrates the case where the mesurements deviate at longer time. This is due to the formation of a porous overlayer which is detached during the mechanical measurements. r_m was proved to give a valuable information about the growth process. It was shown by several authors that r_m followed an Arrhenius law [31, 34, 69, 93, 143]. The activation energy was found in the range 60–100 $kJ.mol^{-1}$. These values indicate that the rate determining step in the deposition mechanism is rather a chemical step than a diffusion one from the solution. The latter should lead to a much lower apparent value (a few tens $kJ.mol^{-1}$). The effects of the solution chemistry (nature of the precursors, complexing agents, pH, see sect. 3), the nature of the substrate (see section 5) on r_m were analyzed and discussed in terms of possible reaction schemes, in a more or less quantitative way [11, 25, 34, 35]. Ortega-Borges and Lincot [34] presented an analytical model allowing to predict the mean growth rate under various conditions, in

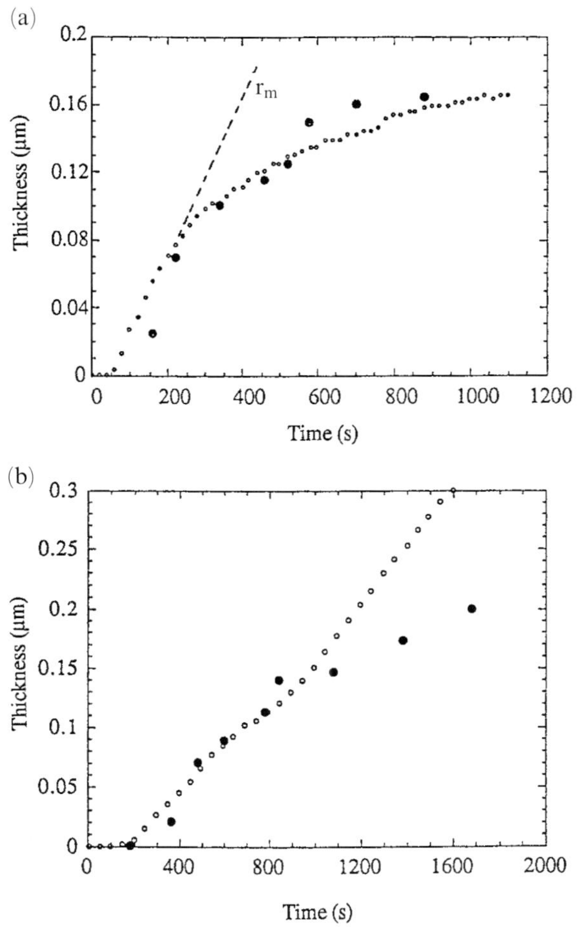

Fig. 7. Typical growth curves of a CdS CBD film measured *in situ* by the quartz microbalance technique (small open symbols). Filled symbols correspond to *ex situ* thickness measurements with a profilometer. (a) Growth of a compact layer with thickness saturation, dotted line shows the initial linear regime, conditions: $[Cd]_T = 0.014$ M, $[TU] = 0.14$ M, $[NH_3] = 1.74$ M, $T = 60$ °C. (b) overgrowth of a porous poorly adherent layer towards long deposition times, same conditions as (a) but with $[TU] = 0.03$ M.

the case of CdS in the ammonia-thiourea system. Rate-limiting surface steps were involved in the proposed reaction scheme. The formation of CdS was stated to take place *via* a surface complex between thiourea and cadmium hydroxide. This model will be discussed more in details in sect. 4.4. *In situ* mass changes m(t) can also be directly monitored with a resolution of a few milliseconds, depending on the data storage capabilities. This approach, developed very recently, allows new insights in the deposition mechanism which will be also presented in this paper [80, 81, 147].

4.2 Quartz Crystal Microbalance In-situ Mass Measurements

QCM was extensively used as a mass sensitive detector in vacuum applications and has become an important tool for monitoring mass changes occuring at the crystal surface when it is immersed in a liquid, as reviewed recently [148]. Sauerbrey provided a description and experimental proof of the linear mass-frequency relation for foreign layers deposited on thickness-shear mode piezoelectric crystals [149]. This relation reads:

$$\Delta f = -\frac{2f_0^2\Delta m}{A\sqrt{\rho_q \mu_q}} \tag{32}$$

where Δf (in Hz) is the measured frequency shift, f_0 the resonant frequency of the unperturbated crystal, Δm (in g) the change in mass attached to the piezoelectrically active surface area A (in cm^2), ρ_q and μ_q the density and the shear modulus of quartz respectively. The derivation of the mass-frequency relation (32) implicitely relies on the assumption that a deposited foreign material exits entirely at the antinode of the standing wave propagating across the thickness of quartz crystal. That is, the foreign deposit is treated as an extension of the quartz crystal. In practice it was verified that Eq. 32 is valid up to loadings approaching 10% of the crystal mass. The constant $K = (-A.\Delta f/\Delta m)$ is theoretically equal to 2.26 10^{-6} f_0^2 $Hz.cm^2.g^{-1}$. For a 6 MHz AT-cut quartz the minimum detectable mass change is of the order of 1 $ng.cm^{-2}$ [144], which roughly corresponds to a tenth of monolayer of carbon atoms. In practice, it is necessary to calibrate the QCM in the operating conditions. It was shown that the average sensitivity measured by silver electrodeposition experiments was strongly dependent on the electrode area: the larger is the active surface area, the closer is the average sensitivity to the theoretical value calculated by taking into account the physical parameters of the quartz crystal [145].

The mass surface density variation Δm can be related to the thichness h of the deposit coating the quartz crystal by considering the relation:

$$h = \Delta m/\rho \tag{33}$$

where ρ is the density of the deposited material. In the case of a discontinuous film an equivalent thickness h_{QCM} can be defined by using (33) [30].

Frequency variations can arise from additional factors as temperature, pressure, solution viscosity, which are generally neglected in usual CBD experimental conditions. Another factor relevant with QCM mass measurements is the effect of surface roughness on the response of the thickness-shear mode resonator in contact with a liquid. It has been demonstrated that for roughness features much less (15%) than the liquid decay length ($\approx$0.25 mm for a 6 MHz quartz in water), the surface may be considered as hydrodynamically smooth and the frequency shift due to the liquid only depends on the density-viscosity product [150]. As features become comparable to or larger than the decay length, new mechanisms can arise, including liquid trapping and compressional wave generation. Enhanced solid/liquid interactions caused by surface roughness will be manifested as an additional mass on the surface. Concerning CBD, there are two situations of interest. First, the surface roughness is due to the quartz substrate itself and does not change significantly during the deposition process. A constant, additive, frequency shift is promoted when passing from air to water solution. From the data in [150], a 15 kHz negative frequency shift is evaluated for an average surface roughness of $R_a = 525$ nm and 3 kHz if $R_a < 10$ nm respectively. The effect is amplified in more viscous liquids. Second, when looking at mass change over a short period of time by comparison to the total deposition time, the morphology of the deposit can be reasonably assumed to be unchanged. On the other hand, it was recently observed by Bruckenstein *et al* that a 5 μm surface finish did not affect the mass sensitivity of the 10 MHz AT-cut crystals used in their silver deposition experiments [151]. Though the question remains open, roughness effects have to be kept in mind as a possible perturbating factor, specially if the morphology of the deposit markedly changes with time.

Within the domain of validity of the Sauerbrey relation, *i.e.* the frequency shift is proportional to the mass change, it is worthwhile to consider the temporal derivative of frequency df/dt. This quantity represents the rate of the piezoelectric substrate mass change. It allows a real-time, *in situ* determination of dissolution and/or film formation reaction rates. This approach was firstly applied to electrochemical corrosion studies during linear potentiodynamic scans, in the case of iron [152] and conductive oxide [153, 154] films. The mass balance from QCM measurements can be compared to mass balance from Faraday's law, providing a direct information about the number of electrical charges consumed per unit of mass [153]. For the temporal derivative to be meaningful, sufficiently high sampling rate (around 1 Hz or more, depending on the potential scan rate) and large signal-to-noise ratio are required. This is readily achieved by smart, commercially available, frequency counters allowing the frequency data array to be directly acquired via an internal memory system. For instance, the gate time can be as low as 1 ms but the random frequency error reaches ± 3 Hz. For a gate time of 0.1 s, the random error goes down to ± 0.06 Hz, lower than the sensitivity of a 6 MHz QCM. For CBD systems, the interest in determining the temporal derivative dm/dt was demonstrated [80] by the formal comparison of dm/dt profiles with models elaborated for transient electrocrystallisation currents [155, 156].

4.3 Combined QCM and Impedance Measurements

Measurement of the interfacial impedance by means of a sine wave of a small amplitude over a wide frequency range has been largely developed for *in situ* kinetic studies in electrochemistry, allowing couplings between various elementary phenomena to be dissociated [157]. The method can also be used for CBD provided conducting substrates are used [143]. Let us consider the CBD formation of a semiconducting layer in depletion conditions [30]. This situation is equivalent from an electrical point of view to the development of an insulating layer on the top of the metal electrode. In a more general case the layer can be considered as partially covering. The corresponding simplified equivalent circuit is shown in figure 8a. Zp accounts for the flow of faradaïc current across the interface and R_s for all the series resistance. C_1 and C_2 represent the respective contributions to the total capacitance ($C = C_1 + C_2$) of the covered substrate area and the uncovered one; θ is the fraction of the covered area. Theses capacitances are expressed respectively as the series combination of C_s, C_i and C_H, which are respectively the substrate capacitance, the dielectric capacitance, and the Helmholtz capacitance per unit area:

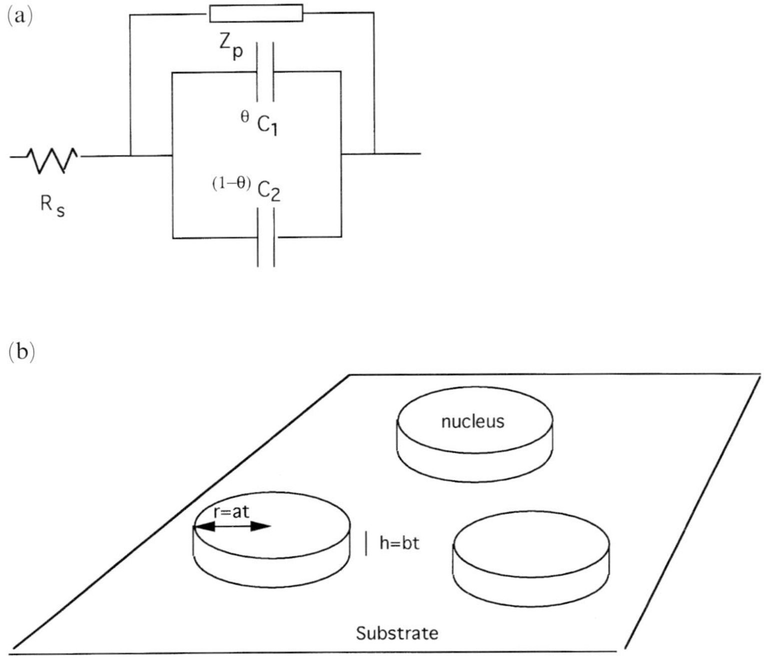

Fig. 8. (a) Simplified equivalent circuit representing the linear response of a conductive electrode partially covered by a surface layer (from [30]). (b) Model used for simulating the growth of the film.

$$\frac{1}{C_1} = \left[\frac{1}{C_s} + \frac{1}{C_i} + \frac{1}{C_h}\right]\theta^{-1} \tag{34}$$

$$\frac{1}{C_2} = \left[\frac{1}{C_s} + \frac{1}{C_h}\right](1-\theta)^{-1} \tag{35}$$

In the case of a metallic substrate, C_s and C_h are much higher than C_i meaning that C_l is close to that of the dielectric film only. In the case of a flat covering layer, its thickness is simply derived by the plane capacitor relation: $h = \varepsilon\varepsilon_0/C_i$ where C_i is deduced from relation (34) and ε is the dielectric constant of the deposited material. When the film does not completely cover the substrate, the measured capacitance depends on the covering properties of the film and is no longer related to the film thickness. The equivalent thickness h_z, defined from the total measured capacitance:

$$h_z = \varepsilon\varepsilon_o\left(\frac{1}{C} - \frac{1}{C_h}\right) \tag{36}$$

is smaller than the real thickness of the layer. The continuous recording of the capacitance signal can be performed by means of a lock-in technique. For these measurements, a large area counter-electrode has to be introduced in the CBD reactor. The frequency range of the superimposed *ac* excitation signal has to be chosen for the equivalent circuit of the interface to be modelled by a simple RC series circuit. A direct measurement of the capacitance is then provided by the quadrature signal.

When the deposition is made on the quartz electrode directly, capacitance measurements can be combined with mass measurements. This has been successfully applied to the study of the growth of CBD CdS layers, deposited from the ammonia-thiourea bath [30]. A quantitative analysis of the experimental results was done by considering a simple growth model shown in figure 8b. It was assumed that (i) the nucleation process was instantaneous, with a density of nuclei N_o, (ii) the nuclei had a circular base, with a radius r increasing with time as $r = a.t$. For a tridimensional growth, nuclei develop as cylinders perpendicular to the substrate surface with a height h(t) varying linearly with time as $h(t) = b.t$: a and b are the parallel and perpendicular rates respectively. Before the coalescence, the surface area covered by the nuclei S(t) is simply given by $S(t) = N_o \pi r^2 = N_o \pi a^2 t^2$. During the coalescence, the covered fraction of the suface $\theta(t)$ is described by the Avrami law [158]:

$$\theta(t) = 1 - \exp\left(-N_o \pi a^2 t^2\right) \qquad (37)$$

From the knowledge of $\theta(t)$, the total capacitance C(t) can be evaluated at any time, using (34) and (35). The equivalent thickness $h_z(t)$ was deduced from (36). In parallel, an equivalent thickness h_{QCM} can be calculated from the total volume of the deposit at time t:

$$h_{QCM}(t) = b.t.\ \theta(t) \qquad (38)$$

This model simplifies in the case of isotropic growth where the perpendicular and parallel rates are equal ($a = b$). This simplification was found to be suitable for describing the kinetics of CBD CdS in [30]. When coalescence is complete ($\theta = 1$), the exponential term in (37) disappears and the predicted deposition rate is hence-forth constant with time (linear regime).

Figure 9 shows experimental results and simulated curves of the equivalent thickness determinations from QCM and capacitance (CAP) where excellent agreement is obtained. At short times the CAP curve is lower than the QCM one. The apparent growth rates are lower than the real (perpendicular) value, the discrepancy being larger for the CAP method due to the fact that the layer does not cover the surface. This corresponds to the coalescence range. Increasing the density of nucleation centers N_o strongly reduces the coalescence sequence, with the limiting case of the 3D-growth of a homogeneous film with a constant rate. This is observed experimentally when changing the solution composition as for instance the thiourea concentration in the case of CdS deposition. The thickness at full coverage decreases down to few tens of nm when thiourea concentration increases. After the coalescence period both measurements are close together indicating the growth of a flat covering film.

A major interest of combined QCM and capacitance measurements is to provide information about the internal structure of the film during the growth. This is exemplified in figure 10a. In this figure, the film growth can be divided into three regions

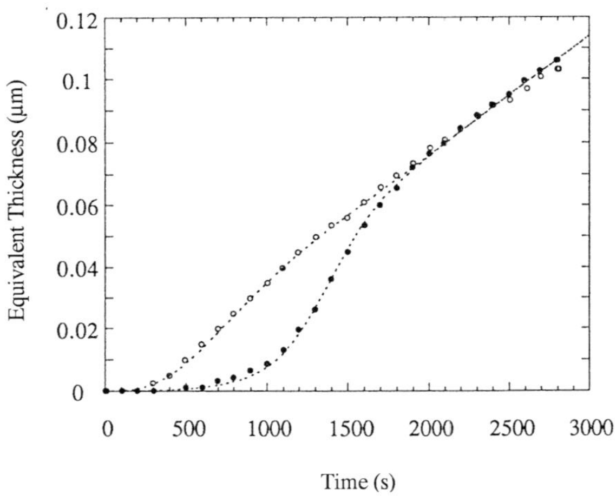

Fig. 9. Illustration of the coalescence phenomenon. Open circles: results from quartz crystal microbalance, filled circles: results from capacitance. Conditions: $[Cd]_T = 0.014$ M, $[ATU] = 0.028$ M, $[NH_3] = 1.74$ M, T = 60 °C. Dotted lines: simulated curves. Parameter values: $C_H = 10\ \mu F.cm^{-2}$, $a = b = 3.9\ 10^{-8}\ cm.s^{-1}$, $No = 6.10^{10}\ cm^{-2}$ (from [30]). With permission of Electrochem. Soc.

(a)

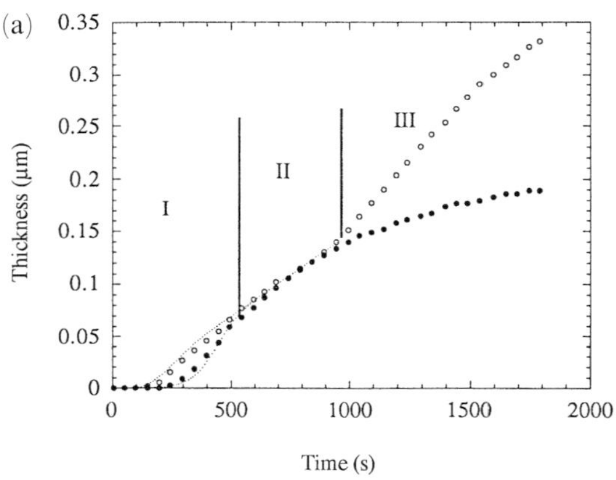

(b)

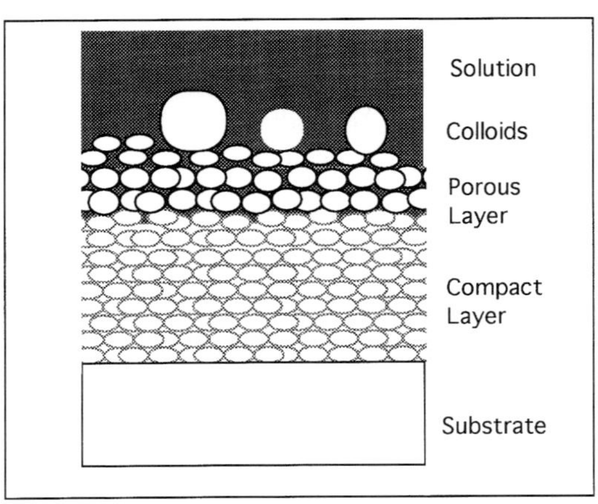

Fig. 10. Illustration of the porous over-layer growth. a) Experimental results. Open circles: QCM measurements, filled circles: capacitance measurements. Conditions: $[Cd]_T = 0.014$ M, $[TU] = 0.03$ M, $[NH_3] = 1.74$ M, T = 60 °C, same as Fig. 7b. (b) Schematic structure of a chemically deposited film. With permission of Electrochem. Soc.

of definite behaviour. First, at short reaction times, QCM and CAP determinations do not coïncide, as predicted by the columnar growth model during the nucleation/coalescence regime (Fig. 9). Second, after the full coalescence is achieved, both determinations are close together. This corresponds to the growth of a homogeneous covering layer, with a growth rate constant or slightly decreasing with time. Third, at longer reaction times, QCM and CAP determinations deviate from each other, the QCM thickness increasing further whereas the CAP thickness tends to flatten. This was explained by the effect of the electrolyte trapped in the outer layer of the deposit which presents a porous character. In these conditions, the QCM thickness is overestimated and the capacitance of the particles in the porous layer do not contribute because of the electrical shunt produced by the penetrating conductive electrolyte. By this way, a change in the structure is evidenced which could be associated to a transition of the growth mechanism from an atom by atom to a colloïdal growth. Figure 10b shows the structure of the CBD films deduced from these experiments. Note that the porous layer formation was also evident by comparing QCM and profilometer measurements (Fig. 7b).

More complicated kinetics can be elucidated by analyzing the dm/dt profile as shown in figure 11 in the case of CdSe layers chemically deposited from an ammonia-

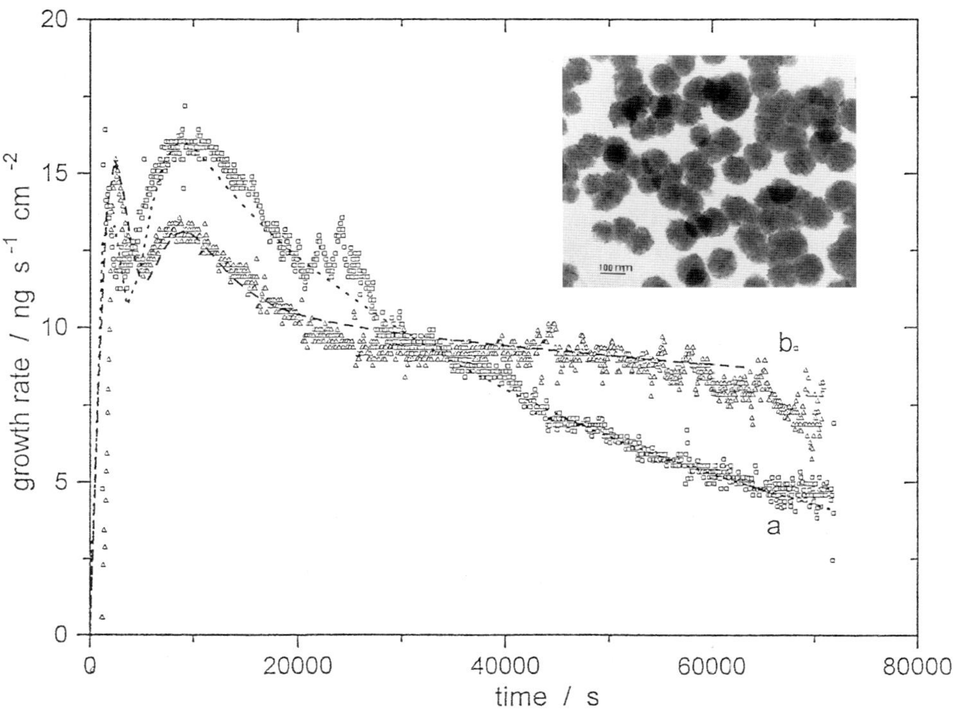

Fig. 11. Comparison between experimental and simulated (dashed lines) dm(t)/dt growth rate of a CdSe film deposited at 45 °C from an ammonia-TEA-selenosulfite bath, without STA (curve a), in the presence of 10^{-5} M STA (curve b). The inset presents a TEM image of a CdSe deposit on a carbon membrane before coalescence showing the formation of flat platelets (from [80]). With permission of Elsevier

triethanolamine (TEA)- selenosulfate bath. Two peaks were evidenced in the dm/dt profile, indicating that the nucleation and growth process occurs *via* two consecutive steps [80]. The first one was interpreted as an instantaneous nucleation of cylindrical nuclei, with bidimensional growth until complete coverage of the substrate. The thickness of this layer was in the range of a few tens of nm. This assumption was supported by low magnification TEM image (inset in figure 11) showing the formation of flat platelets with a surface density of 10^{10} cm^{-2} during this first stage. The second one was interpreted as a tridimensional instantaneous nucleation and growth process occuring on sites randomly distributed on the time-evolving surface, represented by flattened hemispheroïds. A complete analytical model was derived for dm/dt and compared to the experimental profiles. Figure 11 also shows the agreement between QCM experiments and the above two-step model. Additionaly, it is shown in this figure that small amounts of a strongly adsorbing species (silicotungstic acid) decrease the growth of the 3D-growth step but increase the life-time of the chemical bath.

4.4 Origin of the Growth Mechanism: Cluster-by-cluster or Ion-by-ion Process

As illustrated in figure 12, the elementary "brick" for building up a chemically deposited layer can be either a monoatomic (or monomolecular) species (a), or a cluster involving a large number of atoms (b). Mixed growth regime is also possible (c). Case (a) is called generally the "ion-by-ion" process [25, 34]. From the considerations in section 3 it corresponds more likely to the heterogeneous nucleation and growth mechanism. Case (b) is called the "cluster-by-cluster" or colloïdal growth [25, 80, 159]. In that case one can consider that it corresponds to the aggregation on the substrate of particles formed in solution by the homogeneous nucleation mechanism. Experimental evidence of the first one is given by the possibility of growing epitaxial CBD films (PbS, HgS, CdS, CdSe) on monocrystalline substrates (Si, Ge, InP). This point is discussed in detail in sect. 5. Colloïdal growth has been proposed for CdS in the presence of cadmium hydroxide colloïds in solution [159] and also in the case of CdSe to explain the nanostructure of the layer [41, 76]. For CdSe, it was said that the initial nucleation occurs by the conversion of the individual hydroxide clusters to CdSe and further growth of the individual clusters can be considered as an ion-by-ion process. Further thickening of the CdSe film is thought to take place more likely by a

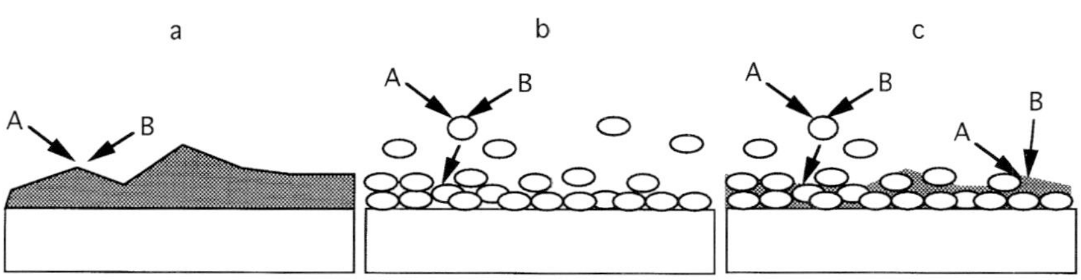

Fig. 12. Schematic representation of the different mechanism for film growth. (a) ion-by-ion growth, (b) cluster-by-cluster growth, (c) mixed growth.

cluster growth mechanism, although parallel ion-by-ion growth is not excluded [41]. In this study the presence of cadmium hydroxide colloids was very well detected by using light diffusion experiments, showing that the transition between ion by ion growth and cluster by cluster growth was corresponding to the cadmium hydroxyde precipitation line. It was further demonstrated that bath conditions favouring colloïd formation are unable to promote CBD epitaxial growth [147]. Time-resolved QCM experiments tend to prove (see after sect. 4.5) that both mechanisms can coexist whereas $Cd(OH)_2$ colloïds are present in solution [147]. On the contrary, in the absence of $Cd(OH)_2$ colloïds in the deposition bath, it is concluded that ion-by-ion growth is certainly valid in agreement with [41].

In the case of CdS, the ion-by-ion process is mainly considered as an explanation of the formation of adherent, specular layers, even in the presence of colloïds in the bath [11]. This is probably the growth mechanism involved in the formation of the compact, adherent, inner CdS layer as discussed in sect. 4.3. At longer reaction times the incorporation of colloids, formed in parallel in solution, gives rise to a porous and less adherent outer layer [30].

The ion-by-ion process has been analysed from *in situ* QCM measurements for the deposition of CdS from ammonia-thiourea solutions, using the measurement of initial growth rates over a wide range of experimental conditions [34]. A general mechanism has been derived for the deposition of a binary compound MX:

(i) Adsorption of precursor M on a specific surface site

$$\mathrm{M} + \mathrm{Site} \underset{k_{-1}}{\overset{k_1}{\Longleftrightarrow}} \mathrm{M_{ads}}$$

(ii) Adsorption of precursor X on Mads

$$\mathrm{X} + \mathrm{M_{ads}} \underset{k_{-2}}{\overset{k_2}{\Longleftrightarrow}} \mathrm{X} - \mathrm{M_{ads}}$$

(iii) Formation of a molecule MX with site regeneration

$$\mathrm{X} - \mathrm{M_{ads}} \overset{k_3}{\rightarrow} \mathrm{MX} + \mathrm{Site}$$

Assuming the conservation of the total number of surface sites N_o, the initial growth rate r_m is given by the equation:

$$\frac{d[MX]}{dt} = \frac{k_1 k_2 k_3 N_0 [M][X]}{(k_{-1}k_{-2} + k_{-1}k_3) + (k_{-2}k_1 + k_1 k_3)[M] + k_2 k_3 [X] + k_1 k_2 [M][X]} \tag{39}$$

with [M] and [X] representing the total concentration of metal and chalcogenide precursors respectively.

This atom-by-atom growth mechanism fits experimental results very well for CdS deposition from QCM investigations in ammonia-thiourea solutions, as shown in figure 13. The rate constants take into account the equilibrium composition of the bath with respect to the concentration of the various cadmium complexes with hydroxide ions and ammonia (see section 3).

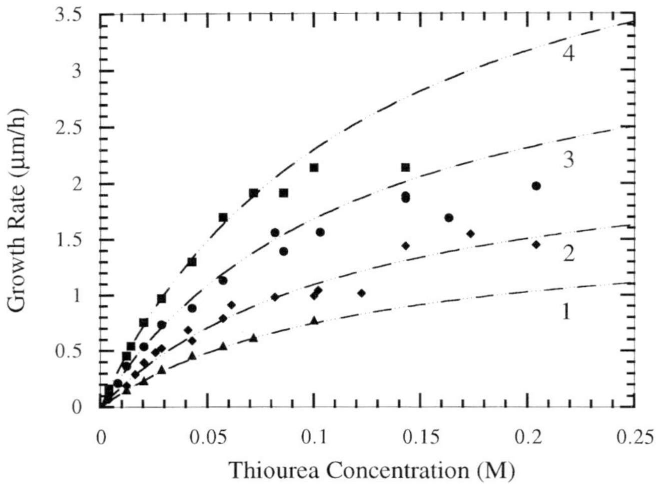

Fig. 13. Influence of thiourea and different cadmium concentrations (1 to 4: from 10^{-3} to 410^{-2} M) on the initial growth rate of CdS films, $[NH_3] = 1.74$ M, T = 60 °C. Points: experimental results, dotted lines: simulated curves [34]. With permission of Electrochem. Soc.

The chemical nature of the surface species has also been considered. Cadmium is reversibly adsorbed as cadmium hydroxide in the first step. The second step corresponds to the adsortion of a thiourea molecule on this species to form a surface complex. The formation of CdS results from the decomposition of this surface complex. An important consequence of this mechanism is that the growth is kinetically controlled by surface reactions, in agreement with the high value of the activation energy and the lack of strong dependence on the hydrodynamic regime. Contrary to the generally assumed mechanism [11] involving the formation of free chalcogenide ions in solution by the hydrolysis reactions (as 22, 27, 28), this mechanism assumes a direct reaction at the surface with thiourea. The adsorption of cadmium hydroxide and its conversion to cadmium sulfide by reaction with thiourea at the surface has been indeed found by Riejke and Bentjen from ex-situ SEM and composition studies [35]. This mechanism also agrees with earlier studies on CdS [135]. In fact it appears that the adsorption of metal hydroxide species*, even in conditions were there is no precipitation in the bulk of the solution, catalyses the decomposition of thiourea and the transformation to sulfide [34]. The similitude between the activation energies of the CBD and the hydrolysis reactions of the precurors is coherent with this mechanism.

*The surface chemistry in basic solutions tends anyway to form hydroxide covered surfaces *via* preferential bonding with metal atoms, even in conditions where the solution in not saturated with the hychoride.

4.5 Transient QCM Experiments

In the previous sections, studies of CBD kinetics were essentially based on the determination of an average growth rate, either independent of time ("linear regime"), or varying with time (complex dm/dt profiles). In a certain way, this is a macroscopic description of the overall CBD process, which has proved to be a powerful tool for analyzing the effects of bath composition or revealing the different contributions in a multistep process. However, such an approach is unable to give insight about the elementary steps involved in the nucleation and growth mechanism at a microscopic level.

A breakthrough was recently brought by the development of time-resolved QCM experiments, that involved an improved signal-to-noise ratio, a sampling interval for data acquisition much lower than 1s, and a well adapted CBD system for this purpose [81, 147]. The latter was CBD CdSe grown from NTA-selenosulfate solutions, intensively investigated by Gorer and Hodes [41]. It was established that there is a certain critical ratio R_c between the cadmium complexing agent (NTA) and cadmium concentration used in preparing the films above which no $Cd(OH)_2$ precipitate is present in the deposition bath. These authors concluded that below R_c the deposition mechanism is initiated on the $Cd(OH)_2$ colloïdal particles adsorbed on the substrate while, above R_c, deposition occurs directly on the substrate by initial CdSe formation, without any mediation by $Cd(OH)_2$ colloids. Simply changing the ratio $R = [NTA]/[Cd]$ by modifying the bath composition allows one to change the deposition mechanism. Time resolved QCM measurements have given new insights in this case also.

Figure 14 shows the evolution of the deposited mass over long reaction times, for $R = 2.2 > R_c$ at two bath temperatures, 70 and 45 °C. At a macroscopic level, a

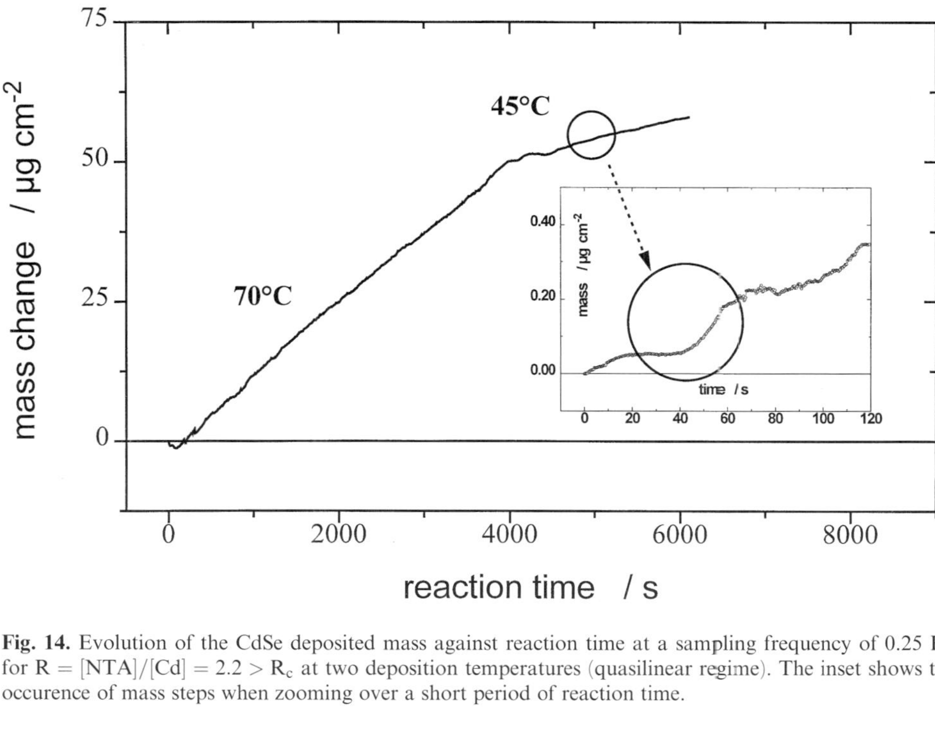

Fig. 14. Evolution of the CdSe deposited mass against reaction time at a sampling frequency of 0.25 Hz for $R = [NTA]/[Cd] = 2.2 > R_c$ at two deposition temperatures (quasilinear regime). The inset shows the occurence of mass steps when zooming over a short period of reaction time.

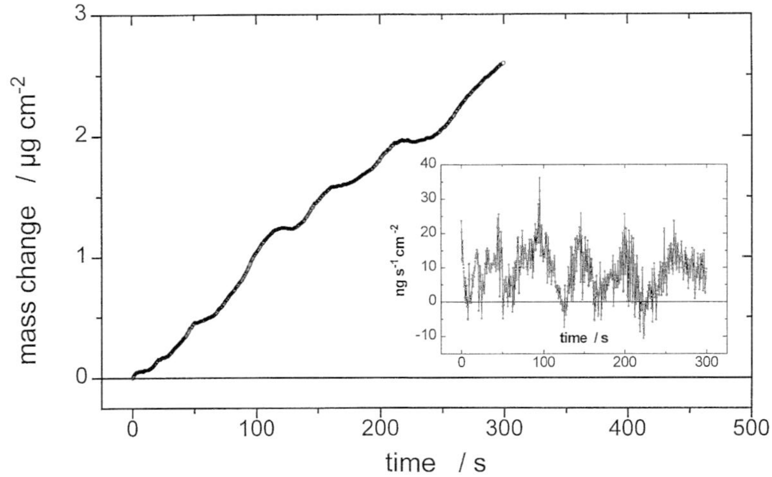

Fig. 15. Evolution of the CdSe deposited mass over a short period of reaction time at a sampling frequency of 10Hz for R = [NTA]/[Cd] = 2.2 > R_c in the presence of 10^{-3} M STA. The presence of steps is illustrated in the inset by the quasi-periodic behavior of the time derivative.

linear variation is observed with a slope decreasing with bath temperature: the film growth is classically characterized by a constant, but temperature dependent, growth velocity. This approach does not differ from that employed for the ammonia-thiourea CBD CdS system, discussed previously [34]. The growth rate follows an Arrhenius behaviour. For most of the bath compositions investigated, in the temperature range 30–80 °C, the activation energy lies between 40–60 kJ.mol-1 a range that is characteristic of a chemically controlled process. However additional information can be gained from a more detailed examination of the time variation of the deposited mass. It is shown by the inset in Figure 14 obtained by zooming the global m(t) curve over a short period of time (2 min). The presence of steps more or less regularly distributed and well separated is evidenced. Figure 15 shows that, in the presence of a very adsorbing species as STA, regular, well defined steps are observed. The regularity can be appreciated by the quasi-sinusoïdal behaviour of the time derivative dm/dt depicted in the inset. The elementary steps in figures 14 and 15 present an average amplitude of 0.1 to 0.3 $\mu g.cm^{-2}$, close to one monolayer of CdSe (1 $\mu g.cm^{-2} \Leftrightarrow$ 1.6 nm). Such patterns have been systematically observed in various conditions of temperature (30–87 °C) and bath compositions ($1.5 < R < 2.4$), both with and without STA. In every case, the profile of an elementary step of mass is well represented by the Avrami law (37), as illustrated by figure 16. That is, every elementary step corresponds to the bidimensional growth of randomly distributed nuclei with coalescence, assuming an instantaneous nucleation. The height of the 2D-steps was found to increase with temperature (figure 17), from about one equivalent CdSe monolayer at 30 °C to a few monolayers at 80 °C. Correspondingly, the time constant τ characterizing the mass transient was strongly dependent on temperature: 1 min at 30 °C and only 1 s at 90 °C (figure 18). The radial growth rate, taken out from the time constant τ, was thermally activated, the activation energy depending on the degree of

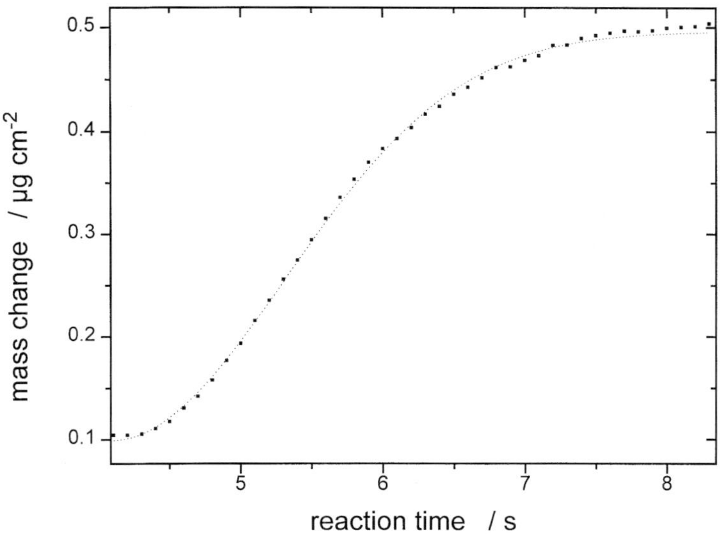

Fig. 16. Profile analysis of an elementary mass step recorded at a 10 Hz sampling frequency, characteristic of a 2D nucleation and growth process. Conditions: $R = 2.4 > R_c$, deposition temperature 50 °C; the dashed line represents a fit according to the Avrami law (Eq. 6) (from [147]). With permission of Electrochem. Soc.

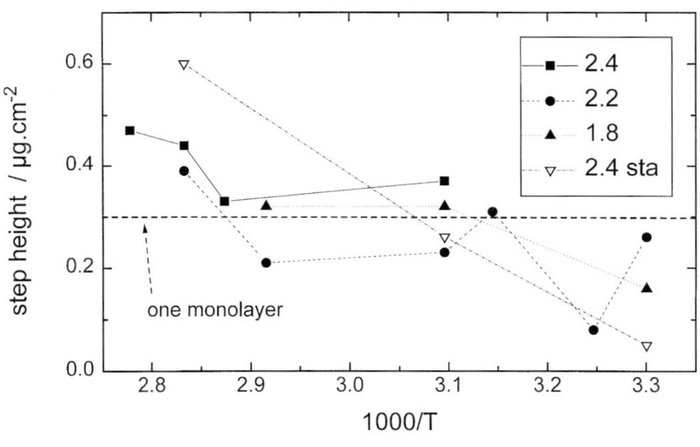

Fig. 17. Dependence of the step height on deposition temperature for various bath compositions. The dashed line represents a mass equivalent to one CdSe monolayer, assuming a thickness (0.606 nm) equal to the lattice parameter of the cubic blende CdSe structure.

complexation (figure 19). For $R = 2.2$, the activation energy is 66 $kJ.mol^{-1}$ (control of the 2D process by a chemical reaction) and significantly lower ($\approx$30 $kJ.mol^{-1}$) at larger $R = 2.4$ (partial diffusion control) or lower $R = 1.8$ (presence of colloïds) degree of cadmium complexation.

These transient QCM experiments bring evidence that, for R > Rc, CBD CdSe growth occurs through a layer-by-layer, ion-by-ion mechanism, also supported by HREM observations showing the formation of platelets constituted of few crystals

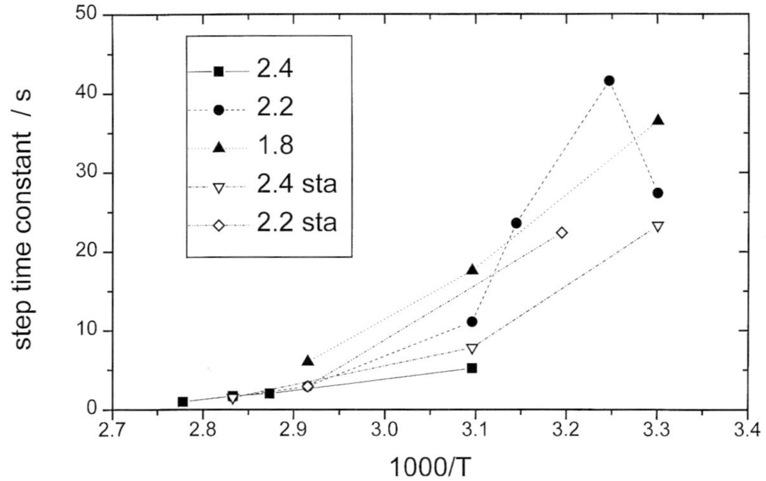

Fig. 18. Dependence of the step time constant on deposition temperature for various bath compositions determined by a step profile analysis according to the Avrami law (Eq. 37).

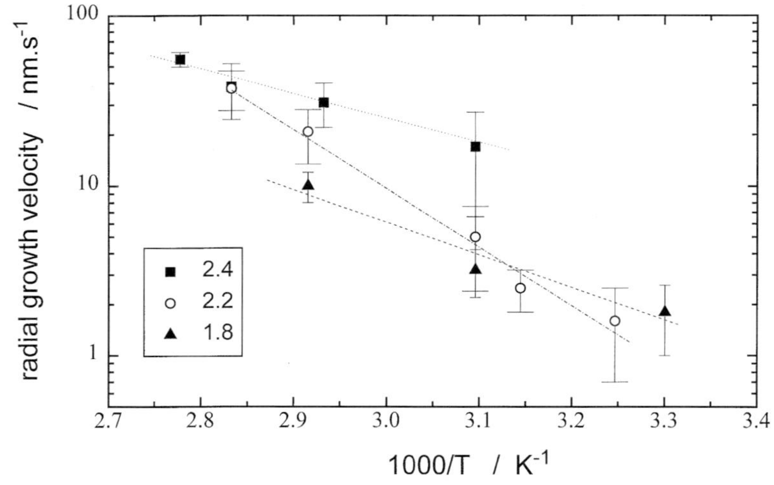

Fig. 19. Dependence of the (2D) radial growth velocity on deposition temperature for various bath compositions.

(see Fig. 22). These results indicate that film growth takes place more favourably at nucleus edges, implying a low surface activity for the area between nuclei and at the top surface of nuclei. They establish that nucleation is the rate limiting step for the growth of CBD CdSe films from a NTA-selenosulfite bath.

For $R < R_c$, when colloïd formation is mediated by the presence of cadmium hydroxide in solution, transient QCM studies show that mass deposition occurs through discrete, sequential events. The latter correspond to much less than one CdSe monolayer and can be interpreted as the rapid bidimensional growth of nuclei by contribution of nanocrystallites already formed in the bath. Such a mechanism becomes progressively disordered and the film growth tends to become irreversibly a tridimensional process at long reaction times.

5 Morphological and Structural Aspects

5.1 Methods of Characterization

The knowledge of chemically deposited films, from a morphological and structural point of view, requires determination of different characteristics: roughness at different scales, nature of the crystallographic structure, existence of textures, nature and density of crystallographic defects like dislocations, stacking faults, twins, etc. such characteristics depend on numerous parameters and are the consequence of the nucleation and growth mechanism.

Scanning Electron Microscopy (SEM) has widespread use [160], allows easy surface observations without special preparation of the samples, and gives resolution in the nanometer range. The determination of grain size from SEM images requires caution because in many cases relations do not exist between the morphology of the deposit and the position of the grain boundaries. The combination of SEM and X-ray analysis gives the possibility to determine the local composition ($\approx$1 μm) of the sample [161].

During the last fifteen years scanning probe microscopies (Scanning Tunneling Microscopy – STM; Atomic Force Microscopy – AFM) have seen tremendous development [162]. It has been demonstrated that these techniques could be used not only in ultra vacuum conditions but also in air and in liquid environments. These advances have opened the possibility to follow *in situ* and at the atomic level the growth or the anodic dissolution of metals and semiconductors [163]. In the liquid phase, these possibilities have been exclusively applied to phenomenon like electrodeposition, underpotential deposition or electroless deposition [164]. The interest of these techniques in the study of chalcogenide semiconductors is illustrated by the STM-AFM observations of CdS monolayers electrodeposited on (111) Au [165] or nanostructures deposited on graphite from Langmuir-Blodgett films [166]. The two processes reveal a six-fold symmetry structure. Unfortunately the *in situ* study of chemical bath deposition by STM or AFM is constricted by the parallel deposition of chalcogenides on substrates and on tips. Nevertheless these techniques could be useful for the *ex situ* study of the first steps of nucleation, specially during epitaxial growth.

X-Ray Diffraction techniques are frequently used for identification of the chalcogenide semiconductor structure. The simplest way of doing this is the direct comparison of the powder diffraction patterns with referenced X-Ray Powder Data File published by the Joint Committee on Powder Diffraction Standards (J.C.P.D.S.). The comparison of the observed intensities of the reflections with those of standard samples can be useful for the identification of a texture. If the crystallites are very small they cause a broadening of the reflections which can be detected at dimensions below 1 μm, and become obvious below 0.1 μm. The mean crystallite diameter can be determined from the linewidth of the XRD spectra by the Scherrer formula [167]. This technique implies a careful determination of the breadth of the reflections that arise from instrumental factors. Specific XRD techniques have been developed for the study of thin film and specially for that of epitaxially grown films [168, 169]. Thus the use of glancing angles with a five circle diffractometer allowed the determination of the epitaxial relations of CdS chemically deposited on InP single crystals [170]. The use of an entrance slit (1 by 0.05 mm) vertically defines a narrow beam and allows one to position a 10 mm diameter sample under a glancing angle. With a glancing angle of 0.6°, the X-ray path in the film is enhanced by a factor 100, so that the X-ray absorption in the films is about 20%, for a 20 nm film thickness. If the goniometer is controlled by a microcomputer, the position and the intensities of selected Bragg peaks is assumed. For investigating crystallinity, two kinds of scans can mainly be used: (i) with a glancing incidence, a Φ scan around an axis normal to the sample surface; (ii) in the condition θ to 2θ around an axis parallel to the sample surface and normal to the X-ray beam.

Electron beams techniques are now frequently developed for the structural determination of materials [171]. The Reflection High Energy Electron Diffraction (RHEED) is a powerful tool which can be used for the study of polycrystalline [172] and epitaxial thin layers [173]. Numerous types of information like crystallographic structure, textures, epitaxial relations, etc, are obtained very simply. Despite the use of grazing angles, the thickness that can be explored is limited to several nanometers. Transmission Electron Microscopy (TEM) is a powerful technique for direct observation of crystalline specimens [174] and a considerable number of structural details like crystallographic defects can be observed when images are combined with selected area diffraction (SAD) patterns. Grain size measurement is one of these possibilities when using bright and dark field images. The main difficulty of the TEM technique arises from the sample preparation which must be thin (thickness <100 nm). Chemically deposited chalcogenides can be directly observed after deposition on thin carbon films supported by gold grids or on polymer films which are subsequently dissolved [175]. Thicker films must be thinned. The Ar^+ ion milling is now a specially interesting technique in the field of semiconductors [176]. The preparation of cross sections is essential for the determination of the interfacial organization of epitaxial layers. High resolution electron microscopy (HREM) is also a widespread technique [177] which gives structural details at an atomic level with very thin (<20 nm) films. The HREM images should be carefully interpreted. It is essential to compare the atomic projection images with simulated images which can be found in the literature for compounds like CdS [178] or CdTe [179]. In most cases TEM apparatus have a resolution near 0.18 nm. Then the atomic column pairs (Cd-S, Cd-Se, Cd-Te), separated by distances between 0.15–0.16 nm, are unresolved. Recently Scanning transmission electron microscopy (STEM) has been developed with electron probes of atomic dimensions. Images formed using elastically scattered electrons show strong atomic (Z) contrast and reveal atomic column location withouth the need of preconceived structure models. With a 0.13 nm probe, Z contrast images have been able to describe the dislocation core near the CdTe/(001)GaAS interface [180]. Image processing is now easely accessible. Images are digitalized using micrographs or on line during the observation. The Fourier Transform analysis gives a local diffraction diagram. The inverse Fourier transform with filtering supplies improved high resolution images [181].

5.2 Crystallographic Structure

Chalcogenides generally adopt the hexagonal Wurtzite or the cubic Zinc Blende structure, both involving tetrahedral coordination of the cation. Most of them present polymorphic structures. Chemically deposited chalcogenides can be obtained in different structures depending on the deposition conditions [11].

In a majority of cases CdS films, analysed by XRD, present a mixed structure with textures where the dense planes (111) in the cubic structure, (0002) in the hexagonal structure, are parallel to the substrate [25, 32, 182–184]. However some of these results are questionable because it is difficult to separate hexagonal and cubic reflections, specially when the peaks are enlarged by a low grain size. The mixed structure can result from the formation of polytype structures and this point will be discussed in 5.4. It seems that nanocrystals present generally a cubic structure [35, 185]. As the crystals become smaller the surface effects become important so, as the fraction of atoms onto the surface increases, the arrangement of them changes in order to minimize the number of dangling bonds and a fcc closed-packed arrangement is favoured [186]. It appears that deposition conditions are able to stabilize one or the other phase. CdS films obtained from the $Cd(NH_3)_4^{2+}$ complex have cubic, hexagonal or a mixed structure, while those obtained from $Cd(CN)_4^{2-}$ and $Cd(EN)_3^{2+}$ complexes exhibit preferentially the hexagonal structure [11]. In solutions consisting of CdX_2, NH_4X, and thiourea, the nature of the anions X, influence the CdS structure [37]. In the presence of iodides an hexagonal structure is found when a cubic structure is obtained with chlorides. I. Kaur et al. [25] have found an influence of solution stirring. CdS films exhibit an hexagonal structure when solutions are vigorously stirred. An ion by ion growth mechnism is then favoured. In the absence of stirring, films are constituted by an aggregation of colloïds which present a cubic structure. The nature of the substrate is also an important parameter in the stabilization of a given structure. Hexagonal CdS is found when deposited on ITO [187], carbon films [188] whereas the cubic structure is obtained on silicon. [189].

The chemical deposition of PbS [56, 60] and ZnS [60, 69, 70] produces films which exhibit only a cubic structure. On the contrary $Cd_{1-x}Zn_xS$ ternary compounds present the hexagonal Wurtzite structure [100]. Chemically deposited PbSe [86], ZnSe [92, 93] and CdSe [41, 78] are generally found with a cubic structure According to A. H. Eid et al. [77] the presence of $Cd(OH)_2$ in the solution could promote the formation of CdSe with an hexagonal structure. Recently H. Cachet et al. [147] found that CdSe films prepared from a solution complexed by sodium nitrilotriacetate could present an hexagonal structure when the temperature is increased up to 80 °C. On the other hand $CdSe_{1-x}S_x$ films where $x > 0.6$ contain a small amount of the hexagonal modification [112].

5.3 Grain Size and Growth Mechanism

Chalcogenide films chemically deposited generally present a microcrystalline structure with typical grain sizes ranging from 2–3 nanometers to 100 nm. The grain size depends on many parameters like composition and temperature of the solution, nature of the substrate etc. [11]. Nevertheless it has been clearly demonstrated, these last years, that the grain size is a direct consequence of the growth mechanism. For example S. Gorer et al. studied the chemical deposition of CdSe [41] and PbSe [86] using complexed solutions containing potassium nitrilotriacetate. The important parameter is the ratio concentration R beween the complexing agent and the metal ion. They observed a critical ratio Rc (see section 4) which corresponds to the transition between the agglomeration of colloïds and an ion by ion growth mechanism. XRD patterns and TEM observations show a sudden increase of the grain size when $R > Rc$. For example at a temperature of 80 °C, where the critical value of Rc is around 2, the CdSe mean grain size is 8 nm if $R < Rc$ and 20 nm when $R > Rc$. H. Cachet et al. [203] have confirmed these results using similar solutions and HREM observations. Figure 20a corresponds to a CdSe deposit on a carbon film obtained with a solution where $R < Rc$; the mean size of individual agglomerated grains is around 5 nm. The Fourier transform diffraction pattern (in the inset), obtained inside one grain, confirms the cubic structure. The inverse Fourier transform of the diffraction pattern, after filtering and enlarging, given in Figure 20b, corresponds perfectly to the projection of atomic columns along [110] of a cubic Blende structure as shown by the simulated image in the inset. These projections are elongated as the resolution of the apparatus is not sufficient to separate Cd and Se atomic columns. It is interesting to compare Figure 20a and Figure 21, which is an image of colloïds recuperated in the solution on a carbon film and dried; the grain size is quite the same ($\approx$5 nm). When $R > Rc$ (Figure 22) monocrystalline and flat nuclei are observed the size of which is found between 20 and 70 nm. These large nuclei present a cubic structure and the HREM image reveals a low density of cristallographic defects. Similar results have been obtained during the CdS deposition from solutions containing cadmium-ammonia complex and thiourea [175]. On the contrary, the chemical deposition of ZnS from ammonia-thiourea solutions corresponds to conditions where ZnS is a little more stable than the hydroxide and it is difficult to obtain a grain size above 5 nm [70].

According to the kinetic studies discussed in section 4, when the growth mechanism corresponds to an aggregation of colloïds, a 3D nucleation is observed. On the contrary, the ion-by-ion growth condition implies a low nucleation rate of 2D nuclei and a layer-by-layer growth. This explains why large monocrystalline grains can be formed. The control of the growth mechanism through the solution composition presents an interest for different applications. The chemical route of colloïdal solutions is particularly well adapted for the preparation of extremely small particles, in the range 1.5 to 5 nm, which present quantum confinement effects. This technique has been used for the preparation of CdS [190], CdSe [41], PbSe [86] and CdTe [191]. On the contrary, it will be demonstrated in section 6 that the ion-by-ion growth mechanism is favorable for the epitaxial growth of chalcogenides on single crystal semiconductors.

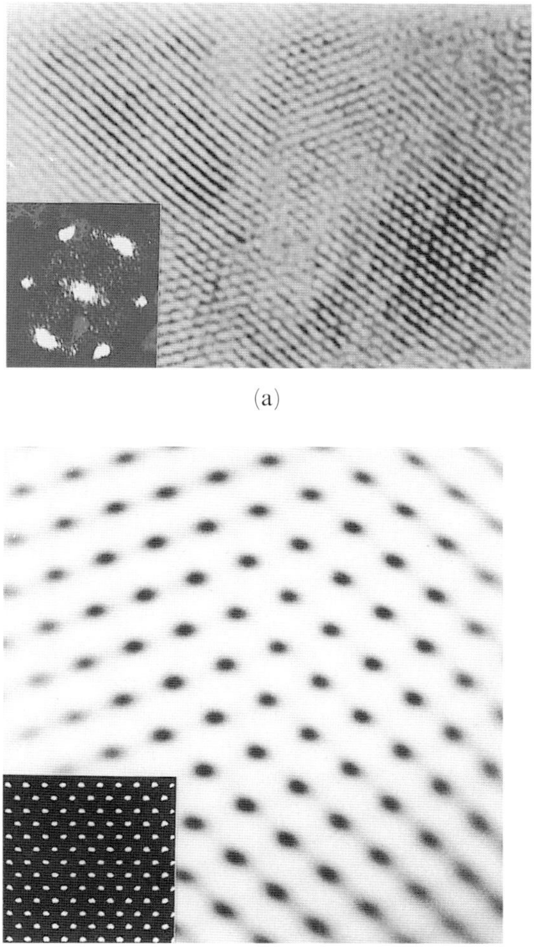

(a)

(b)

Fig. 20. HREM image of CdSe deposited from a solution containing sodium selenosulfate and cadmium sulfate as precursors; complexing agent: NTA with R ([NTA]/[Cd]) = 1.8 (R < Rc) (from [203])
a) Digitalized image and Fourier transform diffraction pattern in the inset.
b) Inverse Fourier transform with filtering of a part of the Fig. 20a; simulated image of [011] CdSe with the blende structure (in the inset). With permission of IUPAC

5.4 Crystallographic Defects and Polymorphism

The ideal situation depicted in 5.3 for the initial step of growth is not valid to describe the crystallographic defects which can be formed in thick films and identified by TEM and HREM observations. CdS and CdSe have been especially studied. Dislocations are observed but in very small amounts [175]. On the contrary extended defects like stacking faults and twins, formed when the initial nuclei coalesce and during the subsequent steps of growth, are frequently detected [175, 192, 193]. The density of defects is highly dependent upon the deposition conditions. For example during the CdS deposition in a standard solution, the stacking fault density is about 10^{11} cm^{-3} (Figure 23a). Decreasing the cadmium concentration to 2.10^{-3} M, while keeping the other parameters constant, increases the stacking fault density by a factor of ten (Figure 23b). The same is observed if the solution contains an excess of thio-

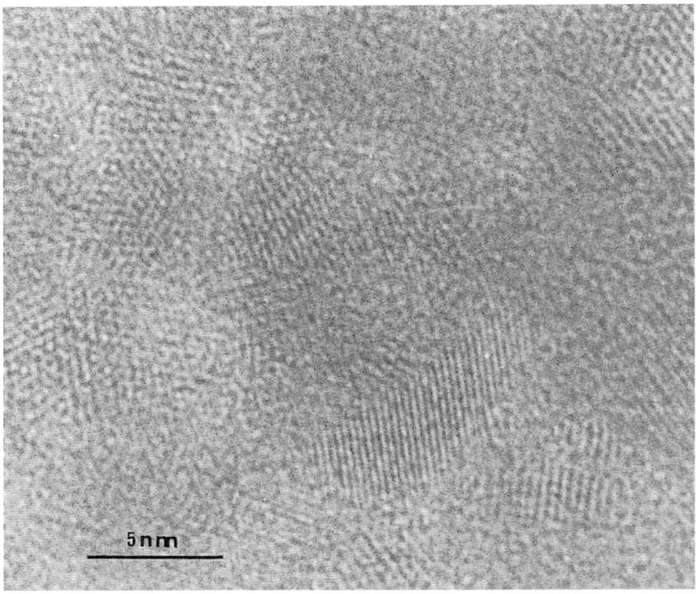

Fig. 21. HREM image of colloïds recuperated in the same solution than that used in Fig. 20.

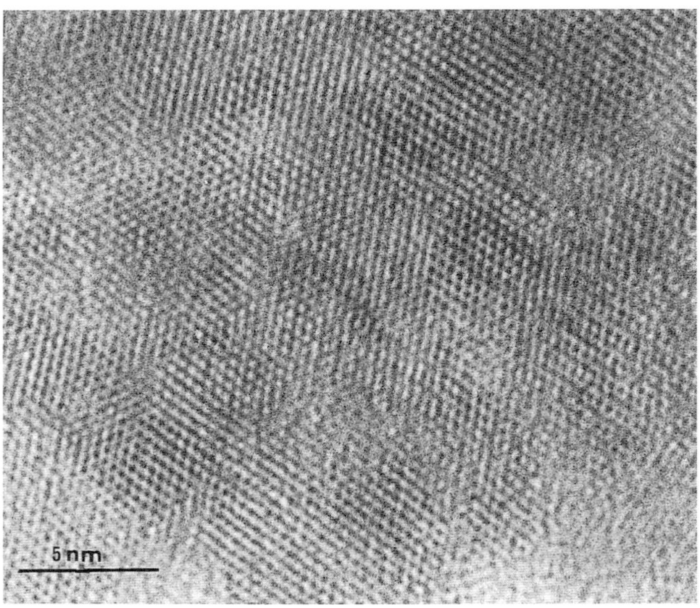

Fig. 22. HREM image of CdSe deposited from a solution containing sodium selenosulfate and cadmium sulfate as precursors; complexing agent: NTA with R ([NTA]/[Cd]) = 2.2 (R > Rc).

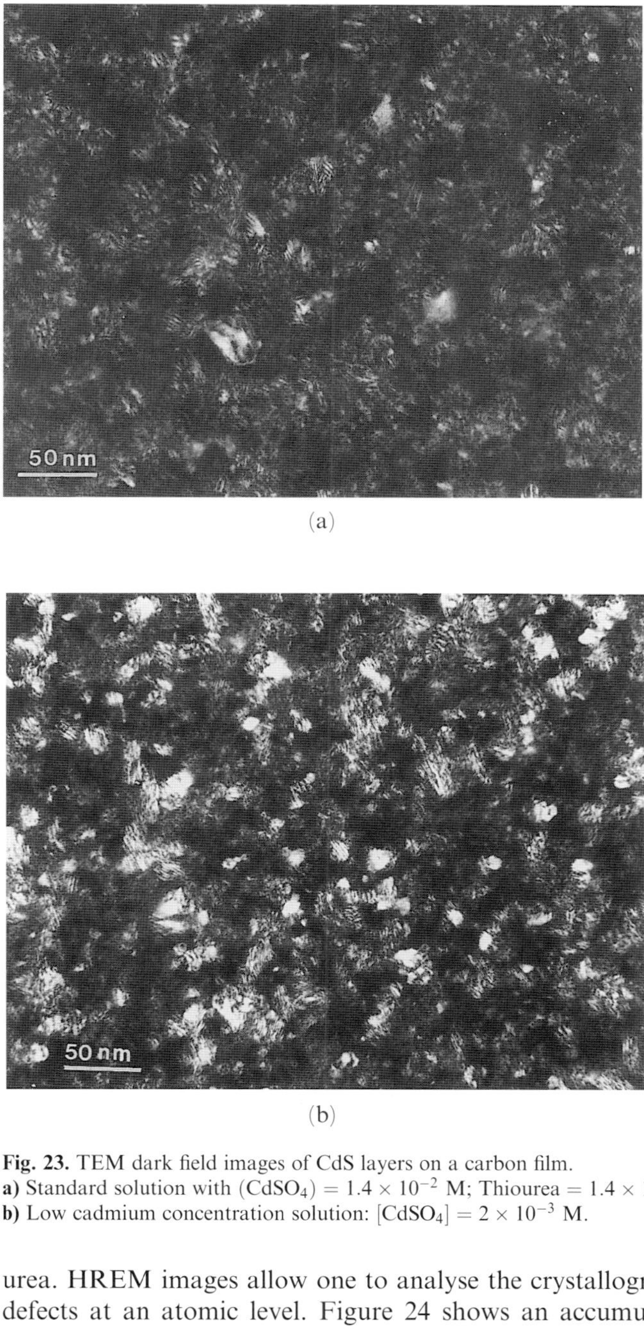

Fig. 23. TEM dark field images of CdS layers on a carbon film.
a) Standard solution with $(CdSO_4) = 1.4 \times 10^{-2}$ M; Thiourea $= 1.4 \times 10^{-2}$ M $[NH_3] = 1.74$ M.
b) Low cadmium concentration solution: $[CdSO_4] = 2 \times 10^{-3}$ M.

urea. HREM images allow one to analyse the crystallographic organization of these defects at an atomic level. Figure 24 shows an accumulation of stacking faults in hexagonal CdS. These defects are formed in the basal planes (0001) of the hexagonal structure and they limit thin lamellae of 3–4 atom layers presenting a cubic structure. Figure 25 gives a representation of a thin cubic lamella in an hexagonal CdS structure; the arrow indicates the position of the stacking fault. Such an organization

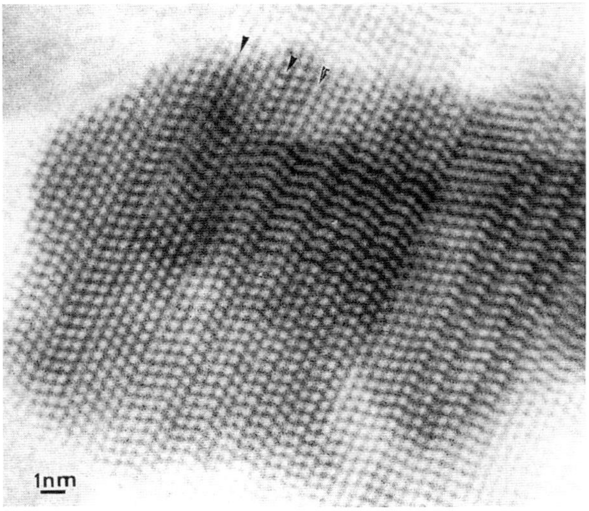

Fig. 24. HREM image of planar defects (see the arrows) in an hexagonal CdSe film. (from [175]) With permission of Elsevier

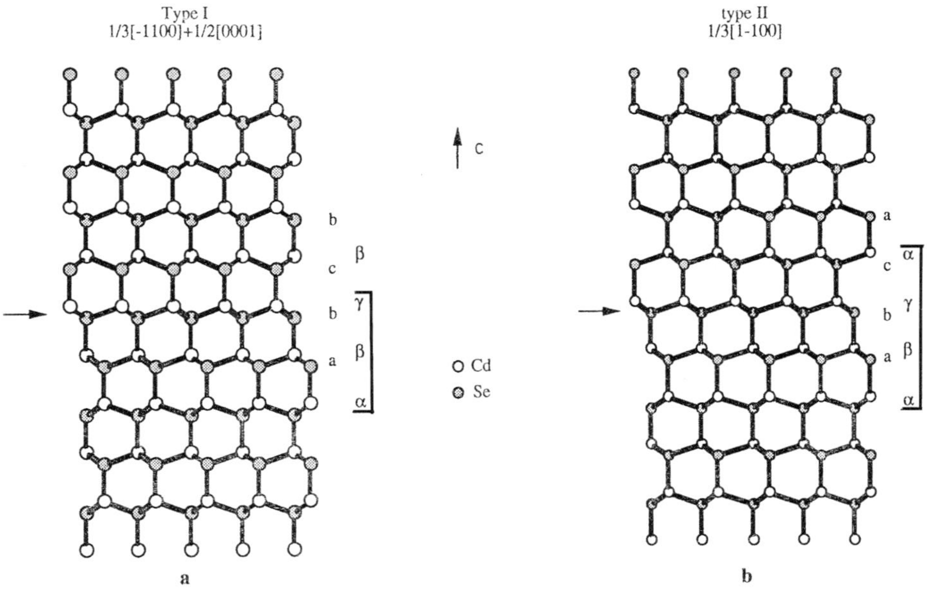

Fig. 25. Arrangement of the hexagonal Wurtzite lattice corresponding to the presence of stacking faults, with 3 or 4 planes presenting the cubic blende modification. (from [175]) With permission of Elsevier

could explain the polymorphism (cubic *vs* hexagonal) which is often reported for CdS (Section 5-2). These structures are also known as polytype structures [195]. According to calculations [194], CdS, CdSe and ZnS can form polytype structures because the energy difference $\Delta E_{W\text{-}ZB}$ between the Wurtzite (W) and the Zinc Blende (ZB) structures is small: 3.1 meV/atom for ZnS and 1.4 meV/atom for CdSe for which the cubic phase is more stable. On the contrary for CdS, $\Delta E_{W\text{-}ZB} = -1.1$ meV/atom and the hexagonal phase is more stable. Polytypes can be defined by the stacking sequence of (111) or (0001) atomic planes with respect to the three possible positions ABC. The cubic structure is noted 3C and corresponds to the sequence ABCABCABC ... The hexagonal structure is noted 3H and corresponds to the sequence ABABABABAB ... The polytype is noted nH or nC, where n represents the number of planes in the unit cell and H the host structure. An example of a short period polytype structure is ABCACBABCACBABC noted 6H. This polytype is simply obtained by a permutation of a sequence BC in the cubic arrangement [195] as indicated by underlining. By further permutations it is possible to reach the hexagonal arrangement. Despite the fact that tens of polytype structures are possible in ZnS [196], only certain arrangements are observed corresponding to the more metastable configurations which have been simulated in numerical models. During the phase transformation process the hexagonality factor h varies from 0 to 1 (or 1 to 0) *via* a sequence of consecutive polytypes (for 4H, the value is 0.5).

As a general rule the cubic-hexagonal transition is simulated by thermal treatments [184, 192, 193, 197–203]. Concurrently the grain size increases. These treatments are generally used to improve the photoresponse in photovoltaic or photoelectrochemical cells. The transition characteristics strongly depend on the annealing conditions and atmosphere. For CdS it starts to take place at about 300 °C in an argon-sulfur atmosphere after 28 hours and does not in argon or air under the same conditions [38]. Some authors claimed that the phase transition was accompanied by an increase of the lattice constant [199] and by the formation of crystallographic defects [200]. TEM and HREM observations have confirmed that chemically deposited CdS annealed at 400 °C contain a high density of planar defects [192, 193]. The cubic-hexagonal phase transition of CdS films deposited on silicon has been also studied by D. Lincot et al. [189]. Two different thermal annealing treatments were performed. The first one was made in argon atmosphere at 400 °C for 0.5 hour (annealing 1). The second one was made in air in the same conditions (400 °C for 0.5 h) but the CdS film was previously dipped for two minutes in a methanol solution at room temperature containing cadmium chloride under saturation conditions (annealing 2). XRD and SAD patterns allow to conclude that after the annealing 1 only a fraction of the cubic CdS had been transformed in the hexagonal modification. On the contrary, after annealing 2, complete vanishing of the cubic spectra takes place. After annealing 1, TEM observations show a slight increase of the grain size (50–70 nm) and a dramatic increase of the density of planar defects. In contrast to annealing 1, annealing 2 produces a spectacular recrystallization with a grain size approaching 90 nm and the density of planar defects is very low. HREM observations provide useful information on the structural organization of the annealed CdS films. Thus a local determination of the structure can be obtained. Figure 26a corresponds to a CdS film after annealing 1. The Fourier transform diffraction pattern (in the inset) corresponds to the major cubic modification. The presence of streaks is related to planar defects. In agreement with the XRD and SAD patterns which reveal a small proportion of the hexagonal modification, some areas present an hexagonal arrangement (Figure 26b); the Fourier transform diffraction diagram (in the inset) corresponds to the hexagonal structure and the HREM image is the atomic column

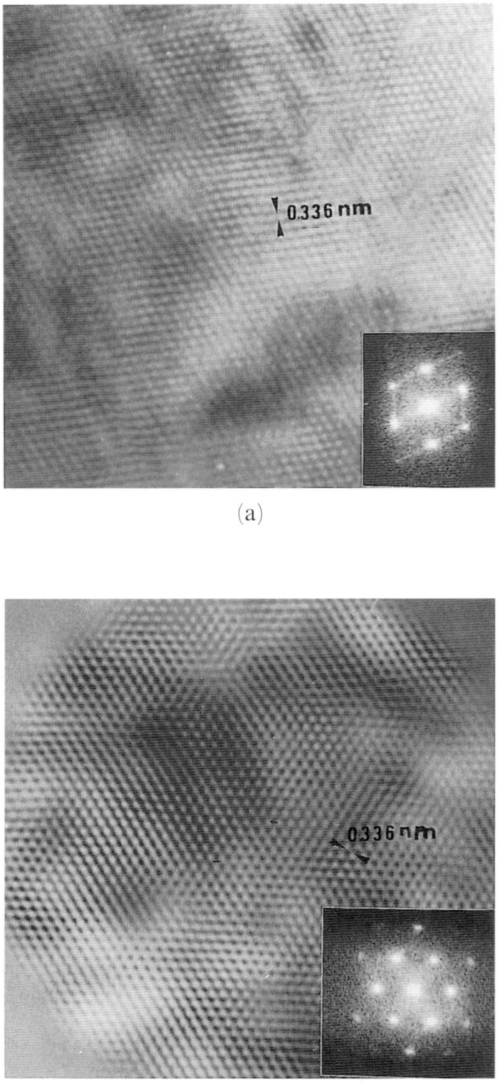

Fig. 26. Digilitalized HREM images of CdS films. Fourier transform diffraction patterns in the insets. (from [189])
a) Cubic structure; note the presence of streaks in the diffraction pattern indicating an high density of planar defects.
b) Hexagonal structure.
With permission of the American Chemical Society

projection on the (0001) basal plane. The main characteristic of the CdS film after annealing 1 is the high density of planar extended defects which appear as a rather regular parallel juxtaposition of thin domains of few nanometers width with rotated atomic directions from one to the other (Figure 27). These defects are formed during the thermal treatment and the existence of hexagonal reflections in the SAD patterns indicate that hexagonal domains are formed in relation to the stacking faults. After annealing 2, large areas inside the grains give HREM images with no stacking faults

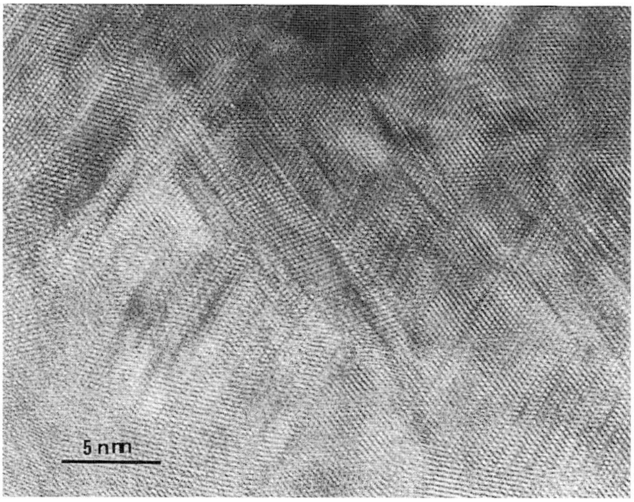

Fig. 27. HREM image of an annealed CdS film showing a stacking fault network. (from [189]) With permission of Taylor and Francis Ltd

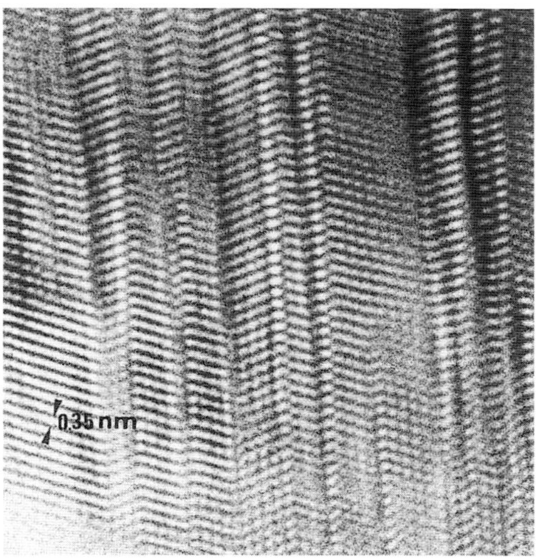

Fig. 28. HREM image of an annealed CdSe film. (from [203]) With permission of IUPAC

or microtwins. Similar results have been obtained during the annealing of CdSe films chemically deposited from selenosulfate solutions [203]. After a 1 hour thermal treatment at 400 °C a partial phase transition cubic-hexagonal is found accompanied by a grain size increase (from 5 to 50 nm). The HREM image (Figure 28) reveals that some proportion of grains contains an important density of planar defects and microtwins.

6 Epitaxial Growth of II–VI Semiconductors

6.1 General Considerations

Direct bandgap II–VI semiconductors have opened up the blue-green region of the spectrum and a range of novel applications in optoelectronics. There are many potential applications for compact laser and Light Emitting Diodes (LED), but it is essential to minimize defects and imperfections which decrease the radiative yield of photons. This explains the interest of the construction of epitaxial structures between II–VI semiconductor layers and bulk semiconductors like Si, GaAs, InP, etc. In particular, quantum wells are formed by thin epitaxial multilayered structures like (Zn, Cd)Se/ZnS. Nevertheless, the choice between bulk semiconductors and the layers deposited or between the multilayers is governed by the lattice mismatch between the two components as the lattice mismatch causes the formation of misfit dislocations. In the optical devices these defects are potential non-radiative centres and at worst they can cause the failure of injection lasers. Figure 29 is a map of energy gap versus lattice constants for a variety of semiconductors; it can be used to select different heterostructures, not only for optoelectronics applications but also for photovoltaïc cells. In the latter application the deposited films are generally polycrystalline and the growth of high-quality epitaxial layers has received little applications.

Several techniques have been reported and, at the present time, the vapor phase deposition processes operating at temperatures around 300 °C are the most used. Thus II–VI compounds films like CdS, CdSe, CdTe, ZnS, ZnSe, and ZnTe have been grown epitaxially on Si, InP, GaAs, GaP, by molecular beam epitaxy (MBE) [204–207], by metal organic chemical vapor deposition (MOCVD) [208–210], or by pulsed laser deposition [211, 212]. Epitaxial deposition from aqueous solutions at low temperatures (< 100 °C) represents another approach. Specific beneficial effects may be also expected due to the simplicity of the process involving low cost investments. On the other hand the low temperature has for consequence the absence of interdiffusion processes around interfaces and the interfacial properties of the solids in contacts with solutions implicate excellent coverage properties at low thicknesses. Different

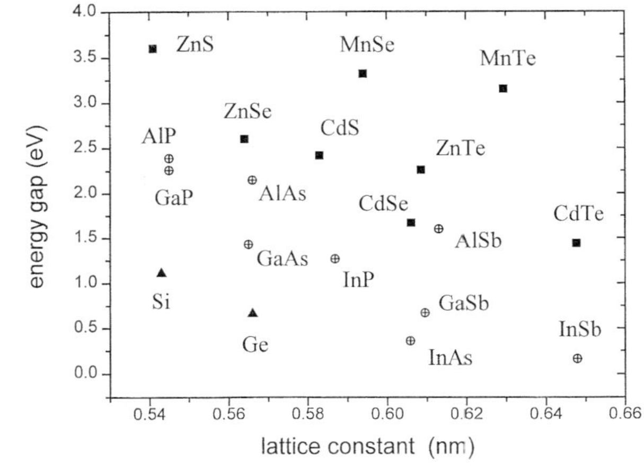

Fig. 29. Lattice constant and energy bandgap for different III–V and II–VI compounds.

techniques using aqueous solutions have been proposed. Some of them imply successive deposition of the two atoms (Cd, Zn, Hg) and (S, Se, Te). Electrochemical atomic layer epitaxy (ECALE) involves underpotential deposition of the elements to form the compound [213]. This method has been mainly used for the epitaxial deposition of a few monolayers CdTe on gold [214] and the evolution of the morphology has been followed by SEM and AFM observations. On the other hand Nicolau et al. [215] have proposed a method based on the successive ionic layer adsorption and reaction (SILAR). They succeeded in the deposition of epitaxial CdS films on Ge, GaAs and InP single crystals. Zinc sulfide films were also deposited on GaAs and InP [215] or Si [216] but they presented a polycrystalline structure. One of the most used technique for the deposition from solutions of II–VI compounds is electrodeposition [217]. Nevertheless few results have been obtained concerning their epitaxial growth. Electrodeposition of epitaxial CdSe quantum dots on gold single crystals has been reported [218, 219]. In the same way epitaxial layers of CdTe [220] and CdSe [221] have been electrodeposited on (111) InP. Using cyclic voltammetry K. Rajeshwar [222] has electrosynthesized CdSe/ZnSe superlattices (non epitaxial). XPS depth profiles have clearly demonstrated the modulation in the Cd and Zn content.

CBO, as CUD, is also well adapted for epitaxial growth. As shown in section 5.2, numerous papers have been published in the field of chemically deposited polycrystalline chalcogenide semiconductors. In comparison with this plentiful literature only a few papers exist concerning the epitaxial growth of these compounds. In this way J. L. Davis and M. K. Norr [55] have shown the epitaxial growth of PbS on (111) Ge using a $Pb(NO_3)_2$-thiourea solution. The epitaxial growth was disclosed by Laue back reflexion X-ray patterns. M. Isshiki et al. [59] confirmed the PbS epitaxy on (111) Ge but they showed that the (100) orientation was more suitable for the epitaxial growth on germanium and indium phosphide. H. Rahnamai et al. [223] tried to obtain the epitaxial growth of PbS on silicon. The RHEED and XRD patterns show in fact a textured growth along ⟨111⟩. N. L. Sharma et al. [106] studied the chemical bath deposition of $Pb_{1-x}Hg_xS$ on (111) Ge and Si. TEM observations, associated with SAD patterns of films floated off the substrates by dipping in 10% nitric acid, show that a polycrystalline growth is mainly found on (111) Si. On (111) Ge some localized regions of the $Pb_{0.67}Hg_{0.33}S$ film give SAD patterns corresponding to single crystals. Nevertheless the orientations are slightly off the [$\bar{1}$11] orientation and depend on the etching of the crystals. G. N. Chaudari et al. [114] tried to obtain the epitaxy of Zn_xSe_{1-x} compounds on (110) GaAs. The use of an exact lattice matching ($x = 0.56$) improves the quality of the films which remain polycrystalline with a [111] zone axis. These last years D. Lincot, M. Froment et al. have shown, by the examples of CdS [224, 225, 246] and CdSe [147], that chemical bath deposition provides an efficient route for preparing epitaxial chalcogenide films on InP, GaP, $CuInSe_2$.

6.2 Influence of the Mismatch on the Epitaxial Growth

This concept was introduced in 1928 by L. Royer [226] who put forward rules of epitaxy, the most important of which is that oriented growth occurs only when it involves the parallelism of two lattice planes that have networks of identical or quasi-identical form and of closely similar spacing. The mismatch m is defined as 100(b-a)/

a, where a and b are the corresponding network spacings in the substrate and the overlayer, respectively. The experiments show [227] that epitaxy occurs only if m is no more than about 15%.

The growth of cadmium sulfide on indium phosphide illustrates the case where there is almost a perfect lattice mismatch ($m \approx 0.3\%$). As discussed in section 4.3, CdS may form two cristallographic structures, one that is hexagonal ($a = 0.413$ nm, $c = 0.671$ nm) and the other fcc cubic ($a = 0.582$ nm). Then the (111) InP face has a surface crystallographic structure identical to the (111) or (0002) planes of CdS. D. Lincot et al. [224, 225] used InP surfaces etched in a 1% solution of Br_2/methanol, then rinsed and finally dipped in sulfuric acid (3 M). The two faces, (111) terminated by In atoms and $(\bar{1}\bar{1}\bar{1})$ terminated by P atoms, were used. The CdS films were chemically grown from a standard solution containing cadmium sulfate (10^{-3} M), ammonium acetate (10^{-2} M), ammonia (0.4 M) and thiourea 5×10^{-3} M), maintained between 30–90 °C. The RHEED patterns of the Figure 30 show that an epitaxial growth is only obtained on the $(\bar{1}\bar{1}\bar{1})$ terminated by P atoms. On the In face polycrystalline CdS is obtained. This indicates the important influence of chemical interactions during initial stages of growth between atoms in the growing film and those at the polar InP faces. The stronger reactivity of this face may be related to the formation of P-Cd or In-S bonds by sulfur-phosphorus substitutions. Such chemical exchange reactions between thiourea and InP have been reported [15]. In section 6.3 it will be demonstrated that, in the case of CdSe, epitaxial growth is obtained on both (111) InP faces. The great significance of surface properties in governing the epitaxial growth is confirmed by the effect of substrate etching. In the absence of such chemical pre-treatments only polycrystalline growth is observed. One can indicate that ammonia is also an efficient etching agent for InP, enabling the removal of surface oxide layers. This leads to *in situ* etching prior to the onset of film growth and may explain why epitaxial quality is improved when the sample is introduced in the solution at room temperature. The accessibility of the atoms of the solution to the atoms of the substrate is then an important condition for the epitaxy. This can explain the dificulty to obtain an epitaxial growth of chalcogenides on silicon because of the Si-H bonds, formed during the etching step, and which are stable in solutions.

X-ray diffraction scans obtained using a five circle goniometer reveal that the CdS films deposited on $(\bar{1}\bar{1}\bar{1})$ InP crystallize in the hexagonal modification. Figure 31 shows Φ scans, taken with a 0.4° glancing angle, relative to two samples prepared under different conditions. Six peaks corresponding to the $(10\bar{1}0)$ reflections are observed which present in the curve a full width at half maximum (FWHM) of 3.6° which is significantly broader than the $(11\bar{1})$ InP (0.9°). The crystallinity of CdS is then not as perfect as that of the substrate. Figure 31b corresponds to a sample prepared under slightly different conditions. The positions of the peaks are at the same location but their intensities are much lower. In this case only a fraction of the film ($\approx 20\%$) is monocrystalline. One important result is that deposition conditions can be found where the polycrystalline content is almost reduced to zero, as in the case of Figure 31a. The perfection of the interface CdS/InP can be studied by the TEM observation of a cross section, in conditions of high resolution (Figure 32). The lattice planes, normal to the (220) InP ($d = 0.207$ nm), are continued in CdS by the $(11\bar{2}0)$ planes ($d = 0.2068$ nm). The good match between the two lattices promotes the absence of interfacial defects.

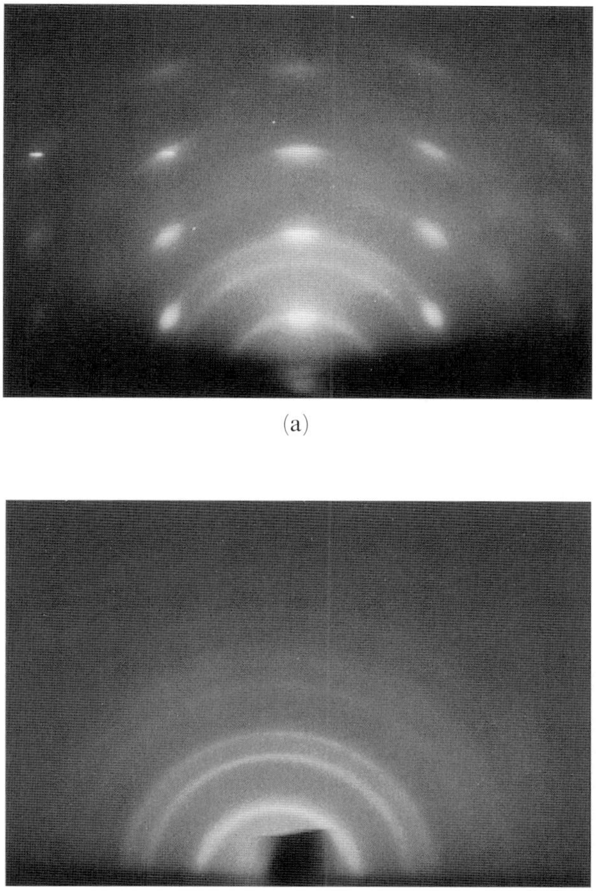

(a)

(b)

Fig. 30. RHEED patterns of CdS chemically deposited on InP single crystals. (from [170])
a) $(\bar{1}\bar{1}\bar{1})$ InP face.
b) (111) InP face.
With permission of Electrochemical Society

The use of $(\bar{1}\bar{1}\bar{1})$ GaP substrates illustrates the effect of an important lattice mismatch on epitaxial growth. GaP has also a cubic structure with a = 0.543 nm and the lattice mismatch with CdS is 7%. The XRD scans with a grazing angle of 0.6° (Figures 33–34) reveal that the epitaxial CdS is a mixture of the hexagonal and the cubic phases in about the same proportion [228]. For the cubic phase (Figure 34) six peaks are observed instead of three as expected owing to the presence of twins in the CdS layer. Twins occur because of two different structural (111) arrangements of the first CdS nuclei which may exist with a 180° rotation. Later, coalescence of these nuclei plays an important part in determining the defect structure (dislocations, stacking faults, twins) of epitaxial films (229). The large mismatch implies a loss of the crystallographic perfection as the CdS peaks have a FWHM of 8° for the hexagonal phase and 7° for the the cubic phase. These values are significantly higher than the

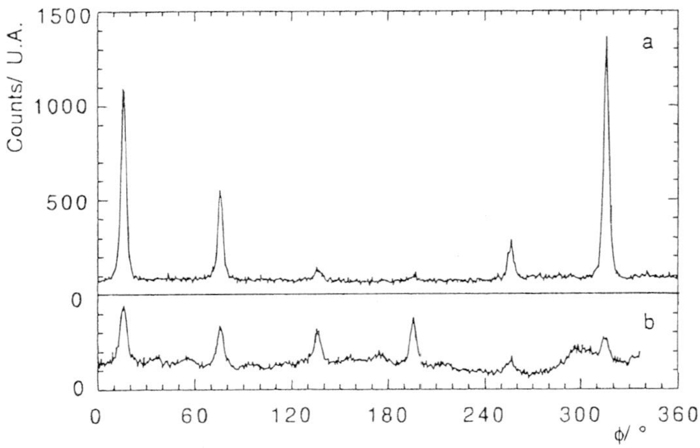

Fig. 31. X-ray diffraction patterns of CdS films (thickness ≈ 25 nm) deposited on $(\bar{1}\bar{1}\bar{1})$ InP. curve a; composition of the solution: $Cd(Ac)_2 = 10^{-2}$ M; $NH_4(Ac) = 2 \times 10^{-2}$ M; thiourea $= 5 \times 10^{-3}$ M; $NH_3 = 4 \times 10^{-1}$ M: T = 85 °C. Curve b: composition of the solution: $[Cd(Ac)_2] = 10^{-2}$ M; $[NH_4(Ac)] = 1.5 \times 10^{-2}$ M; [thiourea] $= 2.8 \times 10^{-2}$ M; $[NH_3] = 1.74$ M; T = 75 °C. (from [170]) With permission of Electrochem. Soc.

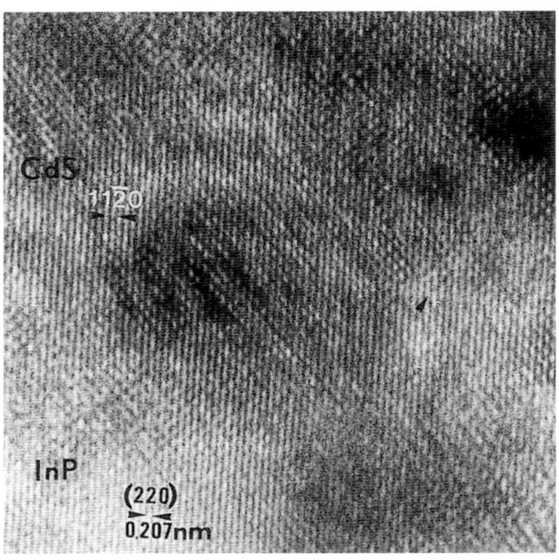

Fig. 32. HREM image of a cross section prepared from a CdS epitaxial layer deposited on $(\bar{1}\bar{1}\bar{1})$ InP with the same solution composition of Fig. 2; T = 60 °C. The arrows shows the interface between CdS and InP. (from [170]) With permission of Electrochem. Soc.

FWHM found during the CdS growth on InP. The TEM and HREM images of a cross-section prepared from a CdS/GaP sample (Figure 35) reveal a high density of planar defects like microtwins and stacking faults. Their origin has been attributed both to the substrate/layer misorientation and to the stresses which are relaxed by atom displacements. Structural models of these defects have been recently proposed in the case of the blende structure [230]. In some places the interface is well resolved and misfit dislocations can be detected (Figure 35b). Some areas show different stacking sequences revealing the presence of a polytype CdS structure [228]. As dis-

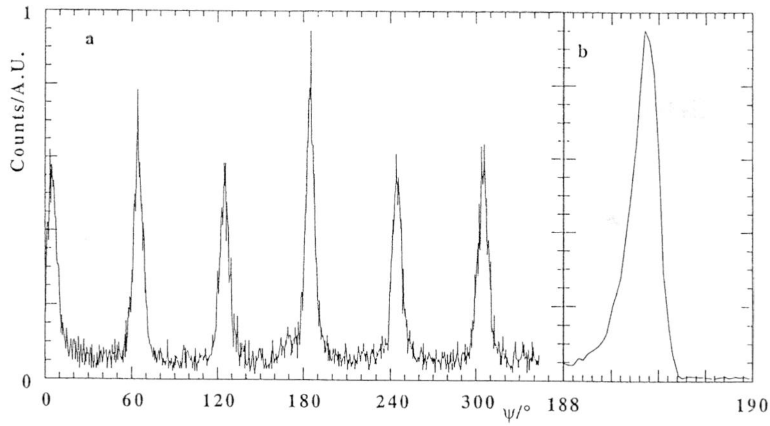

Fig. 33. XRD pattern of an epitaxial CdS film (thickness 64 nm) chemically deposited on (111) GaP. Curve a: (10$\bar{1}$1) reflexions of the hexagonal CdS phase. Curve b (220) reflexion of the GaP substrate. (from [228]) With permission of Editions de Physique

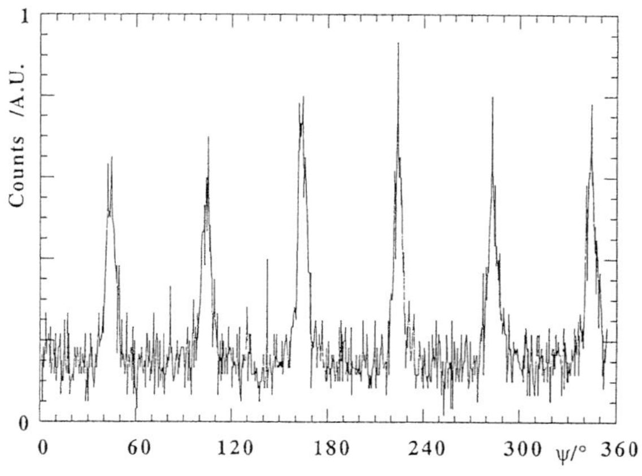

Fig. 34. XRD pattern of an epitaxial CdS film deposited on ($\bar{1}\bar{1}\bar{1}$) GaP. Observation of the six (1$\bar{1}$1) reflexions of the cubic phase (from [228]). With permission of Editions de Physique

cussed in section 4.4 a 4H polytype is concerned with an hexagonality factor of 0.5. On the other hand SEM observations show that the CdS layers present a strong corrugation. In the Figure 36a large density (10^9 cm^{-2}) of pyramids can be observed. According to the literature [231, 232] the stresses which result from a large lattice mismatch can also induce roughness of the epilayers and frequently pyramidal islands are formed.

6.3 Influence of the Substrate Orientation on the Epitaxial Growth

Epitaxial growth of CdS has been also obtained on (100) InP [233] with good quality as demonstrated by the XRD patterns (Figure 37). The FWHM of the four CdS (111) reflexions present a low value (0.6°) compared with that obtained during the CdS growth on ($\bar{1}\bar{1}\bar{1}$) InP (3.6°). Secondly the base line presents a low level. It is

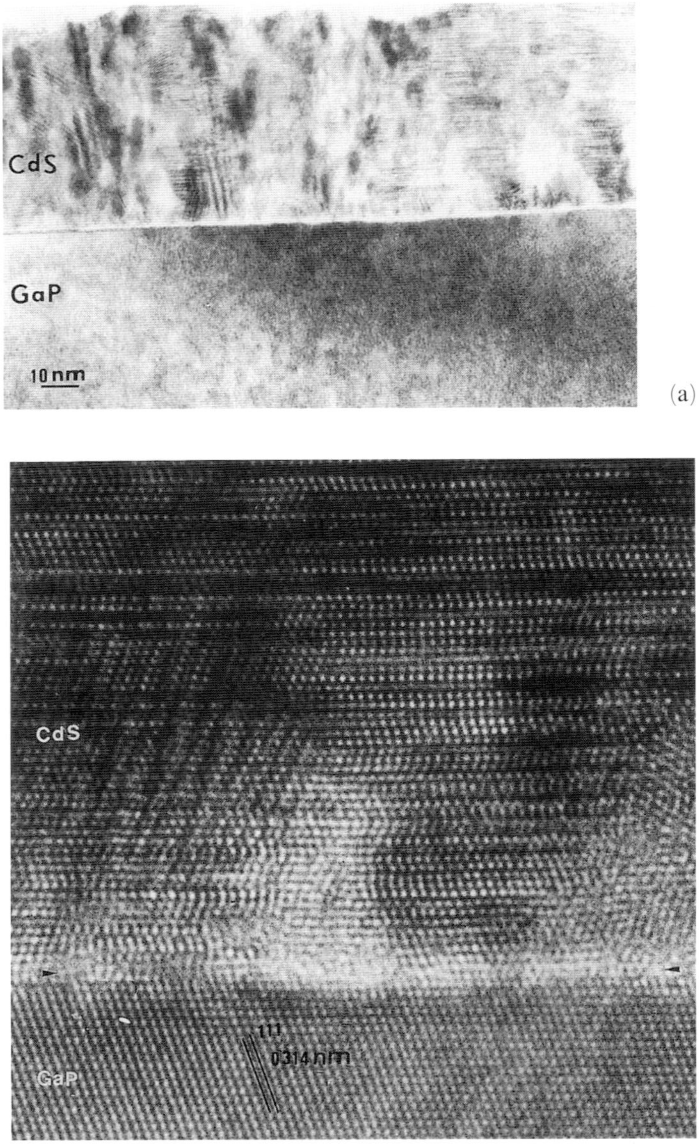

Fig. 35. Observation of a cross sectional thin foil prepared from a CdS/GaP sample (from [228]).
a) Observation of planar defects in a low magnification TEM image.
b) Lattice image of the CdS/GaP interface; the arrows show the position of the interface.
With permission of Editions de Physique

important to remark that contrary to (111) substrates the growth direction does not coïncide with the preferred texture direction of CdS which is ⟨111⟩ or ⟨0002⟩. The epilayer exhibits the cubic structure as it does not exist any epitaxial matching on the (100) face for the hexagonal structure. The HREM observations of the CdS/InP

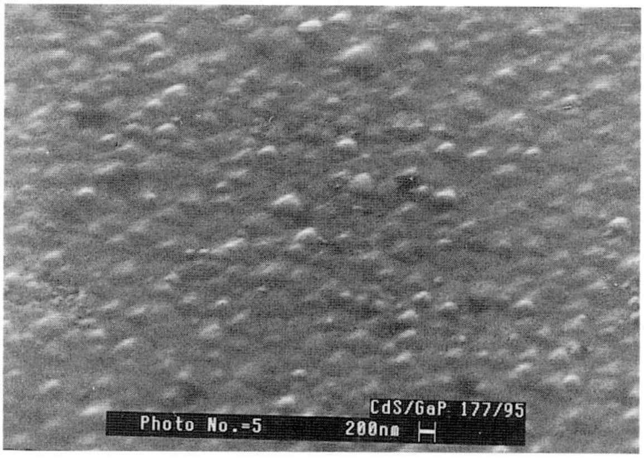

Fig. 36. SEM image showing pyramids at the surface of a CdS epitaxial film grown on $(\bar{1}\bar{1}\bar{1})$ GaP. (from [228]) With permission of Editions de Physique

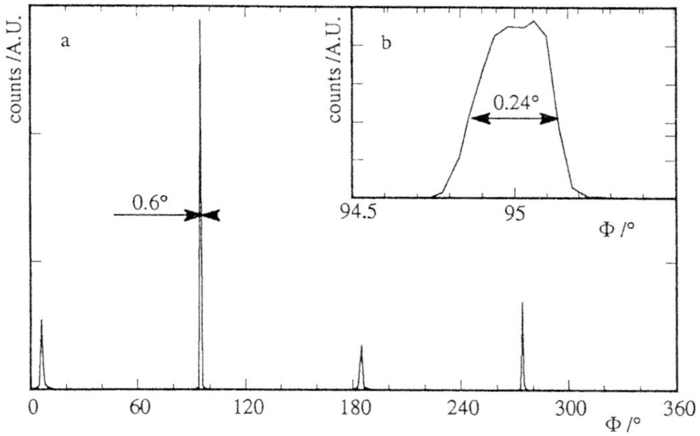

Fig. 37. XRD patterns of an epitaxial CdS film deposited on an (100) InP substrate. Curve a: (111) reflexions of the cubic CdS film. Curve b: (111) reflexion of the InP substrate. (from [233])

cross-sections [233] confirm the good crystallinity of the chemically deposited CdS epilayer and the very low density of planar defects like stacking faults or twins. Contrary to the (111) plane there is no possibility of double-positionning of the first nuclei on (100) and the probability of forming defects during coalescence is reduced. The same conclusions have been deduced during CdS epitaxial growth in the vapour phase. CdS adopts the hexagonal structure when deposited by MOCVD or hot wall epitaxy on (111) GaAs, [234, 235] and the cubic structure when deposited by atomic layer epitaxy on (100) GaAs [236]

6.4 Growth Mechanism and Epitaxy

The growth mechanism of chalcogenide semiconductors synthesized by chemical bath deposition has been discussed in section 4. It has been shown that the solution chemistry governs the growth mode either by aggregation of colloïds or by an atom-by-atom mechanism. It is clear that the first process is inconsistent with an epitaxial growth because it favours a 3D nucleation and growth mechanism called "Volmer-Weber" [227, 237]. In section 3.3, it has been demonstrated that the atom-by-atom mechanism gives rise to a layer-by-layer nucleation and growth mechanism called "Frank-Van der Merwe". This mechanism implies on the other hand a strong binding energy between atoms to be deposited and the atoms of the surface [229]. This mechanism is a necessary condition for epitaxial growth.

As discussed in section 3, S. Gorer and G. Hodes [41] have shown that the growth mechanism of compounds like PbS, CdS, CdSe ... could be easily controlled by using NTA as metal (Cd, Pb) complexing agent. The concentration ratio R between the complexing agent and the metal ion governs the growth mechanism. If R exceeds a critical value Rc, cadmium hydroxide is eliminated and an atom-by-atom growth mechanism is observed. The chemical deposition of CdSe on (111) InP substrates illustrates the relation between growth mechanism and epitaxy [147]. Cadmium selenide films were deposited from solutions described in section 4. A systematic RHEED study of CdSe films deposited on (111) InP (Figure 38) proves that the epitaxial growth occurs only if $R > Rc$. (compare Figures 38a–38b) and is improved if the solution temperature is increased in the range 40 to 90 °C (compare Figures 38b–38d). The XRD scans presented in Figure 39 confirm this effect for CdSe layers deposited at 70 °C (Figure 39a) and 90 °C (Figure 39b). The (220) peaks of the cubic structure have a FWHM of 10° and 7° respectively. Concurently the polycrystalline content decreases. The crystallographic quality of epitaxial CdSe layers electrodeposited on (111) InP is better since the FWHM of the (220) peaks is only 4°. This improvement is probably related to the strong $\langle 111 \rangle$ texture presented by the electrodeposited CdSe layers [221]. When $R \approx Rc$, the epitaxial growth is spoiled because the presence of cadmium hydroxyde traces (compare Figures 38c–38d). An equivalent epitaxial growth is observed both on $(\bar{1}\bar{1}\bar{1})$ and (111) orientations, contrary to the deposition of CdS, the epitaxial growth of which only occurs on $(\bar{1}\bar{1}\bar{1})$. On the other hand the CdSe epitaxial growth on (100) InP is very poor when the chemical deposition of CdS on the same InP orientation produces good epilayers. These observations confirm the high site sensitivity during the atom-by-atom growth. Nevertheless when silicotungstic acid is introduced in the solution, even in conditions where $R > Rc$, polycrystalline CdSe deposits are formed on (111) InP. The strong adsorption of large silicotungstic acid molecules on the InP surface probably weakens the strength of the bonds; then a 3D nucleation and growth mechanism takes place.

7 Application of Chemical Bath Deposition of Chalcogenides

7.1 Photovoltaic Solar Energy Conversion

Chemical bath deposition (CBD) has emerged as a leading deposition method in the area of thin film (CBD) solar cells based on CdTe and on chalcopyrite

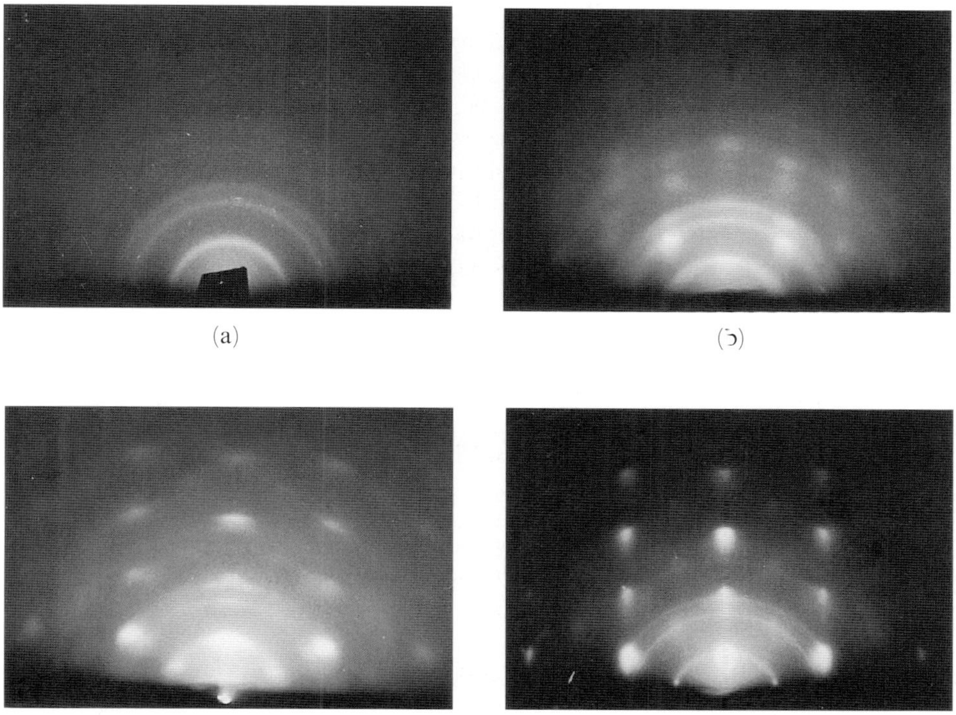

(a) (b)

(c) (d)

Fig. 38. RHEED patterns of CdSe deposited on $(\bar{1}\bar{1}\bar{1})$ InP substrates. Influence of the ratio R ([NTA]/[Cd]) and the temperature on the epitaxial growth; observation in the $\langle 112 \rangle$ azimuth. (from [147])
a) R = 1.75; T = 70 °C (R < Rc); polycrystalline growth.
b) R = 2.2; T = 70 °C (R > Rc); epitaxial growth.
c) R = 2.2; T = 90 °C (R ≈ Rc); poor epitaxial growth.
d) R = 2.4; T = 90 °C (R > Rc): improved epitaxial growth compared to b and c.
With permission of Taylor and Francis Ltd

($CuIn_xGa_{1-x}S_ySe_{2-y}$, $0 \leq x \leq 1$, $0 \leq y \leq 2$, usually called CIS or CIGS) absorbers. The CdTe cell structure consists of glass/SnO_2/CdS/CdTe/contact(C, Au, Ni), illuminated through the glass (backwall cells) while the CIGS cell structure consists of glass/Mo/CIGS/CdS/ZnO, which is illuminated through the ZnO top layer (frontwall cell). All active layers are in the micron thick range or less (1–2 µm for CIS, CdTe, ZnO and 0.05–0.2 µm for CdS). Chemical bath deposition is now widely used for the synthesis of the CdS layer in both structures. CdS acts as an n-type window layer in CdTe cells and as an interfacial buffer layer in CIGS cells.

In the case of CdTe solar cells, CBD CdS is used in combination with electrodeposited [14, 238] or coevaporated CdTe [13, 32]. The process used for evaporated CdTe cells is based on an ammonia bath and has been described in detail [32]. An interesting point is that it seems that it is possible to control the doping of as grown CdS by introducing borate ions in the solution (boron is an n-type dopant in substitution of cadmium atoms). Resistivities of 10 Ω · cm in dark and 2 Ω · cm under illumination were measured. Note that in this process the CdTe layer is deposited at

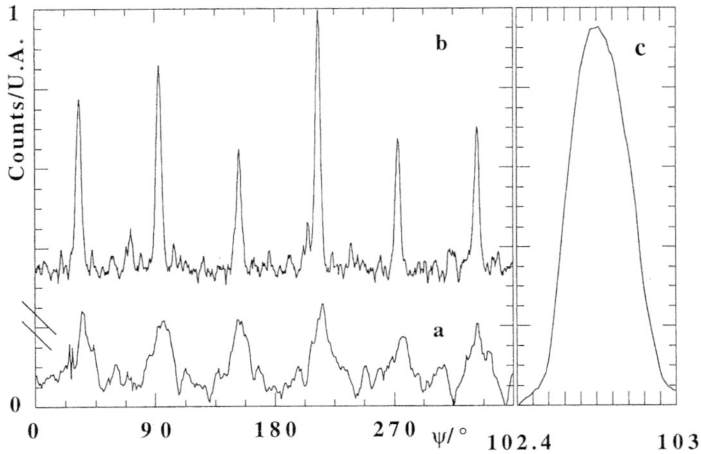

Fig. 39. XRD patterns of CdSe films deposited on $(\bar{1}\bar{1}\bar{1})$ InP; influence of R and T on the epitaxial growth. (from [147])
a) R = 2.2; T = 70 °C; (220) CdSe reflexions; FWHM > 10°
b) R = 2.4; T = 90 °C; (220) CdSe reflexions, FWHM = 7°
c) (220) InP reflexion; FWHM = 0.25°
With permission of Electrochem. Soc.

550 °C, indicating a good stability of the thin (0.2 μm) CdS layer during high temperature processes. Cells made with electrodeposited CdTe are developed by BP Solar, at the pilot production stage. Efficiencies close to 13% are reported on small area cells (about 1 cm^2) and 10% on 30 × 30 cm^2 modules [238]. The fact that the main technologies are both based on solution deposition methods (CBD, ED) is interesting since it demonstrates these techniques are suitable for large area semiconducting device production at low cost. CdS layers are prepared in a batch process with large area multisubstrates (30 × 30 cm^2) processing. The CdS layer thickness is about 60 nm thick deposited on rough tin oxide substrates. This low thickness is a beneficial consequence of the high covering properties of CBD which allows conformal growth on rough substrates [30]. During annealing treatments at 450 °C, beneficial recrystallisation and interdiffusion processes takes place between CdS and CdTe, leading to high junction quality [239]. Basically there is no limitation on the area by the CBD method. Experiments on 30 × 30 cm^2 substrates where only a film of solution of a few mm thick is introduced between the plates show excellent lateral uniformity in thickness (< 5%), and optical properties [240]. For CdS the film thickness on each plate (n in cm), assuming that the thiourea concentration is in excess with respect to the cadmium one, is given by the relation:

$$h = 0.03[Cd(II)]d \qquad (40)$$

where d is the spacing in cm. This shows that for a cadmium concentration of 10^{-3} M a 2 mm spacing is sufficient to deposit 60 nm thick film. The material yield, which is very low (often less than tenths of per cent) in classical CBD reactors because of the low substrate-area-to-solution-volume ratio, is increased with the two plate configuration. For small spacing between the plates (typically 1–2 mm) the material yield in film formation can approach 100% [240], which is very important for waste reduction. This is a beneficial consequence of the fact that the heterogeneous nucleation and growth on the substrates is highly favored in this system.

Thin film solar cells based on Cu(In, Ga)(S, Se)$_2$ also commonly use chemical bath deposited interfacial buffer layers, while CIGS and ZnO layers are deposited by coevaporation, or selenisation and sputtering. Chemically deposited layers of CdS give the best results with record efficiencies at the laboratory scale now approaching 18% (17.7% [241], 17.6 [242]). As for CdTe, upscaling has been demonstrated in various companies in USA (Solarex, Siemens USA), Europe (Siemens, ZSW) and Japan (Showa Shell). High efficiency cells are obtained with CBD-CdS over the entire family of Cu(In, Ga) (S, Se)$_2$ absorbers (from pure $CuInSe_2$ to $CuInS_2$) with rather standardized and simple CBD procedures for CdS (see [241] and [242]) mostly derived from the early work of Kessler et al. [243]. It appears that the deposition process is very flexible in terms of bath chemistry (nature of complexing agent, salt). A typical bath is [Cd(II)] = 10^{-3} M, [TU] = 0.15 M, [NH_3] = 1 M with T = 60 °C [243]. The excess of TU which allows the saturation of surface sites is probably responsible for the good conformal growth and covering properties.

Many investigations are now devoted to understand the positive effect of CBD layers on junction properties, considering the specific effect of interface chemistry [243, 244], and band edge alignment at the interface [245]. These properties may be influenced by the deviations of stoichiometry of CBD CdS layers [44]. Recently it has been shown that CBD CdS can be grown epitaxially on CIS [246], which can explain the good interface quality needed for high efficiency CIGS cells. Electrical properties of CBD CdS may also play a role in the junction formation since these films are not insulating but n-type doped at a level of 10^{16}–10^{17} [245] which is suitable for junction formation with lower doped p-type CIS. Similar values were obtained on other substrates like tin oxide [143] or gold [30] showing that the doping is not related to reactions with the substrate. There are still many open questions concerning the origin of doping and deviations from stoichiometric composition.

Along with studies of CBD CdS, significant efforts are now made in the CIS technology to find alternative layers which will further improve the UV collection with a greater band gap, or which will not contain cadmium. Other materials have been considered, like indium sulfide, zinc sulfide, tin sulfide, zinc selenide. Among these, indium sulfide and zinc sulfide appear to be good candidates. However the best layers are in fact hydroxosulfide films of indium [115] or zinc [117] which lead to efficiencies approching those of CdS layers.

In addition, the use of CBD CdS in cells for spatial applications based on indium phosphide single crystals has been reported [247] in a cell with an efficiency of 17.8%. The structure was InP/CBD CdS/In_2O_3 for which the possibility of epitaxial growth of CdS CBD on InP is a favorable factor.

These recent works show that there are many opportunities for using CBD in large area photovoltaics. In the future it is possible that not only window layers but also absorber layers will be made by CBD (first attempts have been made to deposit $CuInS_2$ [104], $CuInSe_2$ [110] and CdTe [94] by CBD), leading to a completely wet chemical route which is highly relevant to industrial applications as shown for CdTe cells. Promising results have been obtained by Savadogo and Mandal [248] with Schottky barriers on CBD CdSe and Sb_2S_3 which display efficiencies of 7.2 and 5.5% respectively. In the case of antimony trisulfide and cadmium selenide a positive effect of the addition of silicotungstic acid in the bath was shown to form passivating tungten oxide films on the grain surfaces during annealing.

7.2 Optoelectronic Devices

Heterostructures based on III–V compounds (InP, GaAs, GaInAlAs, GaInAsP) are very important in the field of high frequency optoelectronic devices. Interface chemistry control is a key point for device quality and reliability. III–V compounds have highly reactive surfaces leading to an easy formation of vacancies for instance which are detrimental to the electronical characteristics. To solve these problems passivation treatments are needed. Classical treatments are based on vapor phase reactions or other wet procedures like chemical oxidation or ammonium sulfide treatments.

Recently Vaccaro et al. [15, 249] found that chemical bath deposition is also an effective method to passivate metal insulator semiconductor (MIS) devices based on InP. They showed that deposition of a very thin CdS layer from an ammonia bath (2 nm) between InP and SiO_2 lead to spectacular improvement in the electrical properties of the devices. The reason was attributed to a reduction by at least one order of magnitude of the interface state density at the InP surface as compared to the reference structure. They show that the CBD deposition leads to a reduction of the phophorus vacancies densities at the interface and also to elimination of residual oxides.

In subsequent papers [250] they extend the use of CBD to passivate the surface of InAlAs/InGaAs high electron mobility transistors (HEMTs) and metal-semiconductor-metal (MSM) photodetectors with a spectacular improvement of the drain-to-gate current ratio in the first case (8×10^4) and a reduction of dark current by three orders of magnitude in the second case. These results may be a consequence of the fact that contrary to other treatments the device is made by building a new layer on a preserved or even *in situ* chemically treated surface, instead of a surface chemical treatment only which may also perturb the near surface region.

In the area of luminescent devices, chemical deposition seems to be a promising technique. Attempts are made to use transparent CBD layers for contacting light emitting porous silicon. Thanks to the liquid environment, the reactants are allowed to penetrate and fill the pores such as for CdS in p type porous silicon [251]. Luminescence may also be produced by the CBD layer itself such as for CdSe [76, 79, 252]. New applications may arise from using nanocrystalline films where quantum size effects allow to modify the wavelength of the luminescence peak [76, 252].

7.3 Photodetectors

The fabrication of infra-red photodetectors, based on photoconducting properties, has been the first industrial application of chemically deposited films with PbS and PbSe [3] and is still in use for this application. During this time, many candidates for specific photoconductive applications have emerged from chemical bath deposition studies. Materials which display the highest possible conductivity ratio between illumination and dark are needed. In this case contrary to the case of photovoltaic applications the objective is to prepare very resistive films. Chemically deposited films are naturally more likely very resistive. Ternary sulfide CdPbS [97] and HgPbS [105, 106] allow for further adaption of the wavelength range in the infra-red region by changing the alloy composition.

Many investigations are presently carried out on wide gap photodetectors like CdS [253], $Cd_xZn_{1-x}S$ [101], CdSe [254]. Nair et al. [253] report light-to dark-conductivity ratio as high as 10^9 for CdS films (with a conductivity value of 3 Ω^{-1} cm^{-1} under white light 300 $mW.cm^{-2}$). The characteristics depend on deposition conditions, and the decay time varies from a few seconds to a few hours by simply changing the deposition temperature. Similar trends were recently reported with changing the impurity content in the film *via* the thiourea concentration [44]. For $Cd_xZn_{1-x}S$ [101] ratio up to 10^6 are reported. Improvements are observed after doping the films with fluoride, lanthanum and neodynium ions in the bath. Emerging CBD materials are Bi_2S_3 [23], $CuBiS_2$ [102], Sb_2S_3 [62] and Bi_2Se_3 [72] (Bismuth selenide is also interesting for its thermoelectric properties).

7.4 Solar control coatings for architectural or vehicle glazing

Glass covered with specific optical coatings is needed in architecture and in transport. As an example in warm countries, coatings which transmit visible light and reflect infrared light with low thermal emittance are needed. As in photovoltaics, large area processing at low cost is an important aspect, and chemical bath deposition is well suited to satisfy these requirements. Studies are carried out in this direction by P. K. and M. K. S. Nair et al [23, 47, 57, 58] with the deposition of lead sulfide [58] and copper sulfide [47] films. Copper sulfide present excellent characteristics [47]. These optical characteristics result from the semiconducting properties of the films, which possess a direct band gap of about 2.5 eV, allowing the transmission of visible light for wavelengths higher than 0.4 µm while the decrease in transmission from 0.7 µm is due to the absorption from free carriers (the material is degenerate p-type). The films display reflection colors from light purple to greenish blue when the thickness increases from 0.16 to 0.24 µm. Other materials or combination of materials are currently investigated like Cu_3BiS_3 [103], SnS-Cu_xS [255] Cu_xS-PbS [58, 256] in order to improve the characteristics and stability of the films.

Chemical deposition can be anticipated for photothermal conversion applications in architecture. Nair et al. have also proposed the use of cuprous sulfide for this objective [47].

7.5 Photoelectrochemical Applications

The possibility of using chemically deposited films in photoelectrochemistry has been demonstrated for CdSe [74]. The repetition of individual chemical bath deposition sequences allows preparation of thick films up to 3 μm on Ti. After annealing treatment, the films present good photoelectrochemical properties as n-type semiconductors and have been used in a photoelectrochemical cell based on the sulfide polysulfide redox couple. Under illumination (AM1 95 $mW.cm^{-2}$) the photocurrent density was close to 20 $mA.cm^{-2}$ and the open circuit voltage was 0.7 V. Due to a low fill factor, the efficiency was limited to about 6.7%, but significantly higher values should be possible after optimization studies. Photoelectrochemical cells based on nanocrystalline chemically deposited films of CdSe have been also reported [76].

7.6 Other Applications

Chemically deposited non stoichiometric cuprous sulfide films ($Cu_{1.8}S$) have been used as conducting layers as reported by Grozdanov et al. [50]. The films, deposited at 40 °C, present a resistivity of 2.10^{-3} Ω.cm. In addition they present optical transmission values between 50 and 70% in the visible range for a 0.12 μm thick film. These properties have been used for ohmic contacts to ferroelectric films and transparent conducting coatings on polymers. These films can also be used as chemical sensors for Cu^{2+} ions. Note that due to the low deposition temperature polymer substrates can be used [61].

8 Conclusion

In this paper we have tried to present the chemical and mechanistic aspects of chemical bath deposition of chalcogenide compounds as they appear both in the recent literature and also in older studies dealing with hydrolysis of chalcogenide precursors. A better account of these aspects gives clues to understanding the properties of the films such as the dependence of composition on solution composition and competitive precipitation processes, and the dependence of structure on competition between atom-by-atom and colloidal growth deposition mechanisms.

A special focus has been then given to structural properties and the open challenge concerning epitaxial growth of CBD films, which appears as another case of Chemical single Crystal Routes from solutions recently highlighted [257]. It appears that this technique offers many possibilities in this area which have been not yet explored. Important efforts have still to be made to increase the quality of the CBD epitaxial films to the level of vapor phase films, in order to get all the benefits from a unique deposition process at low temperatures. A better control of deposition chemistry in combination with surface chemistry of single crystals is an important aspect.

In the last section we have indicated the main present applications of chemically deposited films. Some of them like photoconductors have been established for a long time while others only emerged recently, as is the case for photovoltaic applications.

We can anticipate that in the near future, chemical deposition will not only concern window layers but also absorber layers, i.e. small band gap semiconductors (Eg = 1–1.5 eV). There are many options for this type of material among sulfide and telluride compounds. Other large area applications or new developments are expected in the area of glass coatings such as large area electrochromic devices prepared by CBD. In addition, applications in microelectronics have recently emerged for III–V devices. It is clear that, owing to intrinsic qualities of the CBD technique in terms of passivating properties, further developments will also appear in this area. It may be the chemically deposited part which will also become a more active functional component, based on specific luminescence or electroluminescence properties, and that coupling with chemical sensors is possible. It is interesting to note that low dimensional structures can be made by CBD, such as quantum sized films where the optical properties are controlled by the grain size of the material. Films with spatially modulated composition and structure (in 2D and 3D) may be new materials efficiently prepared by CBD techniques allowing for instance an improved control of the deposition at the monolayer level, by using flowing solutions for instance [36]. Some weak points which are still associated with CBD concern the thickness range, especially towards thick films and the lack of understanding and control of the electrial properties as compared to their vapor phase homologues.

Looking to the present status of this technique it appears that it is mainly restricted to aqueous solutions and a limited number of precursors especially for the chalcogenide atoms. The use of non aqueous solutions (THF, toluene) with alternative precursors, as reviewed by Rouxel [258] for the homogeneous precipitation of low-dimensional chalcogenides (TiS_2, MoS_2, WS_2, Cr_2S_3, NiTe etc.) may open new routes for film formation by non aqueous-CBD. In addition to chalcogenides, non aqueous solvents also offer the possibility to extend the method to the preparation of III–V compounds films [259].

9 References

1. J. Emerson-Reynolds, J. Chem. Soc. 45, 162 (1884).
2. Y. Okinaka, T. Osaka in: Advances in Electrochemical Science and Engineering, H. Gerisher and C. W. Tobias (ed.), VCH, Weinheim (1994), Vol. 3, p. 55.
3. D. E. Bode in: Physics of Thin Films, G. Hass and R. E. Thun (ed.), (1966), Vol. 3, p. 275.
4. J. Bloem, Appl. Sci. Res. B 6, 92 (1956).
5. R. A. Zingaro, D. O. Skovlin, J. Electrochem. Soc. 11, 42 (1964).
6. A. B. Lundin, G. A. Kitaev, Inorg. Mater. 1, 1905 (1965).
7. G. A. Kitaev, S. G. Mokrushin, A. A. Uritskaya, Colloid J. USSR, 27, 38 (1965).
8. M. J. Mangalam, K. Nagaraja Rao, N. Rangarajan, C. V. Suryanarayana, Brit. J. Appl. Phys. 2, 2, 1643 (1969).
9. G. A. Kitaev , T. S. Terekhova, Russian J. Inorg. Chem. 15, 25 (1970).
10. G.A. Kitaev, T. P. Sokolova, Russian J. Inorg. Chem. 15, 167 (1970).

11. K. L. Chopra, R. C. Kainthla, D. K. Pandya, A. P. Thakoor, Phys. Thin Films 12, 167 (1982).
12. L. Stolt, J. Hedström, J. Kessler, M. Ruckh, K. Velthaus, H. W. Schock, Appl. Phys. Lett. 62, 597 (1993).
13. J. Britt, C. Ferekides, Appl. Phys. Lett. 62, 2851 (1993).
14. J. Barker, S. P. Binns, D. R. Johnson, R. J. Marshall, S. Oktik, M. E. Özsan, M. H. Patterson, S. J. Ransome, S. Roberts, M. Sadeghi, J. Sherborne, A. K. Turner, J. M. Woodcock, Int. J. Solar Energy 12, 79 (1992).
15. K. Vaccaro, H. M. Dauplaise, A. Davis, S. M. Spaziani, J. P. Lorenzo, Appl. Phys. Lett. 67, 527 (1995).
16. C. D. Lockande, Mater. Chem. Phys. 28, 145 (1991).
17. S. S. Dhumure, C. D. Lockande, Thin Solid Films 240, 1 (1994).
18. I. Grozdanov, Semicond. Sci. Technol. 9, 1235 (1994).
19. H. Meherzi-Maghraoui, M. Dachraoui, S. Belgacem, K. Buhre, R. Kunst, P. Cowache, D. Lincot, Thin Solid Films 288, 217 (1996).
20. P. Pramanik, R. N. Bhattacharya, J. Electrochem. Soc. 127, 2087 (1980).
21. R. N. Battacharya, P. Pramanik, J. Electrochem. Soc. 129, 332 (1982).
22. S. Biswas, A. Mondal, D. Mukherjee, P. Pramanik, J. Electrochem. Soc. 133, 8 (1986).
23. M. T. S. Nair, P. K. Nair, Semicond. Sci. Technol. 5, 1225 (1990).
24. L. Huang, R. A. Zingaro, E. A. Meyers, M. T. S. Nair, T. S. Nair, Phosphorus, Sulfur and Silicon 103, 77 (1995).
25. I. Kaur, D. K. Pandya, K. L. Chopra, J. Electrochem. Soc. 127, 943 (1980).
26. W. J. Danaher, L. E. Lyons, G. C. Morris, Solar Energy Mater. Solar Cells 12, 137 (1985).
27. M. T. S. Nair, P. K. Nair, J. Campos, Thin Solid Films 161, 21 (1988).
28. H. Uda, S. Ikegami, H. Sonomura, Japanese J. Appl. Phys. 29, 30 (1990).
29. V. N. Semenov, J. Appl. Chem. 64, 159 (1991).
30. D. Lincot, R. Ortega-Borges, J. Electrochem. Soc. 139, 1880 (1992).
31. J.M. Doña, J. Herrero, J. Electrochem. Soc. 139, 2810 (1992).
32. T. L. Chu, S. S. Chu, N. Schultz, C. Wang, C. Q. Wu, J. Electrochem. Soc. 139, 2443 (1992).
33. B. R. Lanning, J. H. Armstrong, Int. J. Solar Energy 12, 247 (1992).
34. R. Ortega-Borges, D. Lincot, J. Electrochem. Soc. 140, 3464 (1993).
35. P. C. Rieke, S. B. Bentjen, Chem. Mater. 5, 43 (1993).
36. K. Ito and K. Shiraishi, Solar Energy Mater. Solar Cells 35, 179 (1994).
37. T. Nakanishi, K. Ito, Solar Energy Mater. Solar Cells 35, 171 (1994).
38. L. Hernandez, O. de Melo, O. Zelaya-Angel, R. Lozada-Morales, J. Electrochem. Soc. 141, 3238 (1994).
39. K. Ito, T. Tamaru, J. Mat. Sci. Lett. 13, 893 (1994).
40. N. Andriskos, A. J. Karabelas, J. Colloid Interface Sci. 165, 301 (1994).
41. S. Gorer, G. Hodes, J. Phys. Chem. 98, 5338 (1994).
42. P. O'Brien, T. Saeed, J. Crystal Growth 158, 497 (1996).
43. I. O. Oladeji, L. Chow, J. Electrochem. Soc. 144, 2342 (1997).
44. a- A. Kylner, J. Lindgren, L. Stolt, J. Electrochem. Soc. 143, 2662 (1996).
b- D. W. Niles, G. Herdt, M. Al-Jassim, J. Appl. Phys. 81, 1978 (1997).
c- A. Kylner, E. Niemi, 15th E.C. Photovoltaic Solar Energy Conf., (1997), to be published.
45. M. Froment, D. Lincot, Electrochim. Acta 40, 1293 (1995)
46. P. K. Basu, P. Pramanik, J. Mater. Sci. Lett. 5, 1216 (1986).
47. M. T. S. Nair, P. K. Nair, Semicond. Sci. Technol. 4, 191 (1989); ibid 4, 99 (1989).
48. A. J. Varkey, Solar Energy Mater. Solar Cells 19, 415 (1989).

49. I. Grozdanov, C. K. Barlingay, S. K. Dey, M. Ristov, M. Najdoski, Thin Solid Films 250, 67 (1994).
50. I. Grozdanov, C. K. Barlingay, S. K. Dey, Mater. Lett. 23, 181 (1995).
51. G. A. Kitaev, V. I. Dvoinin, A. V. Ust'yantseva, M. N. Belyaeva, L. G. Skornyakov, Inorg. Mater. 12, 1448 (1976).
52. T. Yoshida, K. Yamaguchi, H. Toyoda, K. Akao, T. Sugiura, H. Minoura, Y. Nasaka, Electrochemical Soc. Proc. 97– 20, 37 (1997).
53. P. Pramanik, M. A. Akhter, P. K. Basu, Thin Solid Films 158, 271 (1988).
54. P. Pramanik, S. Biswas, J. Solid State Chem. 65, L45 (1986).
55. J. L. Davis, M. K. Norr, J. Appl. Phys., 37, 1670 (1966).
56. H. Rahnamai, H. J. Gray, J. N. Zemel, Thin Solid Films, 69, 347 (1980).
57. P. K. Nair, M. T. S. Nair, Semicond. Sci. Technol. 4, 807 (1989).
58. P. K. Nair, M. T. S. Nair, A. Fernandez, M. Ocampo, J. Phys. D: Appl. Phys. 22, 829 (1989).
59. M. Isshiki, T. Endo, K. Masumoto, Y. Usui, J. Electrochem. Soc 137, 2697 (1990).
60. L. Huang, P. K. Nair, M. T. S. Nair, R. A. Zingaro, E. A. Meyers, J. Electrochem. Soc. 141, 2536 (1994).
61. P. Pramanik, S. Bhattacharya, J. Mater. Sci. Lett. 6, 1105 (1987).
62. O. Savadogo, K. C. Mandal, Solar Energy Mater. Solar Cells 26, 117 (1992).
63. P. Pramanik, P. K. Basu, S. Biswas, Thin Solid Films 150, 269 (1987).
64. R. D. Engelken, H. E. McCloud, C. Lee, M. Slayton, H. Ghoreishi, J. Electrochem. Soc. 134, 2696 (1987).
65. M. T. S. Nair, P. K. Nair, Semicond. Sci. Technol. 6, 132 (1991).
66. A. Mondal, P. Pramanik, Thin Solid Films 110, 65 (1983).
67. T. G. Leonova, L. F. Bakhturova, V. I. Belyi, S. V. Larionov, N. P. Sysoeva, Inorg. Mat. 27, 1745 (1991).
68. R. Ortega-Borges, D. Lincot, J. Vedel, 11th E.C. Photovoltaic Solar Energy Conf., H. Stephens & Associates, Bedford (1992) , p. 862.
69. J. M. Doña, J. Herrero, J. Electrochem. Soc. 141, 205 (1994).
70. B. Mokili, M. Froment, D. Lincot, J. Physique IV, 5, 26 (1995).
71. P. Pramanik, R. N. Battacharya, A. Mondal, J. Electrochem. Soc., 127, 1857 (1980).
72. V. M. Garcia, M. T. S. Nair, P. K. Nair, R. A. Zingaro, Semicond. Sci. Technol. 12, 645 (1997).
73. R. C. Kainthla, D. K. Pandya, K. L. Chopra, J. Electrochem. Scc. 127, 277 (1980).
74. R. A. Boudreau and R. D. Rauh, J. Electrochem. Soc. 130, 513 (1983).
75. A. Mondal, T. K. Chaudhuri, P. Pramanik, Solar Energy Mater. Solar Cells 7, 431 (1983).
76. G. Hodes, A. Albu-Yaron, F. Decker, P. Motisuke, Phys. Rev. B 36, 4212 (1987).
77. A. H. Eid, S. Mahmoud, J. Mat. Sci. Lett. 11, 937 (1992).
78. O. Savadogo, K.C. Mandal, Mater. Chem. Phys. 31, 301 (1992).
79. J. M. Gracia-Jimenez, G. Martinez-Montes, R. Silva-Gonzalez, J. Electrochem. Soc. 139, 2048 (1992).
80. H. Cachet, H. Essaaidi, M. Froment, G. Maurin, J. Electroanal. Chem. 396, 75 (1995).
81. H. Cachet, M. Froment, G. Maurin, J. Electroanal. Chem. 406, 239 (1996).
82. P. Pramanik, S. Bhattacharya, P. K. Basu, Thin Solid Films 149, L81 (1987).
83. G. K. Padam, Thin Solid Films 150, L89 (1987).
84. C. Lévy-Clément, M. Neumann-Spallart, S. K. Haram, K. S. V. Santhanam, Thin Solid Films 302, 12 (1997).
85. S. Gorer, A. Albu-Yaron, G. Hodes, Chem. Mater. 7, 1243 (1995).
86. S. Gorer, A. Albu-Yaron, G. Hodes, J. Phys. Chem. 99, 16442 (1995).
87. R. N. Bhattacharya, P. Pramanik, Solar Energy Mater. Solar Cells 6, 317 (1982).
88. R. N. Bhattacharya, P. Pramanik, Bull. Mater. Sci. 3, 403 (1981).

89. P. Pramanik, S. Bhattacharya, J. Mater. Sci. Lett. 7, 1305 (1988).
90. G. A. Kitaev, P. T. Sokolova, Russian J. Inorg. Chem. 15, 167 (1970).
91. P. Pramanik, S. Biswas, J. Electrochem. Soc. 133, 350 (1986).
92. C. A. Estrada, P. K. Nair, M. T. S. Nair, R. A. Zingaro, E. A. Meyers, J. Electrochem. Soc. 141, 802 (1994).
93. J. M. Doña, J. Herrero, J. Electrochem. Soc. 142, 764 (1995).
94. R. W. Buckley, 11th E.C. Photovoltaic Solar Energy Conf., H. Stephens & Associates, Bedford (1991), p. 962.
95. M. A. Berchenko, A. I. Belyaev, Russian J. Inorg. Chem. 15, 1034 (1970).
96. R. D. Mukhamed'yarov, G. A. Kitaev, V. M. Markova, V. I. Stuk, Inorg. Mater. 17, 1282 (1981).
97. G. A. Kitaev, V. F. Markov, L. N. Maskaeva, L. E. Vasyunina, I. V. Shilova, Inorg. Mater. 26, 202 (1990).
98. G. K. Padam, G. L. Malhotra, S. U. Rao, J. Appl. Phys. 63, 770 (1988).
99. J. M. Doña, J. Herrero, Thin Solid Films 268, 5 (1995).
100. T. Yamaguschi, Y. Yamamoto, T. Tanaka, Y. Demizu, A. Yoshida, Thin Solid Films 281–282, 375 (1996).
101. a) S. Sushma Shrivastava, S. Bhushan, Indian J. Pure Appl. Phys. 34, 106 (1996).
b) S. Sushma Shrivastava, S. Bhushan, Indian J. Phys. 70A, 487 (1996).
102. L. P. Deshmukh, D. S. Sutrave, B. M. More, C. B. Rotti, K. M. Garadkar, Semiconductor Devices, 421 (1996).
103. P. K. Nair, L. Huang, M. T. S. Nair, H. Hu, E. A. Meyers, R. A. Zingaro, J. Mater. Res. 12, 651 (1997).
104. G. K. Padam, S. U. M. Rao, Solar Energy Mater. Solar Cells 13, 297 (1986).
105. N. C. Sharma, D. K Pandhya, H. K. Sehgal, K. L. Chopra, Mat. Res. Bull. 11, 1109 (1976).
106. N. C. Sharma, D. K. Pandhya, H. K. Sehgal, K. L. Chopra, Thin Solid Films 59, 157 (1979).
107. R. K. Pandey, A. J. N. Rooz, S. K. Kuljarni, Semicond. Sci. Technol. 3, 729 (1988).
108. R. K. Pandey, A. J. N Rooz, R. B. Gore, Semicond. Sci. Technol. 3, 733 (1988).
109. K. C. Sharma, J. C. Karg, J. Phys. D: Appl. Phys. 23, 1411 (1990).
110. R. N. Bhattacharya, J. Electrochem. Soc. 130, 2040 (1983).
111. J. C. Garg, R. P. Sharma, K. C. Sharma, Thin Solid Films 164, 269 (1988).
112. R. C. Kainthla, D. K. Pandya, K. L. Chopra, J. Electrochem. Soc., 129, 99 (1982).
113. L. P. Desmukh, G. S. Shahane, C. B. Rotti, S. G. Holikatti, P. P. Hankare, Semiconductor Devices 388 (1996).
114. G. N. Chaudhari, S. N. Sardesai, S. D. Sathaye, V. J. Rao, J. Mater. Sci. 27, 4647 (1992).
115. D. Hariskos, R. Heberholz, M. Ruckh, U. Rühle, T. Walter, H. W. Schock, Proc. 1st World Conf. Photovoltaic Energy Conversion, (1994), p. 91.
116. D. Hariskos, R. Herberholz, M. Ruckh, U. Rühle, R. Schäffler, H. W. Schock, in: 13th E.C. Photovoltaic Solar Energy Conf., H. Stephens & Associates, Bedford, (1995), p. 1899.
117. K. Kushiya, T. Nii, I. Sugiyama, Y. Sato, Y. Inamori, H. Takeshita, Jap. J. Appl. Phys. 35, 4383 (1996).
118. B. Mokili, Y. Charreire, R. Cortès, D. Lincot, Thin Solid Films 288, 21 (1996).
119. A. N. Nielsen, in: Kinetics of Precipitation, Pergamon Press, (1964).
120. R. J. Hunter, in: Fondations of Colloid Science, Oxford University Press, Vol. 2, (1989).
121. A. E. D. M. van deHeijden and G. M. van Rosmalen, in: Industrial Mass Crystallisation Handbook of Crystal Growth, D. J. Hurle ed., Chap. 7, (1994). p. 315.
122. J. Jolivet, in: De la Solution à l'Oxyde, CNRS Editions, (1994).

123. V. K. La Mer and R. H. Dinegar, J. Amer. Chem. Soc. 72, 4847 (1950).
124. R. Kern, Bull. Miner. 101, 202 (1978).
125. O. Knacke, O. Kubaschewski, K. Hesselmann, in: Thermochemical Properties of Inorganic Substance, Springer Verlag, (1991).
126. A. J. Bard, R. Parsons, J. Jordan, in: Standard potentials in aqueous solutions, Marcel Dekker Inc., (1985).
127. A. Ringbom, in: Complexation in Analytical Chemistry, Interscience, (1963).
128. B. Trémillon, in: Reactions in Solutions, An Applied Analytical Approach, John Wiley and Sons, (1997).
129. G. A. Kitaev, A. A. Uritskaya, L. E. Yatlova, T. S. Terekhova, T. I. Dzyuba, Russian J. Inorg. Chem. 35, 1744 (1990).
130. W. H. Shaw, D. G. Walker, J. Am. Chem. Soc. 78, 5769 (1956).
131. G. Marcotrigiano, G. Peyronel, R. Battistuzzi, J. Chem. Soc. Perkin II, 1539 (1972).
132. G. A. Kitaev, I. T. Romanov, Khim. Technol. 17, 1427 (1974).
133. E. Swift, E. A. Butler, Anal. Chem. 28, 146 (1956).
134. Battistuzzi, G. Marcotrigiano, G. Peyronel, J. Chem. Soc. Perkin II 169 (1975).
135. G. A. Kitaev, Y. N. Makurin, V. I. Dvoinin, Russian J. Phys. Chem. 50, 1828 (1976).
136. G. A. Kitaev, Y. N. Makurin, V. I. Dvoinin, V. R. Mirolyubov, ibid 49, 590 (1975), ibid 51, 51 (1977).
137. L. P. Hammett, in: Physical Organic Chemistry, McGraw-Hill Inc., (1940), p. 340. References cited.
138. G. A. Kitaev, I. T. Pomanov, in russian (1975).
139. R. A. Zingaro, F. C. Bennett, G. W. Hammar, J. Organ. Chem. 18, 292 (1953).
140. N. D. Betenekov, V. P. Medvedev, G. A. Kitaev, Soviet Radiochem. 20, 369 (1978).
141. I. K. Ostrovskaya, G. A. Kitaev, A. A. Velikanov, Russian J. Physical Chem. 50, 956 (1976).
142. G. A. Kitaev, T. P. Bol'shchikova, L. E. Yatkovla, Russian J. Inorg. Chem. 16, 1683 (1971).
143. D. Lincot, J. Vedel, in 10th E.C. Photovoltaïc Solar Energy Conf., H. Stephens & Associates, Bedford (1991), p. 931.
144. M. D. Ward, D. A. Buttry, Science 249, 1000 (1990) and references therein.
145. C. Gabrielli, M. Keddam, R. Torresi, J. Electrochem. Soc. 138, 2657 (1991).
146. S. Bruckenstein, M. Shay, J. Electroanal. Chem. 188, 131 (1985).
147. H. Cachet, R. Cortès, M. Froment, G. Maurin, N. Shramchenko, J. Electrochem. Soc., 144, 3583 (1997)
148. M. D. Ward, in: Physical Electrochemistry. Principles, methods and applications., I. Rubinstein and M. Dekker (ed.), New-York (1995), Ch. 7, p. 293.
149. G. Sauerbrey, Z. Phys. 155, 206 (1959).
150. S. J. Martin, G. C. Frye, A. J. Ricco, S. D. Senturia, Anal. Chem. 65, 2910 (1993).
151. S. Bruckenstein, K. C. Kanige, A. R. Hillman, J. Appl. Electrochem. 26, 171 (1996).
152. H. E. Hager, R. D. Ruediselli, M. E. Buehler, N.A.C.E. 42, 345 (1986).
153. H. Cachet, M. Froment, F. Zenia, J. Electrochem. Soc. 143, 442 (1996).
154. G. Folcher, H. Cachet, M. Froment, J. Bruneaux, Thin Solid Films 301, 242 (1997).
155. M. Y. Abyaneh, M. Fleischmann, J. Electrochem. Soc. 138, 2491 (1991).
156. M. Fleischmann, H. R. Thirsk, in Advances in Electrochemistry and Electrochemical Engineering, P. Delahay (ed.), Interscience, New York (1963), Vol. 3, p. 123.
157. C. Gabrielli, in: Physical Electrochemistry. Principles, methods and applications, I. Rubinstein and M. Dekker (ed.), New-York (1995), Ch. 6, p. 243.
158. M. Avrami, J. Chem. Phys. 7, 1103 (1939); ibid 8, 210 (1940).
159. A. Kitaev, S. G. Mokrushin, A. A. Uritskaya, Colloïd. J. USSR 27, 38 (1965).
160. L. Reimer, in: Scanning Electron Microscopy, Springer Series in Optical Sciences, Physics of Image Formation and Microanalysis, Springer Verlag, (1985), Vol 45.

161. D. E. Newbury, D. C. Joy, P. Echlin, C. E. Fiori, J. I. Goldstein, in: Advanced Scanning Electron Microscopy and X-Ray Analysis, Plenum Press, New York (1986).
162. E. Budevski, G. Staikov, W. J. Lorentz, in: Electrochemical Phase Formation and Growth, VCH, Weinheim, New York (1996).
163. P. Allongue, in: Scanning Tunneling Microscopy of Semiconductor Electrodes. Advances in Electrochemical Science and Engineering, H. Gerisher and C. Tobias (ed.), VCH, Weinheim, New York (1995), vol 4, p. 62.
164. E. Budevski, G. Staikov, W. J. Lorentz, in: Electrochemical Phase Formation and Growth, VCH, Weinheim, New York (1996), p. 75–76.
165. U. Demir, C. Shannon, Langmuir 10, 2794 (1994).
166. P. Facci, V. Erokhin, A. Tronin, Cl. Nicolini, J. Phys. Chem 98, 13323 (1994).
167. H. P. Klug, L. E. Alexander, in: X-Ray diffraction procedures for polycrystalline and amorphous materials, John Wiley and Sons, New York (1974).
168. N. Itoh, Appl. Phys. Lett. 62, 690 (1993).
169. M. A. G. Halliwell, Appl. Phys. A 58, 135 (1994).
170. M. Froment, M. C. Bernard, R. Cortès, B. Mokili, D. Lincot, J. Electrochem. Soc. 42, 2642 (1995).
171. M. H. Loretto, in: Electron Beam Analysis of Materials, Chapman and Hall, London (1994).
172. S. Andrieu, P. Fréchard, Surf. Sci. 360, 289 (1996).
173. J. I. Harris, B. A. Joyce, P. J. Dobson, Surf. Sci. 103, L90 (1981).
174. A. Hirsh, A. Howie, R. Nicolson, D. W. Plashley, M. J. Whelan, in: Electron Microscopy of Thin crystals, Butterworths, London (1985).
175. M. Froment, D. Lincot, Electrochim. Acta 40, 1293 (1995).
176. P. J. Goodhew, in: Thin Foil Preparation for Electron Microscopy, Practical methods in Electron Microscopy, A. M. Glauert ed, Elsevier, Amsterdam (1985), Vol. 11.
177. P. B. Buseck, J. Cowley, L. Eyring, in: High Resolution Transmission Electron Microscopy and Associated Techniques, Oxford University Press, (1988).
178. J. Echigoya, P. Pirouz, J. W. Edington, Phil. Mag. A 45, 455 (1982).
179. R. Hillebrand, H. Hofmeister, K. Scheerschmidt, J. Heidenreich, Ultramicroscopy 49, 252 (1993).
180. S. J. Pennycook, D. E. Jesson, A. J. McGibbon, P. D. Nellist, J. Electron. Microsc. 45, 36 (1996).
181. B. Jouffrey, C. Colliex, in: Electron Microscopy 1994, Interdisciplinary Developments and Tools, Les Ulis (1994), Les éditions de Physique, Vol. 1.
182. S. N. Sahu, J. Mater. Sci. Mater. Electron. 6, 41 (1995).
183. H. H. Hu, P. K. Nair, J. Crystal Growth 152, 150 (1995).
184. O. Zelaya-Angel, A. E. Esparza-Garcia, C. Falcony R. Lozada-Morales, R. Ramirez-Bon, Solid State Comm. 94, 81 (1995).
185. R. J. Bandaranayake, G. W. Wen, J. Y. Lin, H. X. Jiang, C. M. Sorensen, Appl. Phys. Lett. 67, 831 (1995).
186. D. Manno, R. Cingolani, L. Vasarelli, in: Electron Microscopy 1994, Editions de Physique, Les Ulis (1994), Vol. l, p. 977.
187. D. Bhattacharya, M. J. Carter, Thin Solid Films 288, 176 (1996).
188. D. Lincot, R. Ortega-Borges, M. Froment, Phil. Mag. B 68, 185 (1993).
189. D. Lincot, B. Mokili, M. Froment, R. Cortès, M. C. Bernard, C. Witz, J. Lafait, J. Phys. Chem. B 101, 2174 (1997).
190. T. Vossmeyer, L. Katsikas, M. Giersig, I. Popovic, K. Diesner, A. Chemseddine, A. Eychmüller, H. Weller, J. Phys. Chem. 98, 7665 (1994).
191. A. L. Rogah, L. Katsikas, A. Kornowski, Dangsheng Su, A. Eychmüller, H. Weller, Ber. Bunsenges. Phys. Chem. 100, 1772 (1996).

192. M. E. Özsan, D. R. Johnson, S. Oktik, M. H. Patterson, in 12th E.C. Solar Energy Conf., H. Stephens & Associates, Bedford (1994), p. 1604.
193. M. E. Özsan, D. R. Johnson, M. Sadeghi, D. Sivapathasundaram, G. Goodlet, M. J. Furlong, L. M. Peter, A. A Shingleton, J. Mater. Sci. Mater. Electron. 7, 119 (1996).
194. C.-Y. Yeh, Z. W. Lu, S. Froyen, A. Zunger, Phys. Rev. B 46, 10036 (1992).
195. W. F. Knippenberg, Philips Res. Rep. 18, 161 (1963).
196. G. E. Engel, J. Phys: Condens. Matter 2, 6905 (1990).
197. M. T. S. Nair, P. K. Nair, H. M. K. K. Pathirana, R. A. Zingaro, E. A. Meyers, J. Electrochem. Soc. 140, 2987 (1993).
198. O. de Melo, L. Hernandez, O. Zelaya-Angel, R. Lozada-Morales, M. Becerril, E. Vasco, Appl. Phys. Lett. 65, 1278 (1994).
199. P. J. Sebastian, Appl. Phys. Lett. 62, 2956 (1993).
200. O. Zelaya-Angel, L. Hernandez, O. de Melo, R. Lozada-Morales, C. Falcony, H. Vargas, R. Ramirez-Bon, Vacuum 46, 1083 (1995).
201. S. A. Tomas, O. Vigil, J. J. Alvarado-Morales, O. Zelaya-Angel, H. Vargas, A. Ferreira da Silva, J. Appl. Phys. 78, 2204 (1995).
202. R. Lozada-Morales, O. Zelaya-Angel, Thin Solid Films 281, 386 (1996).
203. M. Froment, H. Cachet, H. Essaaidi, G. Maurin, Pure Appl. Chem. 69, 77 (1997).
204. M. A. Herman, Thin Solid Films 267, 1 (1995).
205. D. R. T. Zhan, C. Maierhofer, A. Winter, M. Reckzügel, R Srama, A. Thomas, K. Horn, W. Richter, J. Vac. Sci. Technol. B9, 2206 (1991).
206. K. Sinha, J Menendez, D. W. Niles, H. Höchst, J. Vac. Sci. Technol. B9, 2202 (1991).
207. T. Tadokoro, S. I. Ohta, T. Ishiguro. Y. Ichinose, S. Kobayashi, N. Yamamoto, J. Crystal Growth 130, 29 (1993).
208. W. M. Yim, E. J. Stofko, J. Electrochem. Soc. 119, 381 (1972).
209. M. P. Halsall, J. J Davies, J. E. Nicholls, B. Cockayne, P. J. Wright, J. Crystal Growth 91, 135 (1988).
210. T. F. Kuech, N. R. Perkins, J. Crystal Growth 166, 558 (1996).
211. J. W. McCamy, D. H. Lowndes, J. D. Budai, R. A. Zuhr, Xiao Zhang, J. Appl. Phys. 73, 7818 (1993).
212. W. P. Shen, H. S. Kwok, Appl. Phys. Lett. 65, 2162 (1994).
213. B. W. Gregory, J. L. Stickney, J Electroanal. Chem. 300, 543 (1991).
214. B. M. Huang, L. P. Colletti, B. W. Gregory, J. L. Anderson, J. L. Stickney, J. Electrochem. Soc. 142, 3007 (1995).
215. Y. F. Nicolau, M. Dupuy, M. Brunel, J. Electrochem. Soc. 137, 2915 (1990).
216. S. Lindroos, T. Kanniainen, M. Leskela, Appl. Surf. Sci. 75, 70 (1994).
217. G. Hodes, in: Physical Electrochemistry, I. Rubinstein (ed.), Dekker New York (1995), p. 515.
218. Y. Golan, L. Margulis, G. Hodes, I. Rubinstein, J. L. Hutchison, Surf. Sci. 311, L633 (1994).
219. Y. Golan, G. Hodes, I. Rubinstein, J. Phys. Chem. 100, 2220 (1996).
220. D. Lincot, A. Kampmann, B. Mokili, J. Vedel, R. Cortès, M. Froment, Appl. Phys. Lett. 67, 2355 (1995).
221. H. Cachet, R. Cortès, M. Froment, G. Maurin, J. Solid State Electrochem., 1, 100 (1997).
222. K. Rajeshwar, Adv. Mater. 4, 23 (1992).
223. H. Rahnamai, H. J. G. Ray, J. Nzemec, Thin Solid Films 69, 347 (1980).
224. D. Lincot, R. Ortega-Borges, M. Froment, Appl. Phys. Lett. 64, 569 (1994).
225. M. Froment, M. C. Bernard, R. Cortès, B. Mokili, D. Lincot, J. Electrochem. Soc. 142, 2642 (1995).
226. L. Royer, Bull. Soc. Fr. Mineral. Crist. 51, 7 (1928).

227. J. W. Matthews, in: Epitaxial growth Materials Science Series, Academic Press, New York (1980).
228. D. Lincot, B. Mokili, R. Cortès, M. Froment, Microsc. Microanal. Microstruct. 7, 217 (1996).
229. M. J. Stowell in: Epitaxial Growth, Materials Science Series, J. M. Matthews (ed.), Academic Press, New York (1975), p. 437–492.
230. M. A. Arnold, Z. L. Wang, W. Tong, B. K. Wagner, S. Schön, C. J. Summer, Phil. Mag. A 75, 1209 (1997).
231. A. Ponchet, A. Le Corre, A. Godefroy, S. Salaün, A. Poudoulec, J. Crystal Growth 153, 71 (1995).
232. D. E. Jesson, K. M. Chen, S. J. Pennycook, MRS Bulletin 31 (1996).
233. R. Cortès, M. Froment, B. Mokili, D. Lincot, Phil. Mag. Lett. 73, 209 (1996).
234. J. L. Boone, S. A. Howard, D. D. Martin, G. Cantwell, Thin Solid Films 176, 143 (1989).
235. M. Grün, M. Hetterich, U. Giessen, C. Klingshirn, J. Crystal Growth 141, 68 (1994).
236. T. Tadokoro, S. I. Ohta, T. Ishiguro, Y. Ichinose, J. Crystal Growth 130, 29 (1993).
237. E. Budevski, G. Staikov, W. J. Lorenz, in: Electrochemical Phase Formation and Growth, VCH, Weinheim, New York (1996), p. 1–7.
238. A. K. Turner, J. M. Woodcock, M. E. Özsan, J. G. Summers, in 10th E.C. Photovoltaic Solar Energy Conf., H. Stephens & Associates, Bedford (1991), p. 791.
239. M. E. Özsan, D. R. Johnson, D. W. Lane, K. D. Rogers, in 11th E.C. Photovoltaic Solar Energy Conf., (1994), p. 1600.
240. D. Lincot, B. Mokili, Final Report of Eurocis Joule II project (1994).
241. J. R. Tuttle, A. D. Ward, A. Duda, T. A. Berens, M. A. Contreras, K. R. Ramanathan, A. L. Tennant, J. Keane, E. D. Cole, K. Emery and R. Noufi, MRS Proc., 426, 143 (1996).
242. L. Stolt, K. Granath, E. Niemi, M. Bodegard, J. Hedström, S. Bocking, M. Carter, M. Burgelman, B. Dimmler, R. Menner, M. Powalla, U. Rühle, H. W. Schock, in 13th E.C. Photovoltaic Solar Energy Conf., H. Stephens & Associates, Bedford (1995), p. 1451.
243. J. Kessler, K. O. Velthaus, M. Ruckh, R. Laichinger, H. W. Schock, D. Lincot, R. Ortega, J. Vedel, 6th Int. Phot. Sci. Eng. Conf., (1992), p. 1005.
244. K. Ramanathan, R. N. Bhattacharya, J. Granata, J. Webb, D. Niles, M. A. Contreras, H. Wiesner, F. S. Hasoon, R. Noufi, 26th IEEE PVSC, (1997), to be published.
245. L. Kronik, L. Burstein, M. Leibovitch, Y. Shapira, D. Gal, E. Moons, J. Beier, G. Hodes, D. Cahen, D. Hariskos, R. Klenk, H. W. Schock, Appl. Phys. Lett. 67, 1405 (1995).
246. M. J. Furlong, D. Lincot, M. Froment, R. Cortès, A. N. Tiwari, M. Krejci, H. Zogg, in 14th E.C. Photovoltaic Solar Energy Conf., H. Stephens & Associates, Bedford (1997), p. 1291.
247. S. Saito, Y. Hashimoto, K. Ito, in: Proc. 1st World Conf. Photov. En. Conv., (1994), p. 1867.
248. O. Savadogo, K. C. Mandal, J. Electrochem. Soc. 141, 2871 (1994).
249. H. M. Dauplaise, K. Vaccaro, A. Davis, G. O. Ramseyer, J. P. Lorenzo, J. Appl. Phys. 80, 2873 (1996).
250. K. Vaccaro, A. Davis, H. M. Dauplaise, S. M. Spaziani, E. A. Martin and J. P. Lorenzo, J. Electron. Mater. 25, 603 (1996).
251. M. Gros-Jean, R. Herino, D. Lincot, J. Electrochem. Soc. 145, 2448 (1998).
252. R. Garuthara, G. Levine, J. Appl. Phys. 80, 401 (1996).
253. P. K. Nair, J. Campos, M. T. S. Nair, Semicond. Sci. Technol. 3, 134 (1988).
254. V. M. Garcia, M. T. S. Nair, P. K. Nair, R. A. Zingaro, Semicond. Sci. Technol. 10, 427 (1995).
255. M. T. S. Nair, P. K. Nair, J. Phys. D: Appl. Phys. 24, 450 (1991).

256. P. K. Nair, M. T. S. Nair, Semicond. Sci. Technol. 4, 807 (1989).
257. F. F. Lange, Science 273, 903 (1996).
258. J. Rouxel, in: Advances in the Synthesis and Reactivity of Solids, JAI Press Inc. (ed.), Vol. 2, (1994), pp. 27–91.
259. T. J. Trentler, K. M. Hickman, S. C. Goel, A. M. Viano, P. Gibbons and W. E. Buhro, Science 270, 1791 (1995).

Plasma Engineering

Demetre J. Economou

Plasma Processing Laboratory, Department of Chemical Engineering,
University of Houston, Houston, TX 77204-4792

Contents

List of Symbols

A	area, m^2
$\tilde{A}_\theta$	complex vector potential
B	magnetic induction, T
$\overline{c_i}$	mean thermal speed of species i, m/s
C	capacitance, F
C_{pg}	gas heat capacity at constant pressure, J/kg-K
C_{pSi}	silicon heat capacity at constant pressure, J/kg-K
D_α	ambipolar diffusion coefficient, m^2/s
D_i	diffusion coefficient of species i, m^2/s
Da	Damkohler number
D_+	diffusion coefficient of positive ions, m^2/s
e	elementary charge, $1.602\ 10^{-19}$ C
E	electric field, V/m
E_A	activation energy, J/mole or eV
E_{eff}	effective electric field, V/m
E_0	amplitude of electric field, V/m
ER	etch rate, units vary
f	electron velocity distribution function
$f_0(\varepsilon, t)$	electron energy distribution function, $eV^{-1.5}$
F	force, N
h_B	heat transfer coefficient for wafer backside cooling, J/m^2-K-s
h_T	wafer topside heat transfer coefficient, J/m^2-K-s
H	energy exchange in electron-neutral collision, eV
I_i	current density due to species i, A/m^2
J_i	flux of species i, $1/m^2$-s
$\mathfrak{I}_i$	total (kinetic plus internal) energy of species i, eV
k	Boltzmann constant, eV or J K^{-1}
k_a	attachment rate coefficient, $m^3\ s^{-1}$
k_d	dissociation rate coefficient, $m^3\ s^{-1}$
k_i	rate coefficient for process i, units vary
k_{it}	total ionization rate coefficient, $m^3\ s^{-1}$
k_{sr}	surface reaction rate coefficient, m s^{-1}
k_{vr}	rate coefficient for volume recombination, $m^6\ s^{-1}$
K_i	thermal conductivity of species i, J/m-s-K
Kn	Knudsen number
L	length scale (m), or inductance, H
m_i	mass of species i, Kg
n_i	number density of species i, m^{-3}
N	total gas number density, m^{-3}

N_D	dopant atom density, m^{-3}
N_{Si}	atom density of silicon, m^{-3}
P	pressure, Pa
Pe	Péclet number
q_e	electron energy flux, eV/m^2-s
r	radial coordinate, m
r_e	electrode radius, m
r_w	wafer radius, m
R	resistance (Ω), or universal gas constant, 8.307 J/mole-K
Re	Reynolds number
R_i	reaction rate, units vary
s	sheath thickness, m
t	time, s
T_c	coolant temperature, K
T_e	electron temperature, K or eV
T_g	gas temperature, K
T_i	temperature of species i, K or eV
T_s	substrate temperature, K
u	directed (average) velocity, m/s
u_w	gas velocity at the showerhead electrode, m/s
U_i	internal energy of species i, eV
v	particle velocity, m/s
V	potential, V
$V_{a,b}$	time-average potential of sheath a, b, V
V_f	floating potential, V
V_p	plasma potential, V
x, y, z	Cartesian coordinates, m
Y_s	sputtering yield
z_i	charge number of species i

Greek

α_{acc}	thermal accommodation coefficient
γ	ratio of heat capacities, C_p/C_v
Δ	wafer thickness, m
ΔH_{etc}	heat of etching, J/mole
ΔH_{rec}	heat of recombination, J/mole
ε	energy or electron energy, eV
ε_0	permittivity of free space, 8.8542×10^{-12} F/m
ε_{Si}	emissivity of silicon
κ_g	thermal conductivity of gas, J/m-s-K
κ_{Si}	thermal conductivity of silicon, J/m-s-K
λ	mean free path, m
λ_D	Debye length, m
Λ	characteristic diffusion length, m
Λ_{Cl_2}	free-molecular heat conductivity of chlorine, J/m^2-s-K-Pa
μ	mobility, m^2/V-s
ν	collision frequency (1/s), or kinematic viscosity, m^2/s
ρ	mass density kg/m^3
σ_i	collision cross section for process, i, m^2
σ_{SB}	Boltzmann constant
τ	dimensionless time, or characteristic time, s
Φ	Thiele Modulus
ω	angular frequency, rad s^{-1}

Subscripts

α	ambipolar
a	sheath a
b	sheath b
B	Bohm
c	chemical
e	electrons
f	floating
g	gas
i	species i or reaction i
l	liquid
m	momentum transfer
p	plasma
r	recombination
s	sputtering, surface, substrate
sh	sheath
sr	surface recombination
th	threshold
tot	total
u	energy transfer
vr	volume recombination
$+$	positive ions

Abbreviations

AC	Alternating Current
AFM	Atomic Force Microscope
AMS	Appearance Mass Spectrometry
ARDE	Aspect Ratio Dependent Etching
CCP	Capacitively Coupled Plasma
CVD	Chemical Vapor Deposition
DC	Direct Current
DFT	Density Functional Theory
DSMC	Direct Simulation Monte Carlo
EEDF	Electron Energy Distribution Function
EVDF	Electron Velocity Distribution Function
GEC	Gaseous Electronics Conference
HDP	High Density Plasma
IAD	Ion Angular Distribution
IC	Integrated Circuit
ICP	Inductively Coupled Plasma
IED	Ion Energy Distribution
IR	Infrared
LD-LID	Laser Desorption-Laser Induced Fluorescence
LIF	Laser Induced Fluorescence
LTE	Local Thermodynamic Equilibrium
MC	Monte Carlo
MD	Molecular Dynamics
MERIE	Magnetically Enhanced Reactive Ion Etching
MPRES	Modular Plasma Reactor Simulator
OES	Optical Emission Spectroscopy
PECVD	Plasma Enhanced Chemical Vapor Deposition
PIC-MCC	Particle in Cell- Monte Carlo Collisions
RAD	Radical Angular Distribution

RED	Radical Energy Distribution
RF	Radio Frequency
RIE	Reactive Ion Etching
SOG	Spin-on Glass
STM	Scanning Tunneling Microscopy
TCAD	Technology Computer-Aided Design
TEOS	Tetraethoxysilane
XPS	X-Ray Photoelectron Spectroscopy
ULSI	Ultra Large Scale Integration
UV	Ultraviolet Spectroscopy

1 Introduction

Etching and deposition of thin solid films are central unit operations in the fabrication of integrated circuits (IC) [1], printed circuit boards, magnetic recording heads, and a variety of micro-electro-mechanical devices. A schematic of a cross section of an integrated circuit is shown in Fig. 1 [2]. It consists of patterned layers of metals (e.g., Al/Si/Cu, W, Ti/TiN), semiconductors (e.g., silicon substrate), and insulators (e.g., SOG, TEOS oxide, and silicon nitride passivation layer). A layer may be patterned either by subtractive or additive methods with the aid of a photoresist (a polymer) mask. In the subtractive method (Fig. 2, left) the portion of the layer not protected by the mask is removed by a process known as *etching*. In the additive method (Fig. 2, right), the film is first deposited through a mask by, for example, electrodeposition or physical vapor deposition. The mask is then lifted off leaving behind the desired pattern.

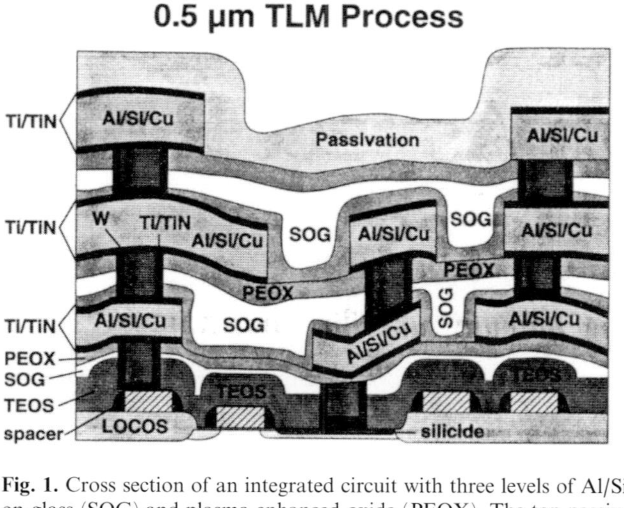

Fig. 1. Cross section of an integrated circuit with three levels of Al/Si/Cu metallization separated by spin-on-glass (SOG) and plasma enhanced oxide (PEOX). The top passivation layer is typically silicon nitride. The silicon substrate is not shown. After [2].

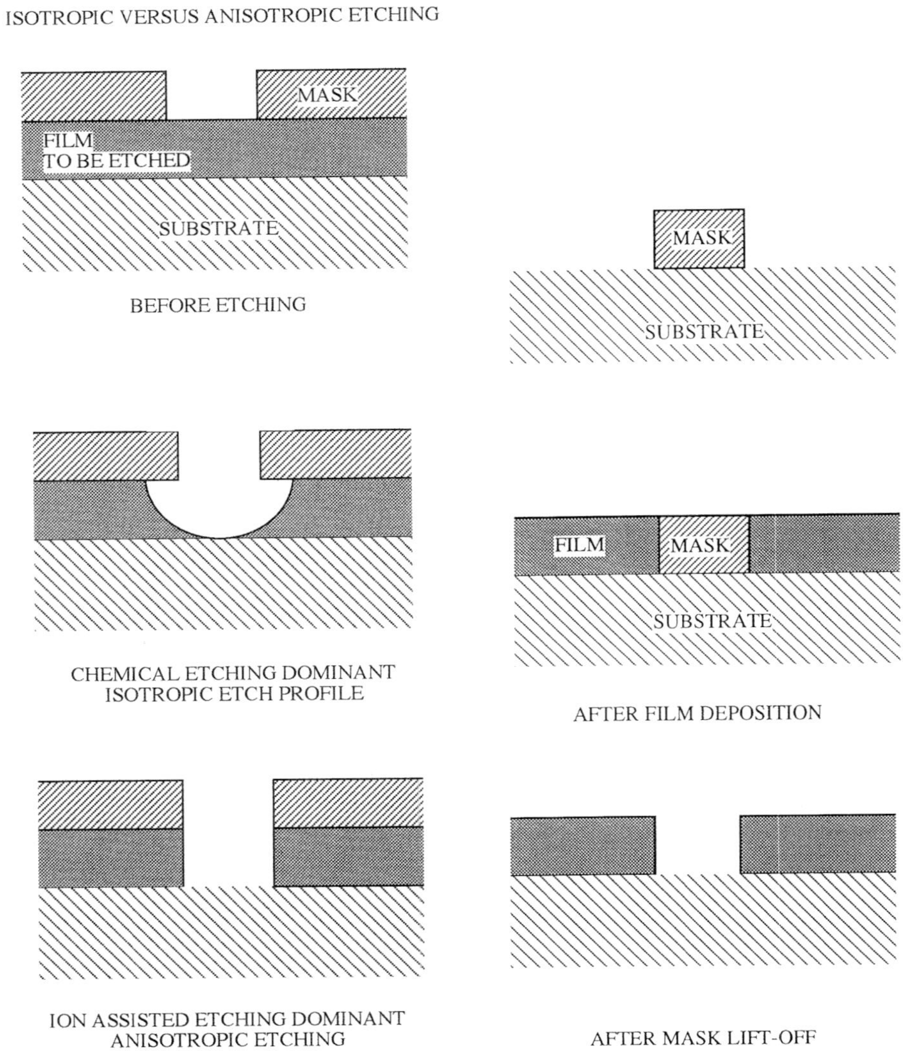

Fig. 2. (left) The subtractive method of patterning a thin film showing isotropic vs. anisotropic etching. (right) The additive method of patterning a thin film.

There are two distinct etching methods: wet and dry. Wet processes make use of a liquid etchant. For example silicon dioxide (SiO_2) can be etched in buffered hydrofluoric acid (HF) to produce (H_2SiF_6) and water. Wet etching offers higher reaction rate and selectivity, but the reaction proceeds at comparable rate in all directions (except for the so-called crystallographic etching of single crystalline materials). This results in a characteristic mask undercut (isotropic etch profile, Fig. 2, left), which limits the resolution of the method to a few microns. Thus, wet etching is advantageous for thicker films of not very fine dimensions. Wet etching is reviewed by Ghandhi [3]

Table 1. Comparison of wet and plasma etching.

	Plasma Etching	Wet Etching
Maturity	Industry standard	Industry standard
Cost	High	Low
Process Control	Fairly easy	Not so easy
Etch Rate	Moderate	High
Selectivity	Moderate	High
Resolution Limit	$<0.25\,\mu$	$>2\,\mu$
Sidewall Profile Control	Moderate	Difficult
Waste Disposal Problem	Low	High

and Vossen and Kern [4]. Dry etching makes use of gaseous plasmas [5–8]. Plasma etching[1] or reactive ion etching (RIE) can provide very high resolution (better that 0.25 μ) by a combination of neutral radicals and ions bombarding the substrate. Wet and plasma etching methods are compared in Table 1. Plasma etching is the method of choice for patterning micron and submicron features in advanced microelectronic device fabrication. Also, the subtractive method (Fig. 2, left) outweighs the additive method (Fig. 2, right) for creating high resolution (submicron) patterns with controlled shapes. In modern semiconductor manufacturing, as many as 30% of all major process steps involve use of a plasma.

Figure 3 is a schematic of a plasma etch process. The case of polysilicon etching in a chlorine plasma is shown as an example. The plasma is generated by applying power between a pair of parallel plates in a low pressure (<1 torr) chamber (Fig. 3a). The Cl_2 feedstock gas is attacked by plasma electrons to produce Cl radicals and Cl_2^+ ions. Radicals diffuse or are convected by gas flow towards the wafer where they adsorb on the surface. Ions accelerate in the sheath naturally occurring over the wafer, and bombard the wafer vertical to its surface (Fig. 3b). The combination of radical and ion bombardment produces $SiCl_4$ product which desorbs and is removed by the gas flow. It is this directional ion bombardment which promotes anisotropic etching of microscopic features (Fig. 3c), whereby the film etches much faster in the vertical as compared to the horizontal direction. At the atomic level, ion bombardment produces a modified surface layer in which the reactant (Cl) is mixed within the silicon lattice (Fig. 3d) to a depth depending on the ion energy ($\sim$10s Å). The energy deposited by ions promotes the formation of products that are either sputtered away or desorb spontaneously in the gas phase. Figure 3 also demonstrates the disparity in length scales encountered in plasma processing. The wafer will soon become 30 cm in diameter, so the reactor scale is of the order of 10s of cm. The sheath thickness ranges from 0.1–10 mm depending on the Debye length and the voltage applied to the electrode. The feature size is rapidly moving in the sub-quarter-micron regime. Finally the lattice has to be described on the Å length scale.

[1] In colloquial terminology, plasma etching usually implies conditions that result in rather weak ion bombardment of the wafer. Reactive ion etching (also referred to as reactive sputter etching) implies intense ion bombardment (see Section 4.1). In this Chapter, the term plasma etching will be used throughout.

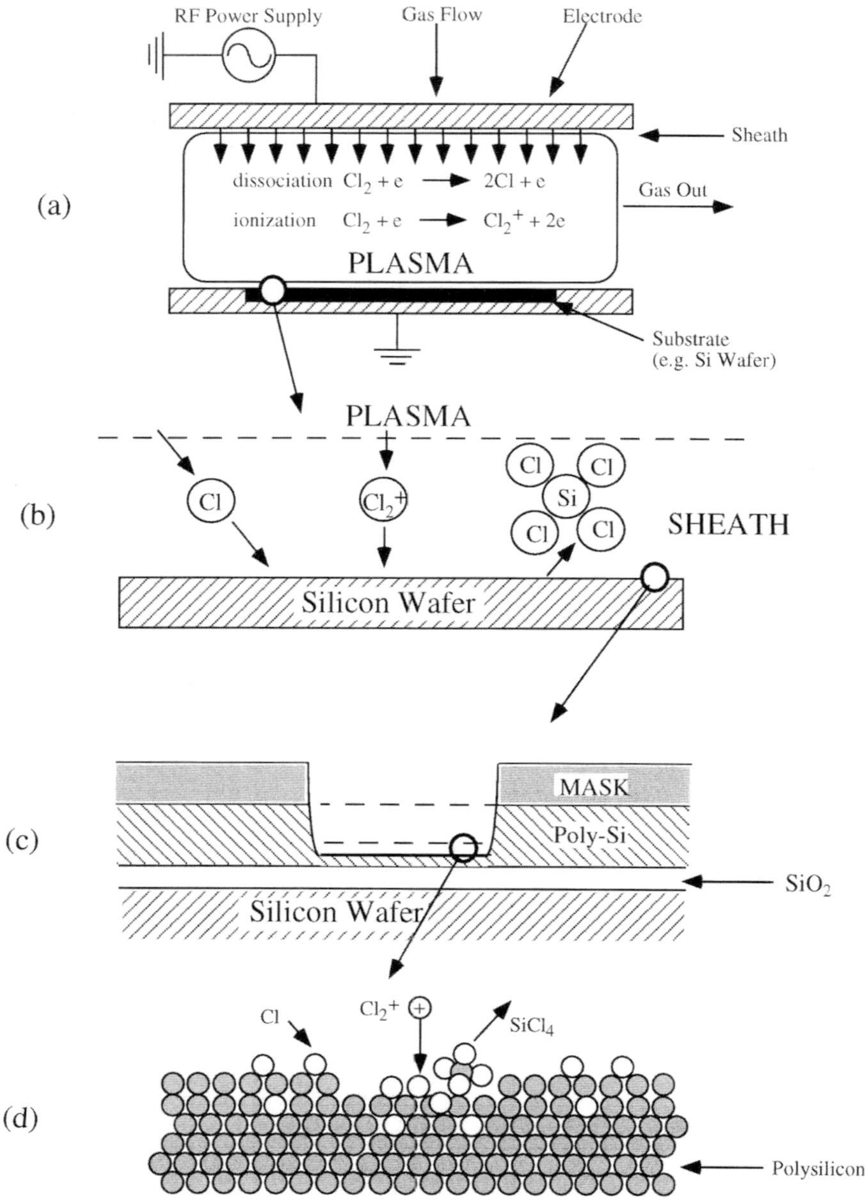

Fig. 3. Panorama of plasma etching using silicon etching with chlorine as an example. This figure also shows the disparate length scales involved from the reactor, to the sheath, to the microfeature, to the atomic scale. Cl radicals and Cl_2^+ ions are generated in the plasma by electron impact of gas molecules (a). Ions accelerate in the sheath and bombard the wafer along the vertical direction (b), thereby inducing anisotropic etching of microscopic features to yield $SiCl_4$, a volatile product (c). Ion bombardment creates a modified layer at the surface where Cl is mixed within the Si lattice (d).

Table 2. Representative gases used for plasma etching and the materials the etch.

Source Gas	Additive Gas	Etchant Specie	Material	Mechanism	Selective Over
CF_4	O_2	F	Si	Chemical	SiO_2/III-Vs
SF_6	O_2	F	Si	Chemical	SiO_2/III-Vs
SF_6	He	F	Si	Ion-enhanced (Low Temp.)	Resist/III-Vs
NF_3	None	F	Si	Chemical	Resist/III-Vs
C_2F_6	O_2	F	Si	Chemical	SiO_2
CF_4	H_2	CF_x	SiO_2/Si_3N_4	Ion-induced	Si
NF_3	Cl_2	F	Si_3N_4	Chemical	Resist
C_2F_6	CHF_3	CF_x	SiO_2	Ion-induced	Si
Cl_2	Ar	Cl	undoped-Si	Ion-induced	SiO_2
Cl_2	C_2F_6	Cl	n-type Si	SP	SiO_2/Resist
Cl_2	BCl_3	Cl_2/Cl	Al	SP	SiO_2/Resist
Cl_2	O_2	Cl/O	Mo, $MoSi_2$, Cr	Ion-induced	SiO_2
Cl_2	None	Cl	III-Vs	Chemical	SiO_2/Resist
CH_4	H_2	CH_x, H	III-Vs	Ion-enhanced	Resist
O_2	H_2O	O, OH	Resist	Chemical	Si, SiO_2

SP = sidewall passivation.

Neutral species do not have any directionality. They can etch equally well in all directions undercutting the mask (e.g., etching of silicon by fluorine atoms). When ion bombardment is necessary for etching to occur (e.g., etching of undoped silicon in chlorine plasma), *anisotropic* (vertical) wall profiles can result (Fig. 2, left). The mask pattern can then be faithfully reproduced into the film. There is a host of gases [9–12] that have been used for plasma etching materials encountered in microelectronics. Table 2 provides only a small sample.

The goals of any plasma etch process are to achieve high etch rate, uniformity, selectivity, controlled shape of the microscopic features etched into the film, and no radiation damage. High *etch rate* is desirable to increase the process throughput (wafers/hr). However, etch rate must be balanced against uniformity, selectivity, and anisotropy. *Uniformity* refers to achieving the same etch characteristics (rate, wall profile, etc.) across the wafer which is projected to be 30 cm in diameter by the year 2000. Uniformity is necessary so that the underlying layer (silicon dioxide in Fig. 3c), for example, is not exposed to potentially harmful plasma radiation in some areas of the wafer while other areas are yet to clear. Also, plasma uniformity is needed to avoid non-uniform charging of the wafer which can lead to electrical damage. *Selectivity* refers to the relative rate of etching of one material with respect to another. Etch processes must be selective with respect to the mask and the underlying film. The mask must not be etched, otherwise the desired pattern will be distorted. Selectivity with respect to the underlying layer is particularly important when that layer is thin or when the process uniformity is not good. For example, a process may call for etching polysilicon over a thin (<50 Å) gate oxide (Fig. 3c). The selectivity of this process must be extremely high ($>50:1$), otherwise a substantial thickness of the oxide may be lost.

The shape of the microscopic features etched into the wafer is of paramount im-

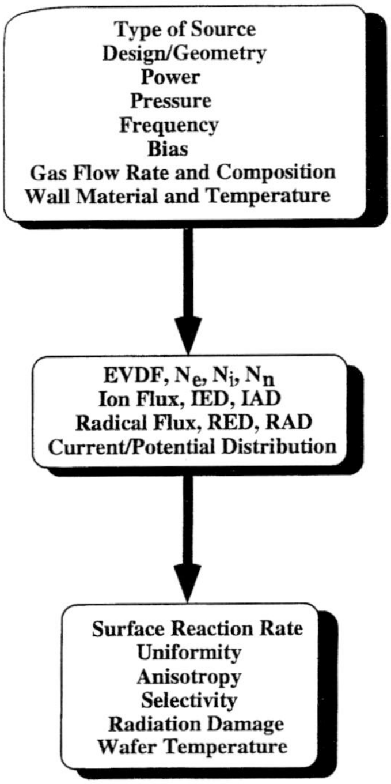

Fig. 4. Representation of the parameter space in plasma etching. The key internal plasma properties (middle) are the bridge between externally controlled variables (top) and the figures of merit (bottom).

portance. Most often anisotropic (vertical) sidewall profiles are required, perhaps with some roundness at the bottom of the feature. *Radiation damage* refers to structural damage of the crystal lattice or more importantly to electrical damage of sensitive devices caused by plasma radiation (ions, electrons, UV and soft X-ray photons) [13]. For example, spatially non-uniform current flowing from the plasma to the wafer can lead to charging and breakdown of thin oxides; or charging of insulating materials within a microfeature can lead to pattern distortion (*notching*, see Section 5.6.1) or high energy ion bombardment can lead to structural damage of the top atomic layers of the etched film.

Manipulation of the plasma chemistry, coupled with the appropriate reactor design, are crucial for meeting the goals of plasma etching. There are many externally controlled variables (process inputs) which can influence the plasma characteristics, and in turn the process output (Fig. 4). Given a reactor type, its dimensions and materials of construction, one can manipulate operating parameters (pressure, power, frequency, etc.) to influence the process output (rate, uniformity, anisotropy, etc.). Plasma process development has been based largely on trial-and-error procedures guided by experience and intuition. A more rational selection of plasma etch chemistry and also of the appropriate plasma reactor design can be based on understanding the fundamentals of the chemical and physical processes taking place in the plasma and on the wafer surface. Such understanding can be obtained by bridging the gap between the process inputs and outputs through key plasma properties as shown in Fig. 4.

In what follows, the fundamentals of plasma engineering are discussed with emphasis on plasma etching. Discussion pertains to the kind of plasmas used in electronic materials processing. Similarities and differences with electrochemical reactor engineering are pointed out along the way, and are summarized in Section 8.

Plasmas are also used for the low temperature deposition of thin solid films, for example amorphous hydrogenated silicon, diamond, and a host of other materials. Since the fundamentals of plasma physics and chemistry are the same for both plasma etching and plasma assisted chemical vapor deposition (PECVD), the latter will only be discussed briefly in Section 6.6. A review of PECVD can be found in [14]. Sputtering is discussed by Chapman [15], and plasma polymerization is covered by Yasuda [16].

2 Plasma Physics

Plasmas of interest are partially ionized gases that contain electrons and ions (positive and negative ions), in a sea of neutrals. This ionized gas must satisfy several conditions in order to be classified as a plasma [17–19], namely (a) the Debye length λ_D (see Eq. 2, below) must be smaller than a characteristic length scale (e.g., plate separation) of the reactor, and (b) there must be a large number of charged particles in a Debye sphere, i.e., $4/3\pi n_e\lambda_D^3 \gg 1$. There is a wide range of plasmas depending on the charge density and temperature [19]. Plasmas used in electronic materials processing are low pressure *glow discharges* with operational and physical characteristics summarized in Table 3. These plasmas are strongly non-equilibrium systems (electrons have much higher temperature than ions and neutrals), and the degree of ionization is small (weakly ionized gases); there are 100 to 100,000 neutrals for every electron in the plasma [6, 15].

Although a plethora of reactor configurations and methods for plasma generation exist (see Section 4), the parallel plate capacitively-coupled reactor (also called *diode*) shown in Fig. 3a is a typical example. A semiconductor wafer rests on one electrode

Table 3. Typical parameter values for RF diode and high density plasma (HDP) reactors [6].

Parameter	RF Diode	HDP	Units
Pressure	10–1000	0.5–50	mtorr
Power	50–2000	100–5000	Watts
Frequency	0.05–13.56	1–2450	MHz
Volume	1–10	1–50	Liters
Magnetic Field	0–100	0–1000	Gauss
Plasma Density	10^9–10^{10}	10^{10}–10^{13}	cm^{-3}
Electron Temperature	1–5	2–12	eV
Heavy Particle Temperature	<0.1	<0.25	eV
Ion Bombardment Energy	100–1000	20–300	eV
Fractional Ionization	10^{-6}–10^{-3}	10^{-4}–10^{-1}	–

and is showered with gas entering through the other electrode. The wafer temperature is too low for reaction to take place in the absence of a plasma. A voltage is now applied between the electrodes. The induced electric field causes gas ionization and a plasma is generated. The electric field imparts a force on charged species (electrons and ions) but not on neutrals. Because the gas is only weakly ionized, charged species collide mainly with neutrals rather than with one another. Depending on the frequency of the applied voltage, plasmas range from DC, to radio frequency (RF), to microwave frequency. DC plasmas are not as common since very often the electrode (and/or wafer) is covered by an insulating material which can't pass DC current. RF plasmas are excited at 13.56 MHz, although other frequencies have also been used. Microwave plasmas are normally excited at a frequency of 2.45 GHz.

The vast disparity in mass between the light electrons and heavy ions and neutrals is responsible for many of the unique characteristics of plasmas: (a) in contrast to ions, electrons can follow the electric field variations with time and thus are able to absorb energy from the field much more efficiently compared to ions [18], and (b) electrons loose very little energy in elastic collisions with neutrals and ions [15]. For example, an electron colliding elastically with an argon atom looses up to $5.5\ 10^{-5}$ times the original electron energy, a very small fraction, indeed. On the contrary, ions loose a substantial amount of energy in collisions with neutrals because their mass is comparable to that of neutrals.

Since electrons pick up most of the energy from the field, and they do not loose this energy efficiently in collisions with neutrals, electrons are "heated" to much higher temperatures compared to ions and neutrals (see Table 3). It is this non-equilibrium that makes plasmas suitable for electronic materials processing. Specifically, the energetic electrons can break bonds (bond energies are a few eV) and initiate radical chemistry in a gas at near room temperature. One would need much higher temperatures to perform the same chemistry by thermal excitation (e.g., thermal CVD [20]) in the absence of a plasma. Thus, it is often said that glow discharge (cold) plasmas are capable of performing high temperature chemistry at low temperatures. As device dimensions continue to shrink, low temperature processing is becoming increasingly important in microelectronics to minimize dopant redistribution and defect generation [21].

2.1 Sheath Formation

Another consequence of the much smaller mass of electrons compared to ions is the formation of a so-called *sheath* over any material surface in contact with the plasma. When a surface is first exposed to the plasma, the surface receives a much higher electron flux compared to the positive ion flux. For a Maxwellian velocity distribution of the particles, the flux due to particle i striking the surface is given by [15]

$$J_i = \frac{n_i \overline{c_i}}{4} = \frac{n_i}{4}\sqrt{\frac{8kT_i}{\pi m_i}} \tag{1}$$

where n_i, T_i, m_i, and $\overline{c_i}$ are number density, temperature, mass, and mean thermal

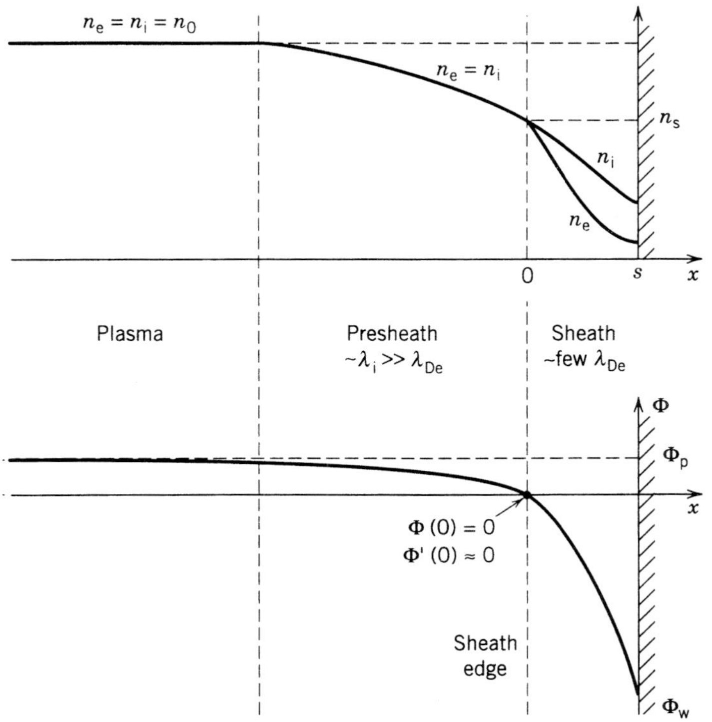

Fig. 5. Establishment of a DC sheath over a wall immersed in an infinite plasma. The sheath is separated from the bulk plasma by a presheath which is of the order of an ion mean free path (λ_i) long. Top shows the electron (n_e) and ion (n_i) density profiles. Bottom shows the potential (Φ) profile. After [6].

speed of particle *i*, and *k* is the Boltzmann constant. Since $T_e \gg T_+$, and electroneutrality holds in the bulk plasma $n_e \approx n_+$ (assuming no negative ions), it follows that $J_e \gg J_+$. Hence the surface charges negatively with respect to the plasma attaining a potential lower than the *plasma potential.* An electric field is thus established that attracts ions and repels electrons so that, at steady state, the net current flowing to a *floating* surface is zero. The surface has then charged to the *floating potential.* The floating potential is obtained by equating the electron and positive ion currents to the wall. The result is [6] $V_f = -kT_e \, ln(m_i/2\pi m_e)^{1/2}$. For argon, the value of the logarithm is 4.7. Since $kT_e \sim 2$–5 eV, the floating potential is ~10–25 V (negative with respect to the plasma potential).

For simplicity, consider a plane wall immersed in an otherwise infinite plasma. Also, assume a time-independent plasma so that a DC sheath develops. The transition from the "bulk" plasma to the wall is shown schematically in Fig. 5, which shows the charge density (top) and potential profiles (bottom) [6]. The bulk plasma is electrically quasi-neutral with almost equal densities of positive and negative charges. The presheath is also quasi-neutral but the density of the charged species decreases from the bulk value. For an electropositive plasma (no negative ions), the electron (and ion) density at the sheath/presheath interface (sheath edge) is 61% of the bulk value [6, 15]. The presheath is of the order of one ion mean free path long and sustains a potential drop of the order of kT_e. It is the presheath that imparts enough energy to the positive ions to reach the so-called *Bohm velocity* $u_B = (kT_e/m_+)^{1/2}$ before entering the sheath [6, 22]. The Bohm velocity is necessary for a stable positive charge sheath to develop. Charge neutrality is violated within the sheath with the positive ion density exceeding the electron density. Because of the existence of net charge, the sheath is a region of relatively high electric field. The sheath thickness over a floating substrate is of the order of the *Debye length* λ_D where

$$\lambda_D = \sqrt{\frac{kT_e\varepsilon_0}{n_e e^2}} \tag{2}$$

The Debye length is the distance over which macroscopic fields can penetrate the plasma. The sheath can be considerably thicker when a negative voltage is applied to the substrate (see Section 5.5). However, for given voltage, the sheath thickness decreases as the Debye length decreases, i.e., as electron temperature decreases or electron density increases. The sheath has an analog in the double layer found next to the electrode in electrochemical systems. The difference is that the electrochemical double layer is much thinner (10–100 Å compared to 0.1–10 mm) because of the much higher charge density and much lower charged species temperature in electrochemical systems.

Formation of the sheath is of critical importance to plasma processing. The strong sheath electric field (normally ~100 times stronger than that in the bulk plasma) is pointing such that positive ions are accelerated towards the wall (Fig. 3b). Since the sheath thickness is much smaller than the lateral dimensions of the electrode, the electric field in the sheath is perpendicular to the macroscopic wafer surface. Hence ions acquire directional energy in the sheath and bombard the wafer along the surface normal. That way, horizontal surfaces which are exposed to ion bombardment etch much faster than vertical surfaces, resulting in anisotropic (vertical) etching (Fig. 2, left). Low temperature processing and anisotropy are the most important attributes of plasma etching from the technological point of view.

2.2 Potential Distribution

Figure 6 shows a typical spatial profile of the potential distribution between the parallel plate electrodes of a DC discharge [10, 15]. In DC discharges of interest, the spacing between anode (right electrode) and cathode (left electrode) is much reduced compared to the classical DC discharge [23], eliminating the *positive column*. The bulk plasma (away from the electrodes) is electrically neutral and sustains only a small electric field. However, as explained above, large potential gradients (hence large fields) develop in the sheath regions near the electrodes. The sheath is much thicker over the cathode which is biased with a large negative voltage (of the order of 1,000 V). The cathode sheath is essentially devoid of electrons. Positive ions accelerate in the cathode sheath and bombard the electrode with high energy ejecting

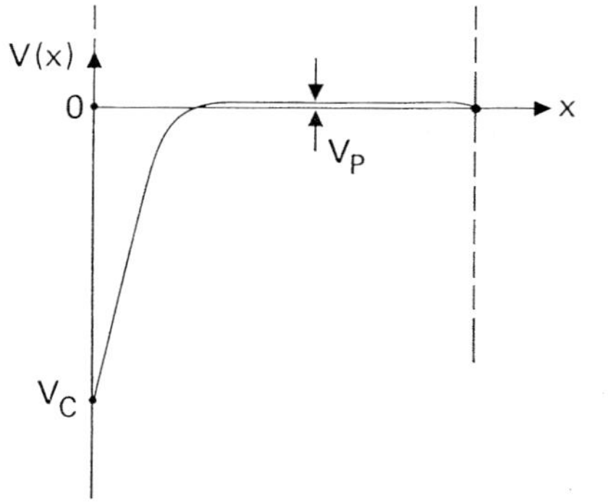

Fig. 6. Potential distribution in a DC discharge sustained between two closely-spaced parallel plates. After [10].

secondary electrons. The sheath potential is such that electrons released from the cathode accelerate through the sheath back into the plasma, where they ionize neutral atoms and molecules. Secondary electron emission by ion bombardment is an important mechanism by which a DC discharge is sustained [15, 23]. In microelectronics, DC discharges are most often used for sputtering of metallic (conductive) films [5, 15].

In the so-called capacitively-coupled plasma (CCP) reactors, AC voltage is applied to the electrode through a capacitor (*blocking capacitor*) which does not allow any *net particle* (DC) current to flow through the electrode. An idealized one-dimensional parallel plate capacitively coupled system is shown in Fig. 7 [6]. The sheaths over the two electrodes are shown as sheath a and sheath b, with time-dependent thickness $s_a(t)$ and $s_b(t)$, respectively. The electron and ion density profiles in the gap are also shown. For excitation frequencies above the ion plasma frequency, $\omega_{p+} = (e^2 n_+/\varepsilon_0 m_+)^{1/2}$, ions do not respond to the instantaneous electric field and the ion density profile is almost time-invariant. Electrons respond, however, and the electron cloud oscillates back and forth between the plates. The sheath potential is such as to repel electrons; hence the electron density drops sharply within the sheath over a distance of a few Debye lengths. In RF CCP discharges, electrons are "heated" primarily by two mechanisms: (a) *collisionless heating* because of the oscillating electric field near the plasma/sheath interface. This can be thought to result because of "collisions" of electrons with an oscillating potential "wall" (the RF sheath edge) [6, 18, 24, 25]. (b) *ohmic heating* because of the electric field accelerating electrons which then collide with neutrals, absorbing power from the field [6, 18]. Collisionless heating predominates at low pressures when electron-neutral collisions are scarse. Secondary electron emission is not necessary to sustain a RF discharges, although it is operative under certain conditions.

Typical spatiotemporal profiles of potential are shown in Fig. 8 [26]. The left electrode is driven by a sinusoidal radio frequency voltage $V_{RF} = V_0 \sin(\omega t)$. The case shown is for an argon discharge with $V_0 = 100$ V and $\omega/2\pi = 13.56$ MHz. The potential distribution is such that the electrodes are bombarded by positive ions during the whole period of the RF cycle, while most electrons are trapped in the plasma. Only electrons with kinetic energy greater than the sheath potential can reach the walls. Electrons "leak" to the walls during a short time in the cycle when

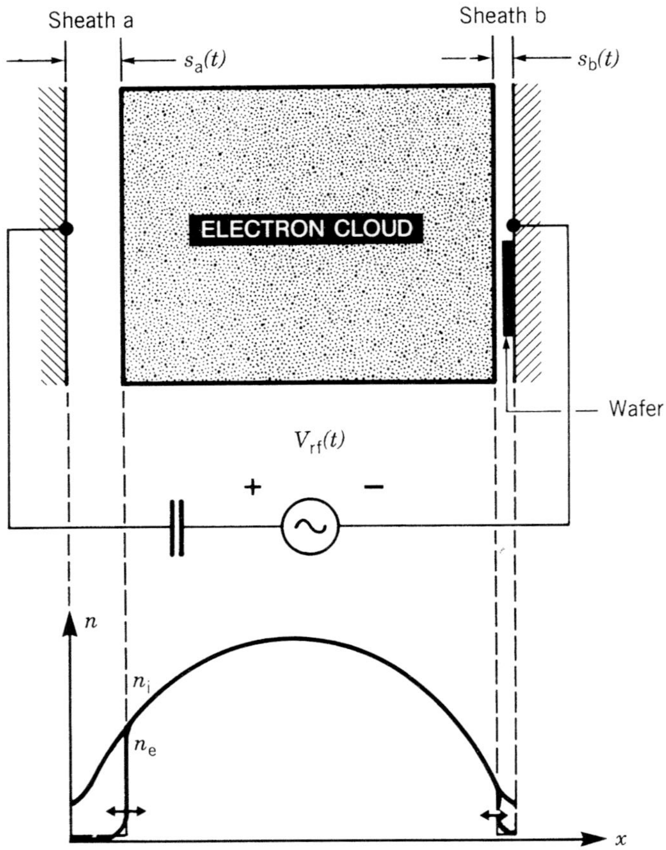

Fig. 7. A RF capacitively coupled plasma between two parallel plate electrodes (diode). The electron cloud oscillates back and forth and is separated from the walls by the sheath. The ion density distribution (n_i) is almost time-invariant if the applied field frequency is greater than the ion plasma frequency. The electron density (n_e) modulates severely in the sheath (bottom). After [6].

the sheath voltage is low; in Fig. 8 this will occur at $\sim\tau = 0.25$ on the left electrode and at $\tau = 0.75$ on the right electrode (the two electrodes are 180 degrees out of phase in this symmetric system). The electron flux is such that when integrated over the RF cycle, the electron current is equal to the positive ion current, since no DC current is allowed to pass through the capacitively-coupled electrode. Each electrode alternates as the cathode and the anode during the RF cycle.

The sheath potential (voltage) is the difference between the plasma potential and the electrode (or wall) potential. The potential of the sheath over the grounded electrode (right electrode in Fig. 8) is equal to the plasma potential. The potential of the sheath over the powered electrode (left electrode in Fig. 8) is the difference between the plasma potential and the electrode potential. The time-average potential of the powered electrode (with respect to ground) is called the *DC self-bias* or *DC bias*. Actually, the DC-bias is the difference of the time-average potential of the sheath

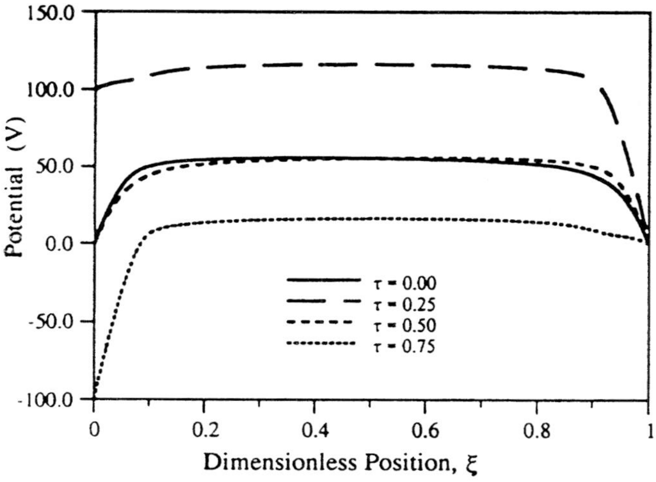

Fig. 8. Spatiotemporal profiles of the potential distribution in a 13.56 MHz argon discharge sustained between two parallel plates (diode). The left electrode is driven by a sinusoidal voltage of 100 V amplitude. The right electrode is grounded. Note that the plasma potential (potential in the bulk away from walls) is more positive than either electrode potential. After [26].

over the grounded electrode from the time-average potential of the sheath over the powered electrode. Since the energy of ions bombarding a wall is directly affected by the potential of the sheath over that wall, the importance of the potential distribution in the reactor is apparent.

Fig. 8 is for a symmetric reactor, i.e., one in which the two electrodes have the same area. Then the time-average potential distribution is symmetric as well, and there is no DC bias developed. In asymmetric reactors the time-average potential distribution looks much like the one in Fig. 6. In fact, assuming capacitive voltage division between the two sheaths, the time-average sheath voltage scales with the inverse electrode area ratio as [15]

$$\frac{V_a}{V_b} = \left(\frac{A_b}{A_a}\right)^n \tag{3}$$

The exponent has the theoretical value of $n = 4$ [27], but a value of $n \approx 1$–2 has been determined experimentally [28]. In any case, Eq. (3) shows that the smaller electrode develops a larger sheath voltage, hence it receives stronger ion bombardment.

The time variation of the potential of the powered electrode (dashed lines) and the plasma potential (solid lines) under different conditions in a high frequency ($\omega \gg \omega_{p+}$) parallel plate discharge are shown in Fig. 9 [29]. Several features are apparent: (a) the plasma potential is always the most positive potential in the system, (b) there is no DC self-bias developing in a symmetric discharge (area of powered electrode equal to that of grounded electrode), or when power is coupled directly to the electrode without a blocking capacitor (DC coupled), (c) with a blocking capacitor (capacitively coupled) the DC bias is negative when the smaller electrode of an

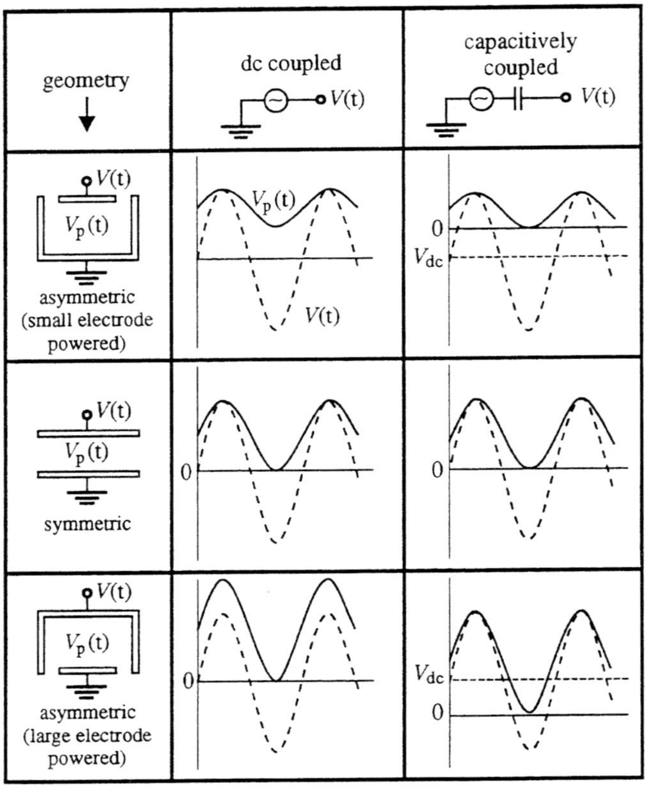

Fig. 9. Time variation of the electrode V(t) (dashed lines) and plasma $V_p(t)$ (solid lines) potentials in a capacitively-coupled (diode) RF discharge for different coupling configurations. After [28].

asymmetric discharge is powered, or positive when the larger electrode of an asymmetric discharge is powered.

At this point it is instructive to contrast the potential distribution in a plasma reactor with that of an electrochemical reactor. In the latter, the potential profile is monotonic with potential drops across the double layers over the cathode and the anode, and another potential drop in the bulk solution. The potential distribution in a plasma reactor is non-monotonic, with the plasma potential being the most positive potential in the system. In addition, the potential drop in the bulk of the plasma (of the order of several kT_e) is a small fraction of the potential drop across the sheath. In electrochemical systems, the potential drop across the bulk solution may be comparable to or even larger than that across the double layer.

2.3 Equivalent Circuit

Equivalent circuits of plasma systems [30, 31] have found use in calculating the electrical interaction of the plasma with the external circuit that powers the plasma. The latter is composed not only of the power supply but also of a *matching network* (this

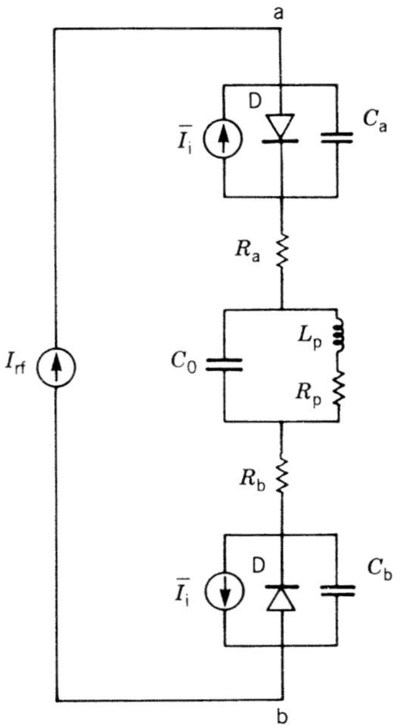

Fig. 10. Equivalent circuit of a RF capacitively-coupled diode. The current source $\bar{I}_i$, which represents ion current, is often replaced by a resistor. After [6].

is used to match the plasma impedance to the 50 Ω internal resistance of the power supply), and stray elements (e.g., stray capacitance to the grounded chamber) [32, 33]. An equivalent circuit of a parallel plate discharge is shown in Fig. 10 [6]. Here the middle block of elements represents the bulk plasma (space-averaged) and the edge blocks represent the sheaths adjacent the two electrodes. The bulk plasma has resistance R_p corresponding to particle collisions with neutrals (ohmic heating), inductance L_p corresponding to particle inertia, and capacitance C_o corresponding to the spacing between the sheath edges (see Fig. 7). The capacitance and inductance of the bulk plasma are normally small and are frequently neglected, leaving only the plasma resistance R_p. Plasma inductance can become important at frequencies approaching the (electron) plasma frequency, $\omega_{pe} = (e^2 n_e/\varepsilon_0 m_e)^{1/2}$ especially in strongly electronegative discharges, where the electron density is orders of magnitude smaller than the ion density. Each sheath is represented by a capacitance (C_a or C_b) that is inversely proportional to the sheath thickness (Fig. 7), a diode to represent the (one-way) electron current flow in the sheath, and a DC current source to represent the ion current. Often, the ion current source is replaced by a resistor. In either case, there is power dissipated by ions in the sheath due to ion collisions with neutrals and/or ions bombarding the walls and transferring their energy to these walls. Resistances R_a and R_b correspond to power dissipation due to collisionless electron heating by the oscillating sheath boundary [6, 18, 24, 25]. It must be emphasized that the elements of the equivalent circuit shown in Fig. 10 depend on discharge conditions. For example the plasma resistance is higher at higher pressures (more collisions) assuming constant electron density, and the sheath capacitance is higher at higher electron densities (thinner sheath).

Equivalent circuits are used to determine important plasma properties (such as electron density, sheath thickness, etc.) from measurements of the current-voltage characteristics of the device (*impedance analysis*, see Section 7.1.7) [32–35]. The equivalent circuit of the plasma reactor has many similarities to that of an electrochemical reactor. For example, in both systems the boundary layers near the wall (sheath or double layer) are represented by capacitors and the bulk is represented by a resistor. The similarity is closer for an AC electrochemical cell.

3 Plasma Chemistry

One of the major complexities in dealing with plasmas is that of plasma chemistry in the gas phase and especially on surfaces. The main reasons are: (a) the existence of a plethora of species neutral and charged, in ground and excited states, (b) the strongly non-equilibrium nature of the discharge, and (c) the lack of knowledge of reaction pathways and rate coefficients. It is worth noting that species at even minute concentrations (e.g., metastables with mole fraction 10^{-5}) can strongly influence the plasma [36]. As another example, extremely small additions of a molecular gas to a rare gas can dramatically change the characteristics of the discharge [37, 38].

Fig. 11 shows the range of kinetic energies and densities of particles typically present in plasma processing reactors [39]. The non-equilibrium nature of the discharge is evident from the fact that different species have different energies (and temperatures). Ions bombarding the cathode (box C) and the secondary electrons

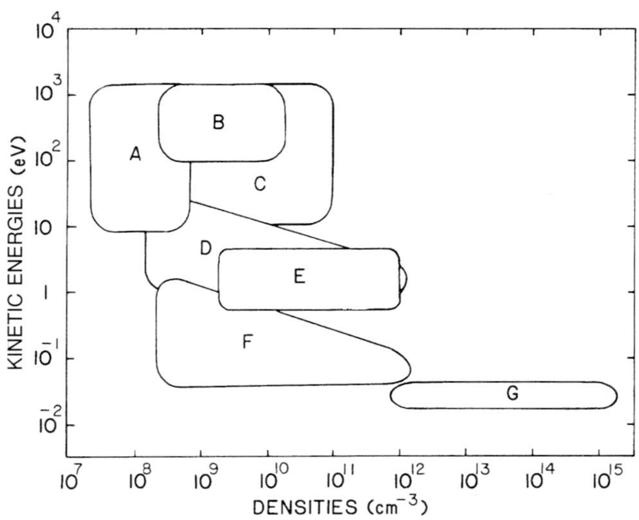

Fig. 11. Ranges of kinetic energies and densities of species typically present in glow discharge plasmas. A=secondary electrons accelerated through the sheath, B=ions backscattered from cathode (most likely neutralized), C=ions accelerated towards cathode, D=electrons in bulk plasma, E=hot ions and neutrals formed in dissociation reactions (Frank-Condon effect), F=ions in bulk plasma, and G=neutral atoms and molecules. After [39].

(box A) they emit (which stream back into the plasma) have the highest energies since they are accelerated through a high voltage sheath. Ions in the bulk plasma (box F) are only accelerated by the much weaker bulk fields and have much lower energies (and temperatures). Neutral species (feedstock gas and reaction products) are orders of magnitude more abundant than charged species and have much lower (thermal) energies (box G). There can be a significant concentration of neutrals (and ions) that are hot (Box E). These result from dissociation or dissociative ionization reactions (see Section 3.1.1) where the excess electron energy is deposited as kinetic energy of the reaction fragments. At low pressures, these fragments do not suffer enough collisions to thermalize [40, 41]. In plasmas of interest the velocity of electrons is much higher than that of positive ions which in turn is much higher than that of neutrals. In addition, electrons have a nearly isotropic velocity distribution, while positive ions have very anisotropic velocity distribution (directed along the electric field). The ordering of electron, ion, and neutral gas temperatures is $T_e \gg T_i \sim T_g$.

Fig. 12 shows a schematic of the steps typically involved in plasma etching [11]. Radicals generated in the plasma by electron impact dissociation of gas molecules (step 1) diffuse or are convected by gas flow (step 2) to the surface where they adsorb (step 3). The adsorbed species (adspecies) react with the surface to from products (step 4). The products then desorb (step 5) and diffuse back into the gas phase (step 6). The surface processes may be strongly influenced by energetic particle bombardment of the surface, including positive ions, electrons and photons. Of these, positive ion bombardment is though to be most important. Negative ions are excluded be-

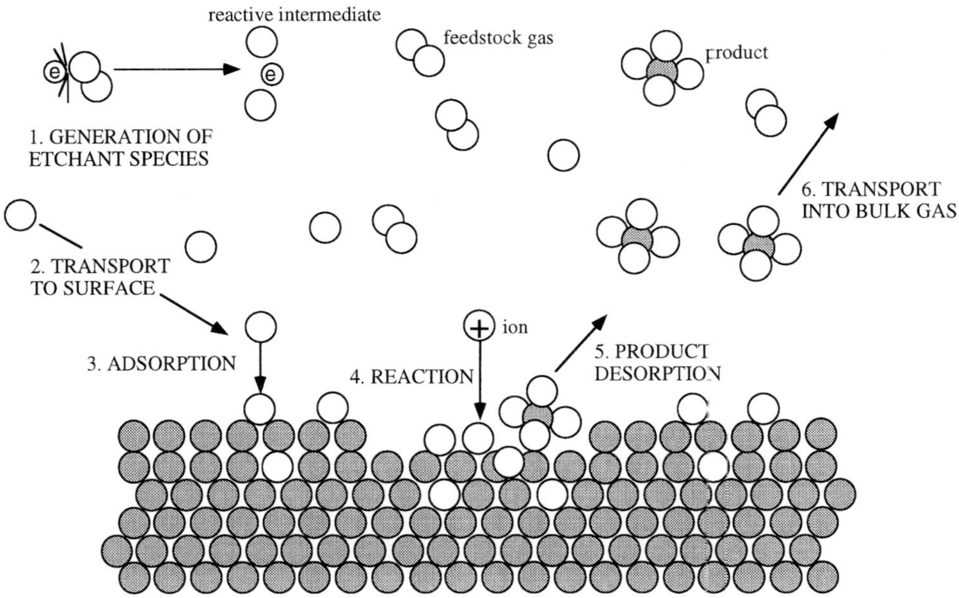

Fig. 12. "Molecular" representation of primary processes occurring in plasma etching. The basic steps are etchant generation (step 1), diffusion to the surface (2), adsorption (3), reaction on the surface to form product (4), product desorption (5), and product diffusion into the flowing gas (6). The surface processes can be strongly affected by energetic ions bombarding the surface. After [11].

Table 4. Representative reactions in argon and chlorine plasmas.

Excitation to Metastables	$Ar + e^- \rightarrow Ar^* + e^-$	R1
Metastable (Two-Step) Ionization	$Ar^* + e^- \rightarrow Ar^+ + 2e^-$	R2
Ground State Ionization	$Ar + e^- \rightarrow Ar^+ + 2e^-$	R3
Metastable-Metastable Ionization	$Ar^* + Ar^* \rightarrow Ar^+ + Ar + e^-$	R4
Metastable Quenching to Resonants	$Ar^* + e^- \rightarrow Ar^r + e^-$	R5
Metastable Quenching to Higher States	$Ar^* + e^- \rightarrow Ar^h + e^-$	R6
Metastable Superelastic Collisions	$Ar^* + e^- \rightarrow Ar + e^-$	R7
Radiative Decay of Resonants	$Ar^r \rightarrow Ar + h\nu$	R8
Excitation to Higher States	$Ar + e^- \rightarrow Ar^h + e^-$	R9
Radiative Decay of Higher States	$Ar^h \rightarrow Ar^* + h\nu$	R10
Momentum Transfer	$Ar + e^- \rightarrow Ar + e^-$	R11
Molecular Chlorine Excitation	$Cl_2 + e^- \rightarrow {Cl_2}^* + e^-$	R12
Atomic Chlorine Excitation	$Cl + e^- \rightarrow Cl^* + e^-$	R13
Molecular Chlorine Radiative Decay	${Cl_2}^* \rightarrow Cl_2 + h\nu$	R14
Atomic Chlorine Radiative Decay	$Cl^* \rightarrow Cl + h\nu$	R15
Molecular Chlorine Ionization	$Cl_2 + e^- \rightarrow {Cl_2}^+ + 2e^-$	R16
Atomic Chlorine Ionization	$Cl + e^- \rightarrow Cl^+ + 2e^-$	R17
Dissociative Attachment	$Cl_2 + e^- \rightarrow Cl^- + Cl$	R18
Dissociation of Molecular Chlorine	$Cl_2 + e^- \rightarrow 2Cl + e^-$	R19
Momentum Transfer to Molecular Chlorine	$Cl_2 + e^- \rightarrow Cl_2 + e^-$	R20
Momentum Transfer to Atomic Chlorine	$Cl + e^- \rightarrow Cl + e^-$	R21
Positive-Negative Ion Recombination	$Cl^- + {Cl_2}^+ \rightarrow 3\,Cl$	R22
Electron-Ion Recombination	${Cl_2}^+ + e^- \rightarrow 2\,Cl$	R23
Wall Recombination of Atomic Chlorine	$Cl + Cl + \text{wall} \rightarrow Cl_2 + \text{wall}$	R24
Volume Recombination of Atomic Chlorine	$2\,Cl + Cl_2 \rightarrow 2\,Cl_2$	R25

cause they are not energetic enough to overcome the sheath potential barrier. It must be emphasized that product *volatility* is a necessary condition for etching to occur. Otherwise, the products block the surface and the reaction stops altogether.

3.1 Homogeneous Chemistry

Table 4 provides a list of the most important types of reactions taking place in plasma etching [9]. Two rather extreme cases are shown. The argon discharge is a typical example of an inert gas plasma, in which negative ions do not play a role (electropositive discharge). Pure argon discharges are not used for etching, but they find widespread application in film deposition by sputtering (Section 6.5). Also, argon is found as a component of gas mixtures used for etching. The chlorine discharge is an example of a molecular plasma that readily forms negative ions (electronegative discharge).

3.1.1 Electron-Impact Reactions

Electrons are the heart of the plasma. Electrons ionize atoms and molecules (reactions R2, R3, R16, R17 in Table 4) to reproduce themselves thus sustaining the discharge by counterbalancing the electron losses. Electrons also dissociate molecules to form highly reactive atoms and radicals which are potential etchants (R19). Electron bombardment of neutrals and ions can excite bound electrons to higher energy levels (R13). As the excited species relax to a lower energy state they can emit photons (R8, R14, R15), giving the plasma a glow characteristic of the gas. Since there are very few electrons in the sheath region, the excitation rate is very low in the sheath. Hence the sheath appears dark (it is also called *dark space*) relative to the bulk plasma glow. In electronegative gases (such as Cl_2 or SF_6), electron attachment to molecules (R18) creates negative ions.

The rate of electron impact reactions is given by [6, 42, 43]

$$R_{ei} = n_e n_i \langle v\sigma(v) \rangle = n_e n_i \int v\sigma(v) f(\mathbf{v}, \mathbf{r}, t)\, d\mathbf{v} \tag{4}$$

where the average $\langle \sigma(v)v \rangle$ is taken over the electron velocity distribution function (EVDF) $f(\mathbf{v}, \mathbf{r}, t)$. Here n_e and n_i are the number density of the electrons and the collision partner, respectively, $\mathbf{v}$ and $\mathbf{r}$ are the electron velocity and position vectors, respectively, v is the magnitude of the velocity vector (speed), and $\sigma(v)$ is the collision cross section, which is a measure of the "effectiveness" of the particular interaction between an electron and a neutral.

Electron collisions can be categorized as elastic and inelastic [6, 15]. Elastic collisions do not alter the internal energy of the heavy particle and result in a small energy loss by the electrons. Inelastic collisions change the internal energy of the collision partner. Inelastic collisions (ionization, dissociation, excitation, etc.) usually have a threshold energy below which the collision can't happen (cross section is equal to zero). In a typical example of an inelastic process, the cross section rises rapidly with electron energy above the threshold energy, reaches a maximum and then declines at high energies. Exothermic collisions may have no threshold, meaning that electrons with very low energy can participate in the collision. A notable example is attachment of electrons to electrophilic gases to create negative ions (reaction R18 in Table 4). In fact, the attachment rate may be highest for thermal electrons, i.e., electrons having "temperature" close to that of the gas. Another example is electron-positive ion recombination (R23). The collision cross section increases as the electron energy decreases. A set of collision cross sections for selected electron impact reactions with molecular chlorine is shown in Fig. 13 [44]. Data on electron collision cross sections for a variety of relevant gases can be found in [45–47].

It is worth noting that the rate of reactions involving electrons depends on the relative velocity between the electron and the projectile. Since the electron velocity is much higher than the neutral velocity, the latter is neglected. Hence the electron-impact reaction rate coefficients are a function of the electron "temperature" rather than the gas temperature. An exception is electron attachment (R18). The corresponding reaction coefficient depends on the gas temperature because the latter affects the rovibrational populations of the molecules and the attachment coefficient can be sensitive to the rovibrational state [45].

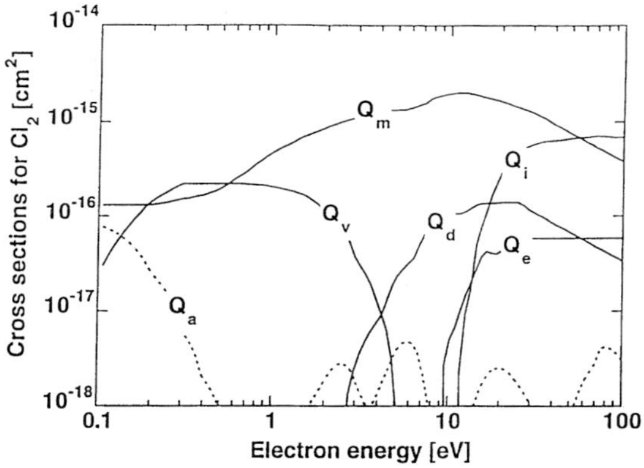

Fig. 13. Electron energy dependence of collision cross section for representative electron impact processes with molecular chlorine. Note threshold energy for endothermic processes (vibrational excitation, ionization, etc.). Q_a, Q_d, Q_e, Q_i, Q_m, and Q_v, correspond to electron attachment (reaction R18 in Table 4), dissociation (R19), electronic excitation (R12), momentum transfer (R20), ionization (R16) and vibrational excitation (not shown in Table 4), respectively. After [44].

3.1.2 Electron Velocity Distribution Function (EVDF)

As shown by Eq. (4), the rate of reactions involving electrons depends on the EVDF, $f(\mathbf{r}, \mathbf{v}, t)$. Determination of the distribution function is one of the central problems in understanding plasma chemistry. The EVDF is defined in the phase-space element $d\mathbf{v}d\mathbf{r}$ such that $f(\mathbf{r}, \mathbf{v}, t)\, d\mathbf{v}\, d\mathbf{r}$ is the number of electrons dn_e at time t located between $\mathbf{r}$ and $\mathbf{r} + d\mathbf{r}$ which have velocities between $\mathbf{v}$ and $\mathbf{v} + d\mathbf{v}$. When normalized by the total number of electrons n_e, it is a probability density function. The EVDF is obtained by solving the Boltzmann transport equation [42, 43, 48, 49]

$$\frac{\partial f(\mathbf{r}, \mathbf{v}, t)}{\partial t} + \mathbf{v} \cdot \frac{\partial}{\partial \mathbf{r}} f(\mathbf{r}, \mathbf{v}, t) + \frac{\mathbf{F}}{m_e} \cdot \frac{\partial}{\partial \mathbf{v}} f(\mathbf{r}, \mathbf{v}, t) = \left(\frac{\partial f(\mathbf{r}, \mathbf{v}, t)}{\partial t} \right)_{col} \tag{5}$$

The right hand side of Eq. (5) is the so-called collision integral which accounts for changes of the EVDF because of collisions electrons undergo mainly with neutrals but also with other electrons and ions. $\mathbf{F} = ez_i(\mathbf{E} + \mathbf{v} x \mathbf{B})$ is the Lorenz force acting on the electrons, where $\mathbf{E}$ and $\mathbf{B}$ are the electric field and magnetic induction, respectively. Eq. (5) is a partial integro-differential equation in seven dimensions (three in space, three in velocity and time), and as such is extremely difficult to solve. In the expressions above, $\mathbf{v}$ is the individual particle velocity not to be confused with $\mathbf{u}$ which is the average velocity of the ensemble of particles.

The EVDF of an equilibrium electron gas follows the Maxwell-Boltzmann distribution (also called Maxwellian, see Eq. 9 below). In equilibrium, there is no net transport of electrons in any direction, i.e., the distribution function is isotropic. Under the influence of an electric field, the EVDF becomes anisotropic. However, for weak enough fields, the degree of anisotropy is very small. The EVDF can then be written as a superposition of an isotropic part f_0 and (a small) anisotropic part, f_1. In this so-called *two-term spherical harmonics expansion* [42, 43, 48], the distribution function is written as $f(\mathbf{v}, t) \approx f_0(v, t) + f_1(v, t) \cos\theta$, where θ is the angle between **v** and **E**, and where the spatial (**r**) dependence of the distribution function has been neglected for simplicity. The two-term approximation is quite popular and has been found to be adequate for a wide range of gases when (a) the electric field to neutral density ratio E/N is relatively small, and (b) the elastic collision cross section is much higher than the inelastic collision cross sections. However, the two-term approximation may be poor when anisotropic electron transport is important. An example is the cathode sheath of DC discharges in which secondary electrons emitted from the wall accelerate in the sheath forming an electron beam.

Often the electron energy distribution function (EEDF), $f_0(\varepsilon, t)$, which is a function of the electron kinetic energy, is used instead of $f_0(v, t)$. The two are related by [48]

$$f_0(\varepsilon, t)\, d\varepsilon = \varepsilon^{-1/2} 4\pi v^2 f_0(v, t)\, dv \tag{6}$$

The EEDF is normalized such that

$$\int_0^\infty \sqrt{\varepsilon} f_0(\varepsilon, t)\, d\varepsilon = 1 \tag{7}$$

Using this normalization, a Maxwellian distribution (Eq. 9) would be a straight line or a semi-logarithmic plot of f_0 (ε, t) vs. ε. Once the EEDF has been determined, the reaction rate coefficient for electron impact reaction i can be calculated by

$$R_{ei} = k_i n_e n_i; \qquad k_i(t) = \sqrt{\frac{2}{m_e}} \int_0^\infty \sigma_i(\varepsilon) f_0(\varepsilon, t)\, \varepsilon\, d\varepsilon \tag{8}$$

which is a working version of Eq. (4), and where, again, the spatial dependence has been suppressed for simplicity.

Fig. 14, left, shows the EEDF in a static (DC) electric field in chlorine gas [50]. Under these conditions the EEDF is a function of the reduced electric field E/N. The distribution is not Maxwellian. This is because inelastic collisions deplete the tail of the distribution. As the field strength increases, the distribution extends to higher electron energies. Using the distribution functions of Fig. 14, top, and the cross section data of Fig. 13, one can calculate rate coefficients based on Eq. (8). Fig. 14, bottom, shows the calculated rate coefficients for electron attachment (reaction R18 in Table 4) and total ionization (sum of reactions R16 and R17) in a chlorine discharge, again in a static field [50]. The rate coefficients increase with E/N. The strong dependence of the ionization rate on E/N is evident. This is characteristic of high threshold processes. The threshold energy corresponds approximately to the activation energy of conventional thermal chemical reactions. The mean electron energy (or en equivalent electron temperature, Eq. 10) as a function of E/N can also be calculated (not shown), and used to express the rate coefficients as a function of $\langle\varepsilon\rangle$ or T_e. Such information is important in plasma reactor modeling (see Section 5.4.1).

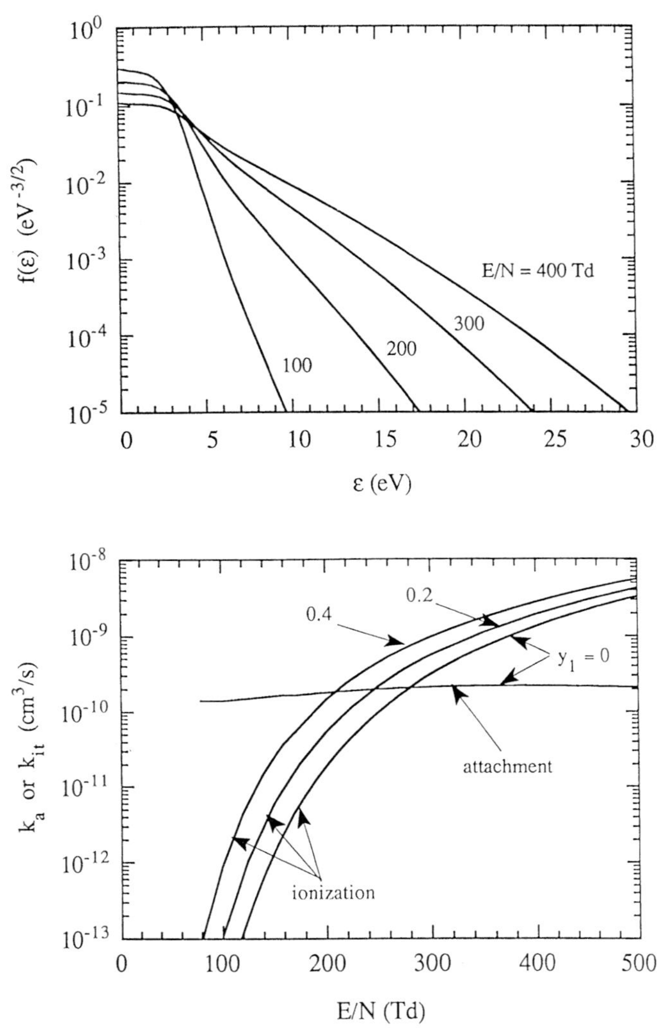

Fig. 14. (top) Electron energy distribution function (EEDF) in molecular chlorine as a function of the electric field to neutral density ratio E/N (1 Td = 10^{-17} V-cm^2). A Maxwellian EEDF would be a straight line on this plot. (bottom) Rate coefficients in chlorine as a function of the electric field to neutral density ratio E/N. k_{it} is total ionization including molecular and atomic chlorine; the atomic mole fraction is denoted by y_1. k_a is the coefficient for attachment to molecular chlorine. The points of intersection of the curves give the value of E/N prevailing in a relatively high pressure (100s of mtorr) self-sustained discharge with the corresponding atomic chlorine content. After [50].

The calculated EEDFs shown in Fig. 14 did not include the effect of electron-electron (e-e) collisions because the degree of ionization was assumed low enough for e-e collisions not to play a role. Under these conditions electron-molecule (atom) reactions predominate. Electron-electron collisions come into play when the degree of ionization exceeds about 10^{-5} [43, 48, 51, 52]. Such collisions tend to drive the distribution towards a Maxwellian. In low density plasmas (Table 3, RF diode column), electron-molecule collisions are more important since the degree of ionization is low ($n_e/N < 10^{-5}$). In high charge density (large n_e), low gas pressure (small N) plasmas (Table 3, HDP column), e-e collisions become more important. However, even when e-e collisions are substantially present, the tail of the distribution may still be depleted of high energy electrons due to inelastic processes.

The Maxwellian distribution is given by

$$f_0(\varepsilon) = \frac{2}{\sqrt{\pi}} \frac{exp(-\varepsilon/kT_e)}{(kT_e)^{3/2}} \tag{9}$$

with the mean (average) electron energy

$$\langle\varepsilon\rangle = \tfrac{3}{2}kT_e \tag{10}$$

It should be noted that an electron temperature is defined properly only when the EEDF is Maxwellian. For non-Maxwellian distributions, an equivalent electron temperature may be defined based on the mean electron energy, by using Eq. (10). Another popular form of the EEDF which can be expressed analytically is the Druyvesteyn distribution [42, 43, 48].

The time response of the EEDF in an AC field is determined by the relative magnitudes of the angular frequency ω of the applied electric field $E = E_0 \sin(\omega t)$ and the momentum (ν_m) and energy relaxation frequencies (ν_u) [53–55]. The collision frequency is given by $\nu_i = \sigma_i v N$. When $\omega \gg \nu_u$, the characteristic time for energy relaxation is much greater than the period of the applied field. As a consequence, the EEDF can't respond to the time variation of the field and does not change appreciably over a cycle. At the other extreme ($\omega \ll \nu_u$) the EEDF can follow the applied field faithfully. Under this condition, the EEDF at any time would be identical to the EEDF obtained in a static field of magnitude equal to the magnitude of the AC field at that particular moment in time. This is the so-called *quasi-steady-state* approximation. Figure 15 shows the time-evolution of the EEDF in chlorine for two excitation frequencies [53]. At 10 MHz (top), the EEDF is modulated strongly since the electron relaxation frequency (especially for the high energy electrons) is higher than the applied frequency. At 100 MHz (bottom), the EEDF is not modulated as much. If one were to increase the frequency to 2.45 GHz, modulation of the EEDF would cease altogether. It should be noted that ν_u is an increasing function of electron energy, for the energy range of interest. This is the reason, the tail of the EEDF is modulated more strongly than the low energy part in Fig. 15.

Many investigators have used to effective field approximation [42, 53] to describe the EEDF under high frequency excitation, when time modulation is absent. In this static approximation, the high frequency field is considered equivalent to an effective DC field of magnitude

$$E_{eff} = \frac{E_0/\sqrt{2}}{\sqrt{(\omega/\nu_m)^2+1}} \tag{11}$$

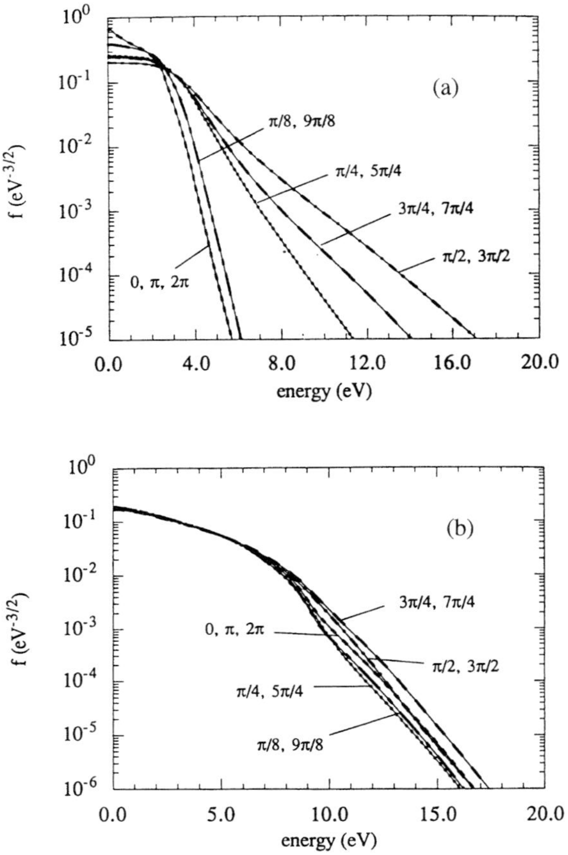

Fig. 15. Time dependence of the EEDF in chlorine under an applied AC field with 10 MHz (top) and 100 MHz (bottom) excitation frequency. The peak value of the field was $(E/N)_0 = 500$ Td. After [53].

As noted above, solving the Boltzmann equation is problematic because of the multidimensionality of the problem. A promising approach to calculating the electron distribution function in low pressure plasmas is the so-called *non-local* approach to electron kinetics. This was proposed by Bernstein and Holstein [56] and popularized by Tsendin [57], who initially suggested this approach for the positive column of a DC discharge. Since then, the non-local approach has been applied to a variety of low pressure gas discharge systems [58].

The non-local approach assumes that the distribution function is almost isotropic (i.e., the two-term approximation applies) and that the isotropic part f_0 is not modulated in time. The Boltzmann equation is then reduced to a time-invariant but space-dependent equation for f_0; f_1 is recovered as a derivative of f_0. For the majority of electrons in the discharge, which are trapped by the self-forming electrostatic field, a further simplification is applicable. At low enough pressures, the energy relaxation length for these electrons is large compared to the reactor dimensions. Since electron heating is a slow process compared to the time scale of electron spatial displacement, the total energy of the electrons (kinetic plus potential) is approximately constant in space. One can then space-average the Boltzmann equation and derive an equation for f_0 as a function of the *total* electron energy, that does not depend explicitly on the spatial coordinates [59]. This is a tremendous simplification of the Boltzmann equation in multidimensional plasmas. However, the spatial dependence has to be retained for electrons which have high enough energy to overcome the sheath potential and reach the walls (free electrons), or for electrons that have an energy relaxation length smaller than the reactor dimension. For monatomic gases (e.g., argon) this includes electrons with energy above the first excitation potential of the atom. In molecular gases (e.g., chlorine, sulfur hexafluoride, carbon tetrafluoride) the electron energy relaxation length of trapped electrons is much shorter that in atomic gases because of a plethora of relatively low threshold energy (few eV) collisions (e.g., vibrational excitations), see Fig. 13. Therefore, as the gas pressure increases, the non-local approach is expected to break down much sooner for molecular gases as compared to atomic gases. Also, for molecular gases, the traditional two-term approximation of the EEDF may not be applicable because the inelastic collision cross sections are comparable to or even exceed the elastic collision cross sections.

3.1.3 Heavy-Particle Reactions

Common types of chemical reactions among heavy particles are also shown in Table 4 (More can be found in [6]). An important reaction is radical recombination in the gas phase according to

$$\mathrm{A} + \mathrm{B} + \mathrm{M} \rightarrow \mathrm{AB} + \mathrm{M} \tag{12}$$

Reaction R24 in Table 4 is an example. The rate coefficient k_{vr} has an Arrhenius dependence on gas temperature,

$$k_{vr}(T_g) = A(T_g)\, exp\left(-\frac{E_A}{RT_g}\right) \tag{13}$$

Another important kind of heavy-particle reaction is that between ions and molecules or atoms. Typical examples are charge exchange and ion-molecule recombination

$$\mathrm{A}^+ + \mathrm{B} \rightarrow \mathrm{A} + \mathrm{B}^+ \tag{14}$$

$$\mathrm{A}^+ + \mathrm{B} \rightarrow \mathrm{AB}^+ \tag{15}$$

Charge exchange between a molecule and the daughter ion (symmetric charge exchange) is particularly efficient. Charge exchange is often responsible for transforming a fast ion and a slow neutral into a fast neutral and a slow ion in the sheath, thereby reducing the energy by which ions bombard the wafer surface. Ion-molecule (atom) recombination is responsible for generating molecular ions in atomic gas plasmas.

Finally, ion-ion neutralization reactions such as

$$A^+ + BC^- \rightarrow AB + C \tag{16}$$

are important as negative ion sinks in the gas phase, since negative ion loss to the walls is negligible. These reactions are often dissociative (see reaction R22 in Table 4). Data on heavy particle reactions of interest can be found in [60–62].

Thermal dissociation reactions are not important in plasma systems because of the low gas temperature. Dissociation of molecules is induced by energetic electron bombardment instead. In contrast, thermal dissociation is a major mechanism for radical generation in thermal CVD systems [20].

3.2 Surface Chemistry

Surface reactions are of ultimate importance since the goal of plasma processing is surface modification [7]. One can distinguish between several types of processes [12] which can occur on the surface of a solid exposed to a plasma (Fig. 16). The surface is bombarded by neutral radicals and molecules, positive ions, electrons, and photons.

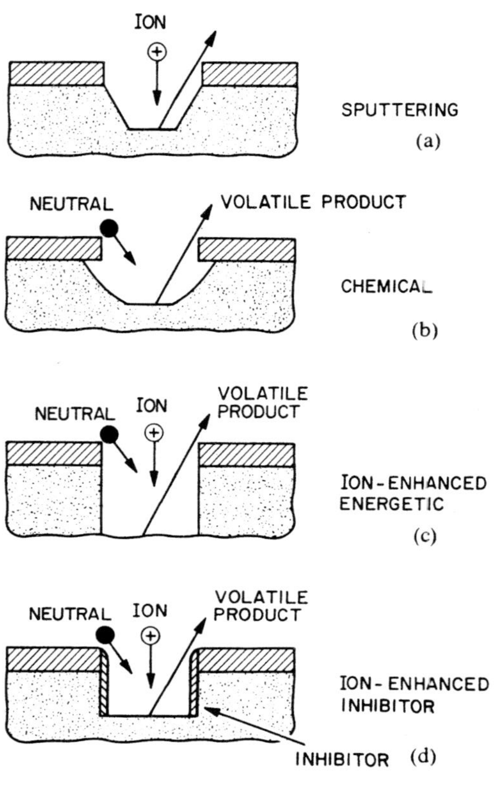

Fig. 16. The four basic mechanisms of etching by a plasma. (a) sputtering, (b) chemical etching by the action of neutrals only, (c) ion-induced etching; reaction occurs only on surfaces undergoing ion bombardment, and (d) ion-enhanced etching; neutrals could etch even in the absence of ion bombardment; the latter increase the etch rate. In order to avoid etching of the sidewalls by neutrals in case (d), the plasma chemistry is selected so as to "coat" the sidewalls by an extremely thin layer which inhibits sidewall etching. Ion bombardment clears the inhibitor film allowing etching to proceed anisotropically. After [12].

Of these, neutral specie and positive ion bombardment is generally considered most important. *Sputtering* (Fig. 16a) refers to the process of ejection of surface atoms induced by ions transferring momentum to the surface [15]. Sputtering can be thought of as atomic scale sandblasting. This is a physical rather than a chemical mechanism, and as such has very low selectivity. For ion energies of interest, the etch rate due to sputtering is given by [63, 64]

$$ER_s = Y_s \frac{J_+}{\rho_s} = \alpha(\varepsilon_+^x - \varepsilon_{th}^x)\frac{J_+}{\rho_s} \tag{17}$$

where $x = 0.5$, Y_s is the sputtering yield (atoms of substrate removed per incident ion), J_+ and ε_+ are the ion flux and energy, respectively, and ρ_s is the density of the substrate material; α is a constant characteristic of the material ($\alpha = 0.0337$ eV$^{-0.5}$ and 0.0139 eV$^{-0.5}$ for Si and SiO_2, respectively), and ε_{th} is the sputtering threshold energy ($\approx$20 eV and 18 eV for Si and SiO_2, respectively) [64]. The latter is the minimum ion energy required for sputtering to occur at all. The sputtering yield for some solid/ion combinations can be found in [15, 61]. Plasma etching systems are designed to minimize sputtering. Ion bombardment can also result is ejection of electrons (secondary electrons) from the surface. The secondary electron emission yield (electrons per incident ion) depends mainly on the type of surface and the ion energy (see also [15, 61]).

In *chemical etching* (Fig. 16b) neutrals react with the surface spontaneously to yield product. An example is etching of silicon with F atoms. Rapid surface fluorination leads to SiF_4 which desorbs into the gas.

$$\mathrm{Si} + 4\mathrm{F} \rightarrow \mathrm{SiF_4} \tag{18}$$

Ion bombardment is not necessary for this reaction to occur. The reaction rate in the absence of ion bombardment is given by Eq. (58) of Section 6.1. Generally, chemical etching has very high selectivity.

Instead of etching, radicals can recombine on the surface (reaction R24 in Table 4). This may be an important loss mechanism of potential etchant species, especially at low pressures. Recombination can occur between two radicals adsorbed on the surface at adjacent sites, (Langmuir-Hinshelwood mechanism [65]). The product of recombination desorbs into the gas.

$$\mathrm{A.S} + \mathrm{B.S} \rightarrow \mathrm{AB(g)} + 2\,\mathrm{S} \tag{19}$$

This mechanism has been postulated for recombination of Cl atoms on alumina (both A and B in Eq. 19 are Cl atoms). Another mechanism is for a gas-phase radical to abstract another radical adsorbed on the surface (Rideal mechanism [65]). The latter case can be shown schematically as,

$$\mathrm{A(g)} + \mathrm{B.S} \rightarrow \mathrm{AB(g)} + \mathrm{S} \tag{20}$$

This mechanism is followed in recombination of O-atoms on quartz (both A and B in Eq. 20 are O atoms).

Ion bombardment can influence one or more of the surface reaction steps in plasma etching. For example, ion bombardment disrupts the crystal lattice generating "active" sites where neutrals can adsorb (step 3 in Fig. 12), or ions promote the reaction of adsorbed species with the surface (step 4), or ions clear the surface from adsorbed products (by sputtering) thereby exposing "fresh" surface to the incoming etchant flux. In *ion-induced etching*, reaction between a species with the surface occurs only in the presence of ion bombardment. An example is etching of undoped silicon with chlorine. Figure 17 shows the result of a well-defined experiment that demonstrates the principle [66]. A piece of silicon wafer at ambient temperature can be bombarded under Ultra High Vacuum conditions by a thermal beam of Cl_2 molecules and an energetic beam ($\sim$1000 eV) of argon ions (Fig. 17, top). No etching occurs when the surface is bombarded by chlorine alone (not shown). Dissociative adsorption of chlorine on the silicon surface occurs, but surface reaction leading to the volatile $SiCl_4$ compound can't occur under these conditions [67]. When the surface is bombarded by the ion beam alone, a small etch rate is observed, corresponding to sputtering of the surface. When the chlorine beam is also turned on, so that the surface is bombarded by molecules and energetic ions simultaneously, the etch rate is much higher than the sputtering rate. The ion beam has now induced a reaction of chlorine with the surface leading to $SiCl_x$ product formation. At the molecular level, ion bombardment induces mixing of the adsorbed gas (Cl) with the crystal by the so-called *collision cascade* [68]. The depth of the mixed layer depends on the ion energy, and typically extends several atomic layers into the crystal lattice. Energy deposited by the ions in the mixed layer favors the formation of weakly bound species ($SiCl_x$, $x = 1$–4) that are either sputtered away or desorb spontaneously into the gas phase. The formation of a surface layer is discussed further in Section 6.1.

In *ion-enhanced etching*, ion bombardment accelerates the reaction of the neutral with the surface. Without ion bombardment that reaction would still occur but at much reduced rate. An example is shown in Fig. 17, bottom. When the silicon surface is exposed to XeF_2 alone, a small etch rate (chemical etching) is observed. Apparently, fluorination of the silicon surface leading to volatile SiF_4 does not require ion bombardment. Also, when the surface is exposed to argon ions alone, a small sputtering rate is observed. However, simultaneous exposure to the neutral and ion beams leads to an etch rate much higher than the sum of the chemical etching and sputtering rates. More details on the etching of silicon and also of oxide can be found in Section 6.

In general, the total etch rate can be expressed as the sum of three components: physical sputtering, spontaneous (chemical) etching and ion-induced or ion-enhanced etching [64, 69–71].

$$ER_{tot} = ER_s + ER_c + ER_+ \tag{21}$$

All three processes can occur simultaneously when a surface is exposed to a plasma. The relative importance of one mechanism over the others depends on the material system (surface and gas), the ratio of neutral to ion flux, and the ion energy. In general, conditions are selected such that ion assisted chemistry dominates. In cases that chemical etching dominates, a mechanism for wall passivation (Section 3.3) is necessary to avoid mask undercut.

In some cases ion bombardment has no effect on etching. A notable example is

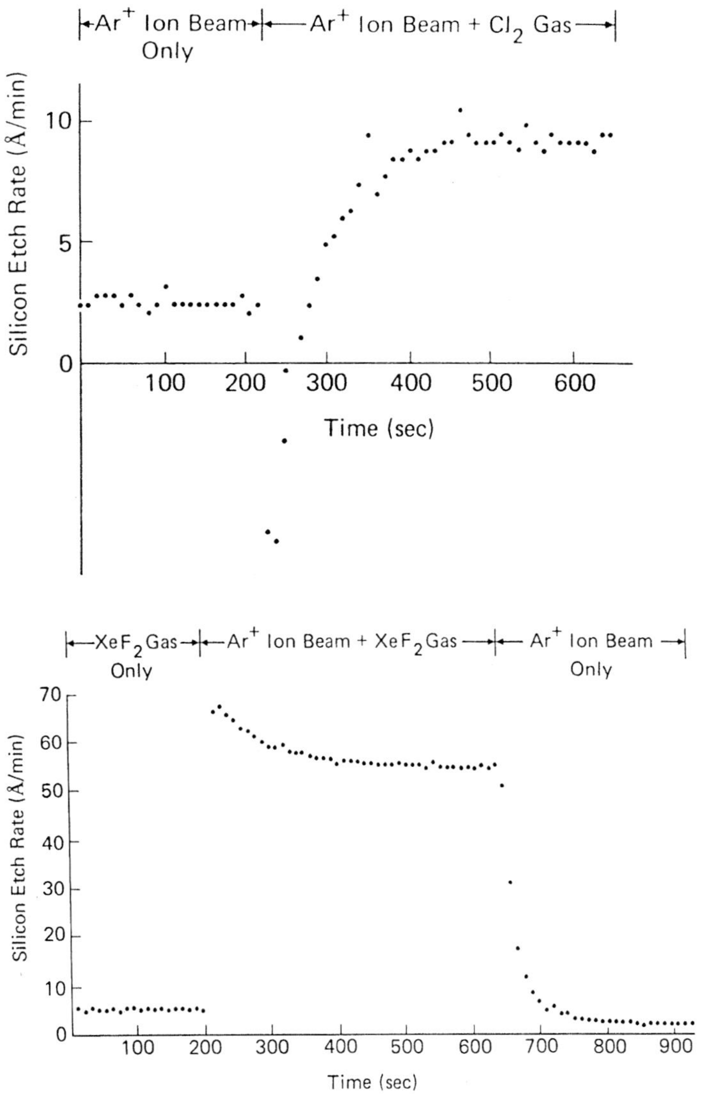

Fig. 17. Demonstration of ion-induced etching (top). Chlorine gas alone does not etch silicon. Ion bombardment alone yields a low sputtering rate. Simultaneous exposure to ions and neutrals results in an enhanced etch rate (much higher than the sputtering rate). Demonstration of ion-enhanced etching (below). XeF_2 can etch silicon (chemical etching) even in the absence of ion bombardment. Ion bombardment alone yields a low sputtering rate. Simultaneous exposure to ions and neutrals results in greatly enhanced etch rate (much higher than the sum of chemical etching and sputtering). After [66].

etching of aluminum in chlorine containing plasmas [72]. The etch rate is independent of ion energy since the only function of the ions in this case is to keep the surface "clean" by sputtering away any adsorbates (e.g. oxygen forming aluminum oxide) that can potentially block etching. In still another situation, ion bombardment can retard etching as in the case of copper etching by chlorine [73].

3.3 Anisotropy Mechanisms

Anisotropic etching refers to the situation where reaction proceeds only along the bottom surface of a microscopic feature, or more precisely, only along surfaces that are not perpendicular to the macroscopic wafer surface. Anisotropic etching is based on the premise that energetic ion bombardment is directed along the normal to the wafer surface. Of course, the finite ion temperature can result in a spread of ion velocities around the surface normal [41, 74, 75]. Plasma etching processes are designed to minimize this angular distribution of ion flux. Also, the ion energy must be optimized in the sense that it should be enough to cause the desired surface reaction but not too high to cause undesired sputtering or radiation damage. In many cases that optimum lies in the range of 50–300 eV, depending on the system.

There are basically two mechanisms to achieve anisotropy (Fig. 16). When gasification of the solid by neutral species is not spontaneous, etching can occur only on surfaces exposed to ion bombardment (ion-induced etching). Thus, anisotropy can be readily achieved (Fig. 16c). However, when the chemical reaction between neutrals and the surface is spontaneous, a "wall passivation" mechanism is necessary to achieve anisotropy (Fig. 16d). Wall passivation means deposition of thin polymeric films or "inhibitors" that block etching of the surface by the neutrals. The passivation film does not form on surfaces exposed to ion bombardment, allowing etching to proceed unimpeded on these surfaces. The deposition of passivation films is controlled by judicious selection of the plasma chemistry [76] and wafer temperature. Oftentimes, the photoresist (polymer) used as mask contributes to the formation of passivation films. Spontaneous chemical etching can be stopped if the wafer is chilled to low enough temperatures to quench the reaction. This is another way to promote anisotropic etching.

A possible mechanism of formation of a sidewall passivation film in the case of silicon etching in a $HCl/O_2/BCl_3$ plasma is shown in Fig. 18 [77]. $SiCl_xH_y$-type byproducts are sputtered away by ion bombardment from the bottom of the trench. A portion of the sputtered flux strikes and sticks on the sidewalls on the trench. Oxygenation of the byproducts on the sidewalls results in a silicon dioxide type of film that resists etching. The sidewalls do not receive any appreciable ion bombardment and hence, depending on conditions, a rather thick inhibitor film may be formed.

4 Plasma Reactor Configurations

4.1 Low Plasma Density Reactors

These have the operational characteristics shown under the "RF diode" column of Table 3. An example of a low plasma density system is the capacitively-coupled diode shown in Fig. 3a. The substrate electrode can be made larger to hold many wafers,

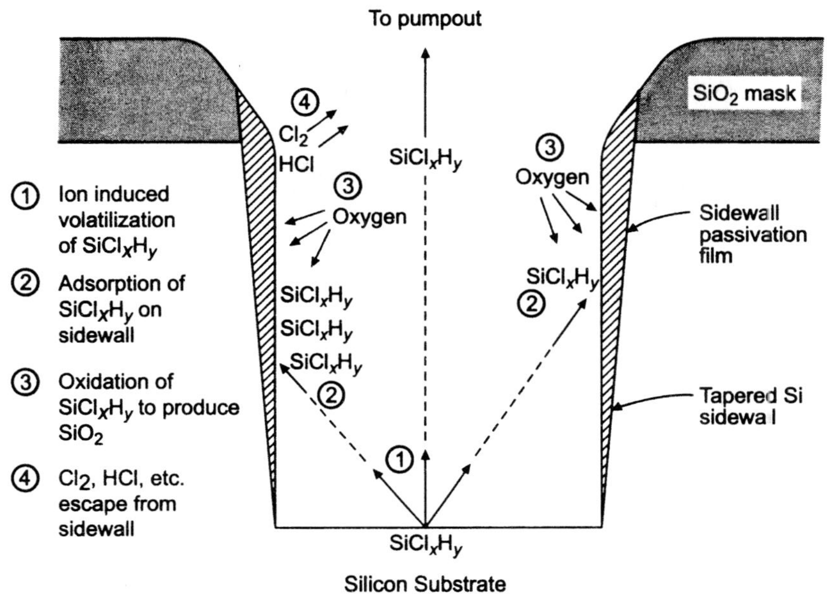

Fig. 18. Model of formation of sidewall passivation film during RIE of silicon in a $HCl/O_2/BCl_3$ plasma. Material sputtered away from the feature bottom (1) redeposits on the sidewall (2) where it is oxidized (3, 4) to form a passivation film. After [77].

but due to the tendency of increasing wafer size, single wafer tools (processing one wafer at a time) are favored for improved uniformity and better process control. Fig. 3a shows a symmetric configuration (electrodes of equal area), that is normally operated at relatively high pressures (>100 mtorr). As a result of collisions in the sheath [78], the energy of ions bombarding the wafer is rather low (see also Fig. 33, bottom). In colloquial terminology this is referred to as *plasma etching* configuration. A variant of this diode system is shown in Fig. 19a, in which the counterelectrode is the (grounded) chamber itself. This is a strongly asymmetric system (large difference in electrode areas). As a result the sheath over the smaller electrode holding the wafer attains a larger voltage (Eq. 3). In addition, the operating pressures are relatively low (<100 mtorr) resulting in intense ion bombardment. This is referred to as *reactive ion etching* (RIE) or reactive sputter etching configuration. The addition of a magnetic field parallel to the wafer holder (Fig. 19b) reduces electron losses and hence enhances the plasma density. At the same time the sheath potential is reduced minimizing unwanted sputtering. The resulting configuration is called magnetically enhanced reactive ion etching (MERIE) [5]. Oftentimes, the magnetic field is rotated to improve etch uniformity. In another variant of the diode reactor (so called *triode*) a third electrode is added to partially decouple plasma density and ion bombardment energy.

The volume-loaded *barrel* reactor shown in Fig. 19c, is not used for applications that require anisotropic etching. A typical example is photoresist stripping. High

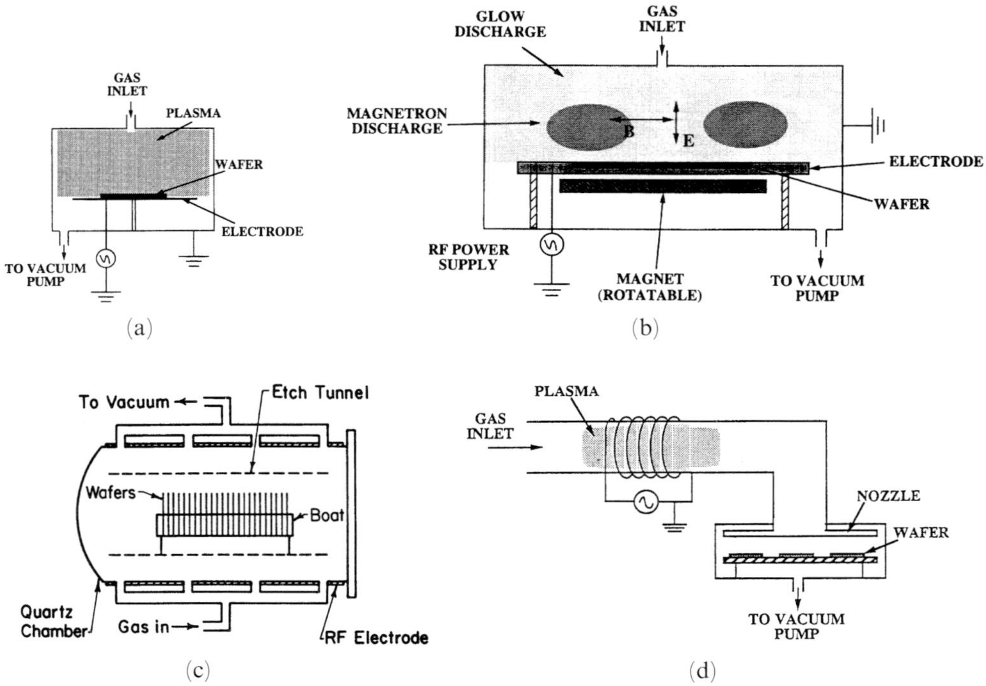

Fig. 19. Schematic of common plasma reactor configurations besides the one shown in Fig. 3a. Reactive ion etching (RIE) (a), magnetically enhanced reactive ion etching (MERIE) (b), barrel (c), and downstream etching (d) reactors. In the barrel and downstream reactors etching is purely by neutral radical species (chemical etching). After [11].

throughput is one of the advantages of the volume-loaded tools. The plasma is generated in the annulus between the quartz barrel and the etch tunnel, a perforated metal cylinder. The wafers are loaded on a quartz boat which is placed inside the tunnel. Reactive radicals formed in the plasma diffuse through the perforations of the etch tunnel and then diffuse in-between the wafers where they react [79, 80]. When the plasma is confined in the annular space outside the tunnel, the wafers are not subjected to ion bombardment, but the wafers can still be exposed to plasma radiation. There are applications that require the wafers to be far away from the plasma so as to avoid electrical damage of devices built onto the wafers. In such cases a *downstream etching* reactor is used (Fig. 19d). The plasma is generated at a remote location. Reactive radicals formed in the plasma are transported by gas flow downstream to the wafer chamber where etching occurs. If the plasma is far enough from the substrate, charged particles recombine before reaching the wafer. Thus the wafer is exposed to neutral species alone. This configuration is used only for applications that do not require anisotropic etching, for example, photoresist stripping [81] and selective nitride etching over oxide [82].

4.2 High Density Plasma (HDP) Reactors

A disadvantage of the diode CCP reactor is that the plasma (electron and ion) density can't be controlled independently of the ion bombardment energy. As the applied RF voltage amplitude is increased to increase the plasma density, so does the sheath potential and consequently the ion bombardment energy. Excessive ion energy is not beneficial as it can lead to unwanted sputtering and heating of the wafer. Furthermore, etching of modern sub-quarter-micron device structures requires extreme directionality of the impinging ions. This can only be achieved by avoiding ion collisions in the sheath. Finally, etching of wafers with ever increasing diameter demands a uniform plasma over large areas. To satisfy these requirements, high (charge) density, low (gas) pressure plasmas have been developed. Typical characteristics of these high density plasmas (HDP) compared to conventional RF diodes are shown in Table 3. Salient features of HDP reactors are: (a) (quasi) independent control of plasma density and ion bombardment energy can be achieved by decoupling plasma generation from substrate (wafer) bias, (b) high plasma density (small Debye length) and low gas pressure (long mean free path) result in collisionless sheath that promotes ion directionality, and (c) low gas pressure facilitates diffusion, promoting uniformity over large diameter substrates. Examples of HDP are inductively coupled plasma (ICP), helical resonator, electron cyclotron resonance (ECR), and helicon plasmas (Fig. 20) [6, 11, 83]. A common feature of these reactors is the separate zones for plasma production and wafer placement. These two zones can be 10s of cm apart. Plasma is generated in the upper zone and diffuses into the lower part of the chamber enclosing the substrate. A set of optional multipole permanent magnets (not shown) can surround the substrate chamber to minimize plasma losses to the walls, thus improving plasma density and also plasma uniformity.

Inductively coupled plasma (ICP) reactors (Fig. 20a) are particularly attractive because their design is relatively simpler and they are easily scaleable to large diameter substrates [84, 85]. In ICPs, the plasma is excited in a cylindrical chamber (r, z, θ) by a solenoidal or planar (stovetop-type) coil powered at radio frequencies, for example 13.56 MHz. The coil current induces a time-varying magnetic field which in turn induces an azimuthal (in the θ-direction) electric field that couples power to the plasma, i.e., heats the plasma electrons. For common excitation frequencies (less than the plasma frequency), the electromagnetic fields are absorbed by the plasma within the *skin depth*. For typical conditions, fields penetrate a few cm into the plasma. The power is deposited non-uniformly in the shape of a toroid (see also Fig. 25). Because of the low pressure, however, species diffusion is facile and the plasma fills the whole reactor. In the absence of any capacitive coupling from the coil, the plasma potential is relatively low ($\sim$20 V) thus minimizing unwanted sputtering of the reactor walls. Capacitive coupling from the coil can result in larger plasma potentials. The wafer platen can be biased independently by a separate RF power supply in order to control the energy of ions bombarding the wafer [86]. As in the case of capacitive coupling from the coil, larger (but controllable) time-dependent plasma potentials are then established.

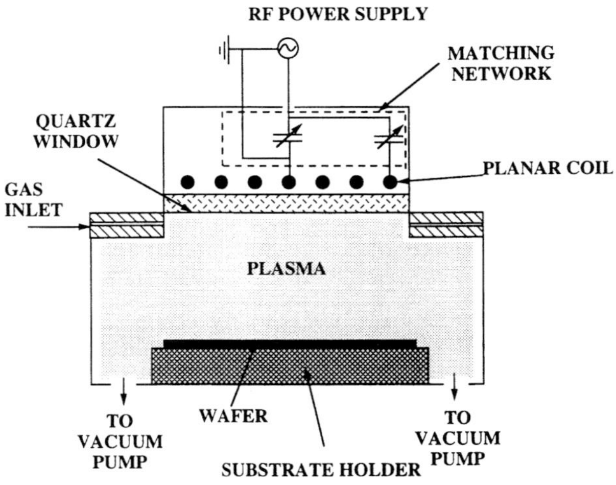

(a)

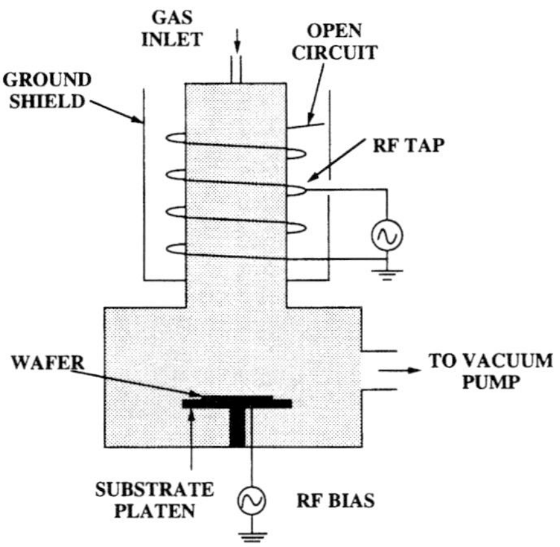

(b)

Fig. 20. High density plasma (HDP) reactor configurations. Inductively coupled plasma (ICP) (a), helical resonator (b), electron cyclotron resonance (ECR) (c) and helicon (d) reactors. After [11].

5 Plasma Modeling and Simulation

5.1 Problem Statement

The problem statement can be summarized by Fig. 4. Given a reactor type and configuration, geometrical dimensions and materials of construction, as well as a set of operating parameters (inputs including plasma power, gas pressure, excitation frequency, substrate voltage or power, and feedstock gas composition and flow rate),

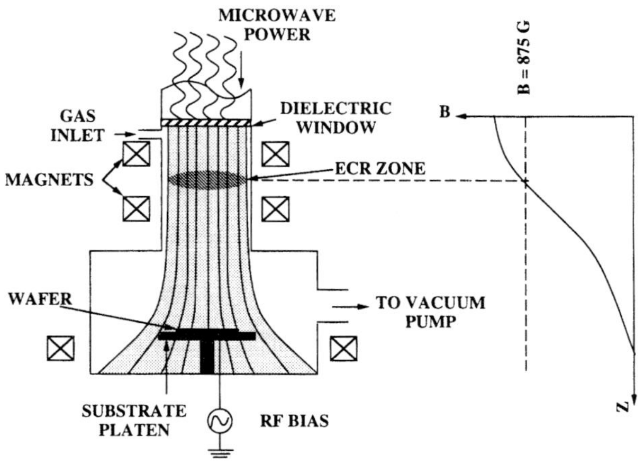

(c)

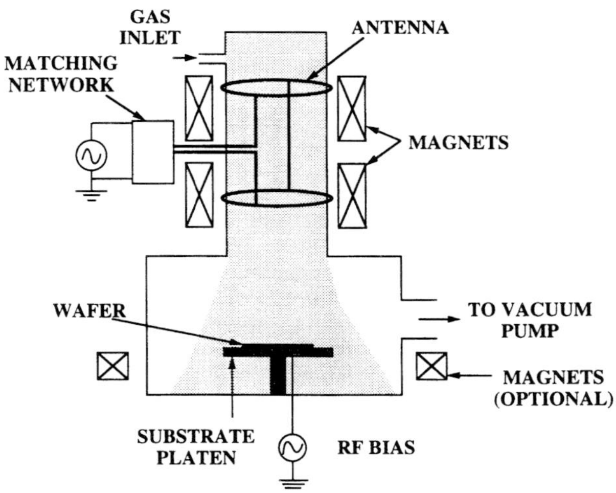

(d)

Fig. 20. (*continued*)

determine the following key plasma properties: The electron velocity distribution function (EVDF), the space and time variation of electron, ion, and neutral species densities and velocities, the flux and energy distribution of ions and neutrals bombarding the electrodes and their uniformity across the electrodes, and the potential and current distribution in the system. These variables and their spatiotemporal variation provide insight into the plasma reactor behavior. When the reactor is loaded with a wafer, one is in addition interested in the outputs (figures of merit) including the etch (or deposition) rate, uniformity, anisotropy, selectivity, radiation damage, and wafer temperature. One can also pose the inverse problem which is more difficult to solve, that is: given a material to be etched (or deposited)

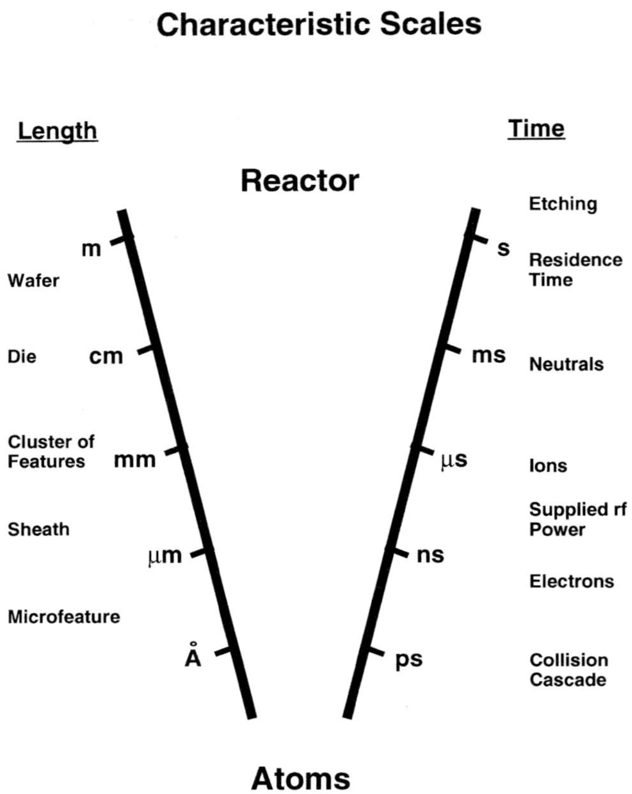

Fig. 21. Disparity in length and time scales in plasma processing.

and specifications on the rate, uniformity, anisotropy, and selectivity, determine the reactor configuration, dimensions, and operating parameters to achieve the task.

5.2 Disparate Length and Time Scales

One of the central issues in modeling the behavior of plasma processes is the disparity in length and time scales (Fig. 21). Length scales range from atomistic, to microscopic (microfeature width), to mesoscopic (sheath, cluster of features) to macroscopic (reactor, wafer). Even if one focuses on reactor scale simulations, the presence of the extremely thin sheath in high density plasmas introduces a formidable range of length scales from 10s of microns to 10s of cm. The range of time scales is also extremely wide, from ps for the collision cascade, to the ns response time of electrons, to μs for ions, to 10–100s of ms for heavy species chemistry and gas residence times, to mins for the duration of etching processes. Similar issues arise in electrochemical engineering as well as in many other fields of science and engineering. In pattern definition by electrochemical etching or electrodeposition, for example [87, 88], one needs to describe behavior at the progressively diminishing scales of reactor, workpiece (e.g., circuit board), cluster of features, individual feature, films possibly covering the surface, and the reaction(s) at the interface between film and underlying material.

One can extend the picture to include a process composed of several unit operations, or even the whole factory. When the whole range of scales is included, the problem becomes one of describing the behavior of an entire plant starting from molecular principles. This problem is unsolved at the present time, but given the rapid advancements in modeling and simulation at each individual scale, and the growth in computational power, one can envision that the overall coupled problem will be solvable within the next few decades.

Until recently, work done to connect the disparate length scales solved the problem in a sequential manner. For the plasma etching problem for example, Economou and Alkire solved for the macroscopic (reactor) scale [69] and used the results as a boundary condition for solving the sheath equation (mesoscopic scale) [78]. In turn they used the sheath simulation results as a boundary condition for the microscopic (feature) scale [89]. Feedback from the feature to the reactor was not considered. Lately, some progress has been made in self-consistent coupling (with feedback) of the disparate length scales of reactor (10s of cm) and microfeature (<1 µm) evolution in CVD [90, 91]. Gobbert et al. [90] developed a multiscale simulator that links the reactor and microfeature scales through a mesoscopic scale ($\sim$ few mm) model. The mesoscale corresponds to the size of a die on the wafer and contains clusters of microfeatures. Both reactor and mesoscale models were solved based on the continuum approximation ($\mathrm{Kn} = \lambda/\mathrm{L} \ll 1$) using a finite element method, while the feature scale model was solved based on ballistic transport (no gas phase collisions). The calculation iterated among the different scale models until a converged solution was obtained. Thus, the effect of changing surface area, for example, due to microfeature shape evolution during CVD, could be accounted for in the simulation. Rodgers and Jensen also developed a multiscale simulator for CVD [91]. Feature scale Monte Carlo simulations were used to develop a "surface reactivity" map as a function of location, $\varepsilon(r)$, which was used in the flux boundary condition for the reactor scale model. In this continuum finite element model the wafer surface was taken as planar. Information was exchanged back and forth (Fig. 22) between the models until convergence. Again, the effect of changing surface topography during CVD could be accounted for.

5.3 Plasma Simulation Approaches

Modeling and simulation of plasma systems has emerged as a tool for enhancing one's intuition about the physicochemical processes occurring in the plasma, for understanding the complex spatiotemporal plasma dynamics, and for assisting in the design of new reactors or the optimization of existing ones [92–96]. A comprehensive plasma model should account for at least the following phenomena [97]:

1. Mass, momentum, and energy transport of charged and neutral species coupled with plasma chemistry.

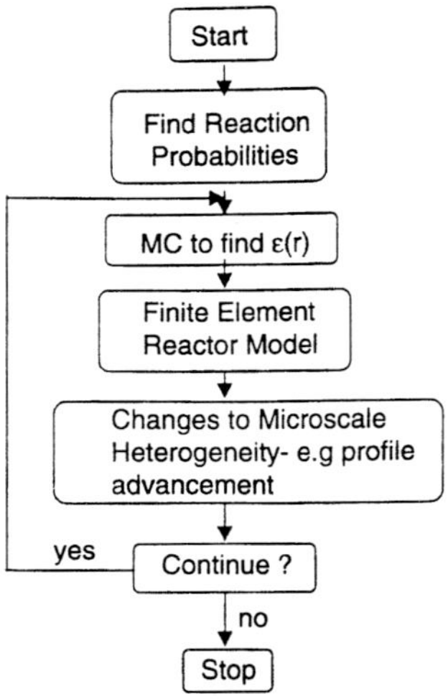

Fig. 22. A methodology for two-way coupling between the microscopic (feature) and macroscopic (reactor) scales in CVD. After [91].

2. The electric and magnetic field distribution in the reactor.
3. The variation of electron velocity distribution function with operating conditions.
4. Species transport in the sheath, and the resulting angular and energy distribution of ions, electrons, and any hot neutrals bombarding the surface.
5. Surface chemistry, especially ion-assisted reactions, recombination, or polymerization that might occur.
6. Feature profile evolution.
7. Heat transport in the semiconductor wafer.

Because of the strong interaction and coupling of the phenomena enumerated above, performing a comprehensive plasma simulation which couples all of these phenomena is an extremely challenging task, and is still an unsolved problem. One way to attack this problem is to break it down into smaller pieces, separating the length and time scales. An approach is shown in Fig. 23, top. The reactor scale simulation includes the bulk plasma and the sheath. Wafer heat transport should also be included here, although this item has not received attention in the literature, except for isolated cases [98]. Wafer heating is important as it can affect the rate of recombination, chemical etching and polymerization on the surface. In many cases, bulk plasma and sheath are solved together, i.e., the same equation set is applied to the whole reactor. This approach is especially convenient for low plasma density systems in which the sheath thickness is an appreciable fraction of the reactor length scale [97], but has also been practiced, to a limited extend, in HDP reactors [99, 100]. However, the most common approach for HDP systems (in which the sheath is extremely thin), is to solve the bulk plasma and the sheath separately [101–103], and then apply a "splicing" procedure to connect the two [102, 103]. The reactor scale model (or the sheath model if done separately from the bulk plasma) provides boundary conditions (species fluxes, energy and angular distributions) to the feature scale model. A surface chemistry model should describe the reactions between neutral and charged particles with the surface, including ion-assisted chemistry, recombina-

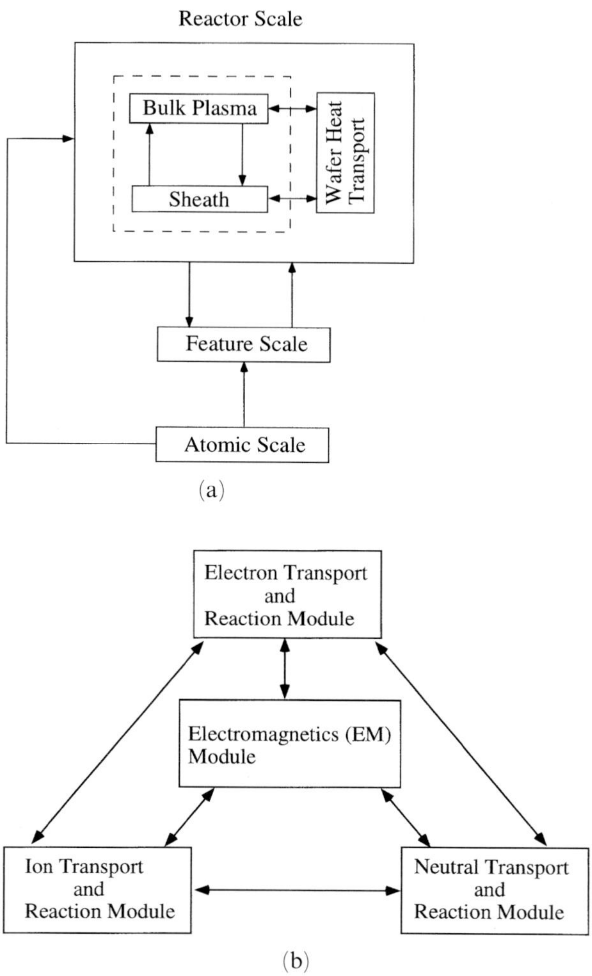

Fig. 23. A methodology for modeling plasma reactors by braking the problem into smaller parts: (top) From the reactor, to the feature, to the atomic scale. (bottom) Modules used for reactor scale model.

tion, polymerization, or even secondary electron emission. This model can be phenomenological (Langmuir-Hinshelwood kinetics) or one based on molecular simulation approaches (Atomic Scale, see Section 5.7). It provides boundary conditions for the feature evolution model as well. An atomic scale model can also be used to calculate rate coefficients for plasma chemical reactions, and these can be fed to the reactor scale model. Although published works on plasma modeling tackle these models in isolation from one another (for example no coupling between reactor and feature scales), efforts to construct an integrated plasma reactor model are continuing.

The reactor scale model is further split into "modules" (see Fig 23, bottom) to separate the disparate time scales of electron, ion, and neutral transport. This is essentially an equation splitting approach. Calculation of the EVDF by solving the Boltzmann Eq. (5) is then part of the electron transport module. The EVDF determines the space- and time-dependent electron energy and transport properties, as well as the electron-particle reaction rate coefficients (Section 3.1.2) for given electric and magnetic fields and plasma gas composition profiles. The electromagnetics (EM) module solves for the electric and magnetic field distributions in the reactor using the Maxwell equations. The neutral and ionic species concentrations are obtained from the corresponding modules. Information is cycled among the modules until a converged solution has been obtained.

5.4 Reactor Scale Simulations

Plasma reactor simulations range from zero-dimensional (well-mixed) to three-dimensional. Well mixed [104–106] and one-dimensional models (including plug flow models [107, 108]) are best for sorting out the complicated gas and surface chemistry to arrive at a reduced reaction set for use in multidimensional simulations. Two-dimensional simulations can address the important aspect of reaction uniformity across the wafer radius. Three-dimensional simulations are useful for studying azimuthal asymmetries in the reactor due to non-axisymmetric power deposition, or non-axisymmetric gas inlets and pumping ports [109, 110].

Simulations of plasma reactors of the type used in the microelectronics industry started to appear only in the 1980s. A number of two-dimensional plasma reactor simulations were reported [69, 111, 112] focusing on the transport and reaction of neutrals only (neutral transport and reaction models). The electron density was *assumed* to have a uniform or Bessel function profile, and the electron energy was not calculated as a function of space and time in the reactor. The radical source terms (by electron-impact dissociation, for example) were estimated and the conservation equations for mass, momentum, and energy transport of neutrals were solved to obtain the fluid velocity profiles, gas temperature and the concentration distribution of radicals; this is a computational fluid dynamics problem. Charged particle transport was not considered, and the effect of plasma gas composition (different than the feedstock gas composition) on the plasma properties was not accounted for. Aydil and Economou [50] improved on these works by using a well-mixed model for the discharge physics to obtain the electron density and electron impact reaction rate coefficients. These were used as input to a two-dimensional transport and reaction model to calculate etch uniformity. They also extended the model to include wafer heating effects [98].

Up until the early 1990s, simulations that solved for the radio frequency (RF) plasma dynamics (so-called glow discharge models) were confined to one spatial dimension (1-D) [36, 113–126]. In addition, most of these simulations did not solve for the transport and reaction of neutrals, i.e., the effect of gas excitation and/or dissociation on the plasma characteristics was not accounted for. This can sometimes be a severe limitation since even minute quantities of excited species can alter the discharge properties [36]. *Self-consistent* RF plasma simulations which solve for the coupled effects of charged and neutral species transport and chemistry have only been reported within the past several years in 1-D [36, 121, 123], 2-D [97, 100, 127–132] and 3-D [109, 110]. Multidimensional simulations are particularly useful since they can address the important issue of plasma uniformity and the spatiotemporal plasma dynamics along both the radial and axial direction. However, most 2-D simulations still do not include neutral transport and chemistry and consider noble gases (argon and helium) only [133–138], not reactive gas plasmas. In view of the above discussion, multidimensional self-consistent plasma reactor simulation is of very young age.

There are three kinds of glow discharge simulations: fluid, kinetic and hybrid. *Fluid* simulations use moments of the Boltzmann equation describing species density, momentum and energy conservation [48, 92, 122]. They require some assumptions regarding the species distribution function to achieve closure of the equations. *Kinetic* simulations, including Particle-In-Cell with Monte Carlo Collisions (PIC-MCC) [139–141] or Direct Simulation Monte Carlo (DSMC) [41, 142, 143] yield the particle distribution functions as an output of the simulation. They are considered more accurate than fluid simulations at low pressures when the species mean free path is comparable to or longer than a characteristic length scale (Kn > 0.2) or for highly non-equilibrium situations. However, there is evidence that fluid simulations can perform well even at low pressures for which their assumptions must be scrutinized [144]. Kinetic simulations are computationally intensive as compared to fluid simulations. *Hybrid* simulations have been developed [59, 121, 131] in an attempt to preserve the accuracy of kinetic simulations and at the same time reduce the computational burden.

5.4.1 Fluid Model Equations

A fluid (or continuum) model of a plasma reactor treats the system as three interpenetrating interacting fluids, i.e., electron, ion, and neutral fluids. The fluid equations are obtained from the Boltzmann Eq. (5) by first multiplying that equation by ϕ, where ϕ is a constant or a function of the velocity $\mathbf{v}$, and then integrating over the velocity space to obtain the "average" of ϕ. The resulting equations are called moments [48, 92, 122, 129]. Kinetic simulations such as DSMC and PIC/MCC solve for the species distribution function by integrating the Boltzmann equation using pseudoparticles. On the other hand, fluid simulations solve equations for "average" quantities such as the directed velocity [48].

When $\phi = 1$, the species i number density continuity equation (where i can be electrons $i = e$, positive ions $i = +$, negative ions $i = -$, or neutrals $i = n$) is obtained as the zeroth moment,

$$\frac{\partial n_i}{\partial t} + \nabla \cdot (n_i \boldsymbol{u}_i) = \sum_j R_{ij} \tag{22}$$

where n_i and $\boldsymbol{u}_i$ are the density and directed velocity of species i, respectively. R_{ij} is the rate of production or loss of species i due to the volumetric reaction j.

The first moment is the (vector) momentum equation which can be derived by setting $\phi = m\mathbf{v}$,

$$\frac{\partial}{\partial t}(n_i m_i \boldsymbol{u}_i) + \nabla \cdot (n_i m_i \boldsymbol{u}_i \boldsymbol{u}_i) = -\nabla P_i + n_i \boldsymbol{F}_i - n_i m_i \boldsymbol{u}_i \nu_{mi}, \tag{23}$$

where ν_{mi} is an effective momentum exchange frequency of species i and P_i is the partial pressure given by $P_i = n_i kT_i$, and T_i is the species temperature. Eq. (23) assumes that the pressure tensor is isotropic which appears to be a good approximation in the absence of strong magnetic fields. In Eq. (23), the terms represent (in order from left to right) local acceleration, convective acceleration (inertia), motion due to pressure gradient, motion due to the presence of force fields (e.g., electric field), and momentum exchange due to collisions with the background species.

When the terms on the left hand side of Eq. (23) are negligible (see [122] for a discussion of this point), and after setting $\mathbf{F} = ez_i\mathbf{E}$, where z_i is the charge number of species i (positive or negative), one obtains

$$\boldsymbol{u}_i = \frac{ez_i\mathbf{E}}{m_i\nu_{mi}} - \frac{\nabla(n_i kT_i)}{n_i} \tag{24}$$

Since the flux for each carrier is $\mathbf{J}_i = n_i\boldsymbol{u}_i$ one can write

$$\mathbf{J}_i = z_i\mu_i n_i\mathbf{E} - D_i\nabla n_i \tag{25}$$

where μ_i and D_i are the particle's mobility and diffusivity respectively, and the species temperature has been taken outside the differential as if the temperature were constant. Eq. (25) is known as the drift-diffusion approximation and is often used in place of the full momentum Eq. (23) for simplicity [113–115, 133, 135, 137]. The drift-diffusion approximation has been questioned at low pressures and in the sheath where species inertia may not be negligible [92]. Eq. (25) does not contain a term corresponding to convective flux by bulk gas flow $(n_i\boldsymbol{u}_g)$ where $\boldsymbol{u}_g$ is the gas flow velocity. This term is normally small compared to the other two terms in Eq. (25). Actually the gas velocity should have been included in the last term of Eq. (23) since the momentum loss of the ion fluid, for example, is proportional to the *relative* velocity between ions and neutrals [48]. It was neglected because $u_+ \gg u_g$. Under typical conditions, the drift flux dominates over the diffusion flux in Eq. (25) for positive ions, while the two fluxes nearly balance each other (and are in opposite direction) for electrons and negative ions. Comparing Eqs. (24) and (25) one obtains,

$$\mu_i = \frac{e}{m_i\nu_{mi}} \tag{26}$$

$$D_i = \frac{kT_i}{m_i\nu_{mi}} \tag{27}$$

Combination of Eqs. (26) and (27) results in Einstein's relation,

$$\frac{D_i}{\mu_i} = \frac{kT_i}{e} \tag{28}$$

Although electrons are mobile enough to respond to the variations of the electric field at 13.56 MHz (i.e., $v_{me} \gg \omega$, and the electron inertia can be neglected), ions are massive and can't follow the field faithfully. Recognizing this fact, Richards et al. [115] introduced an equation for an "effective" field to which the ions (positive and negative) respond.

$$\frac{\partial \mathbf{E}_i^{eff}}{\partial t} = v_{mi}(\mathbf{E} - \mathbf{E}_i^{eff}) \tag{29}$$

That way, the drift-diffusion Eq. (25) can be used for both electrons and ions, except that the actual electric field is replaced by the effective field, in the case of ions. Of course, when the full ion momentum Eq. (23) is used, Eqs. (25) and (29) become immaterial.

The second moment of the Boltzmann equation is the (scalar) energy conservation equation which can be derived by setting $\phi = m\mathbf{v} \cdot \mathbf{v}/2 = \Im$, where $\Im$ is the total (kinetic plus internal) energy. For particle i, $\Im_i = m_i u_i^2/2 + U_i$ where U_i is the internal energy

$$\frac{\partial(n_i \Im_i)}{\partial t} + \nabla \cdot (n_i \boldsymbol{u}_i \Im_i) = -\nabla \cdot P_i \boldsymbol{u}_i + n_i \boldsymbol{F}_i \cdot \boldsymbol{u}_i + \nabla \cdot K_i \nabla T_i - \sum_j R_{ij} H_{ij} \tag{30}$$

Here K_i is the thermal conductivity and H_{ij} is the energy loss of species i due to collision j. In order from left to right, the terms in Eq. (30) represent the time rate of change of total energy, convective transport of energy, rate of work done by pressure forces, energy exchange with the force field, energy transport by conduction, and energy loss due to collisions.

An equation for the thermal energy U_i (For a Maxwellian distribution $U_i = 3/2kT_i$) can be obtained by taking the dot product of $\boldsymbol{u}_i$ with the momentum Eq. (23), and subtracting the resulting equation from the total energy Eq. (30) [92]. However, a simplification is frequently made in plasma simulations, namely that $U_i \gg m_i u_i^2/2$, and $\Im_i$ in Eq. (30) is simply replaced by U_i. The resulting equation, written here for electrons ($i = e$), reads,

$$\frac{\partial}{\partial t}(3/2 n_e k T_e) + \nabla.\mathbf{q}_e + e\mathbf{J}_e.\mathbf{E} + 3v_{me}\frac{m_e}{m} n_e k(T_e - T_g) + \sum_j R_{ej} H_{ej} = 0 \tag{31}$$

where the substitution $\mathbf{F} = -e\mathbf{E}$ has been made, and the total electron energy flux is

$$\mathbf{q}_e = -K_e \nabla T_e + \tfrac{5}{2} k T_e \mathbf{J}_e \tag{32}$$

The thermal conductivity of a monatomic "gas" is

$$K_i = \frac{3}{2} k D_i n_i \tag{33}$$

In Eq. (31), the electron energy loss term has been decomposed into losses due to elastic and inelastic collisions (last two terms on left hand side). The term containing the dot product of electron flux and electric field provides for electron heating by the field. This would correspond to ohmic heating in electrochemical systems. In RF plasma systems, in addition to ohmic (Joule) heating, non-Joule heating is also important, especially at low pressures (< 100 mtorr) [6, 18, 24, 25]. The ability of the fluid model to capture this essentially non-collisional heating has been scrutinized in the literature [145]. Turner [146] has proposed a modification to the fluid equations to account for this effect. Some authors write Eq. (31) in terms of the thermal energy U_e, i.e., they don't make the substitution $U_e = 3/2kT_e$ [122, 137]. This way, they have an equation in terms of the mean electron energy instead of one in terms of electron temperature. The two formalisms are equivalent.

When solving a fluid equation for the (average) electron energy, an assumption has to be made regarding the EEDF. In the case of Eq. (31), the EEDF has been implicitly assumed to be Maxwellian. The electron energy determines the rate coefficients of electron impact source terms that enter the species mass balance (Eq. 22), and also the electron transport properties (momentum exchange frequency, mobility, diffusivity) that enter the electron momentum equation (either Eq. 23 or Eq. 25). These coefficients and transport properties can be expressed as a function of the average electron energy [53–55, 92, 129] based on the solution of a DC Boltzmann equation. Instead of solving for the average energy, one can solve the (time-dependent) Boltzmann equation for electrons and calculate the EEDF (a kinetic approach), in essence replacing the electron energy equation. This essentially hybrid (kinetic EEDF, fluid for the rest) approach has been used extensively in the literature [59, 121, 131].

Finally, the Poisson equation links the potential field with the charge density

$$\nabla^2 V = -\frac{e\sum_i z_i n_i}{\varepsilon_0} \tag{34}$$

Depending on the approximations made, different sets of the equations shown above are used by different authors. Most often, the drift-diffusion approximation is made for both electrons and ions [113–115, 126, 127, 133, 135, 137]. Other authors solve the full momentum equations for either electrons or ions (using drift-diffusion for the other species) [99, 130, 134] or for both electrons and ions [116, 125, 136]. An energy equation for ions is normally not solved. Almost always, it is assumed that ions are at the same "temperature" as neutrals. At very low pressures (or more accurately for high Kn) this is not a good assumption. Most workers also assume a neutral gas temperature. Others solve an energy balance for the neutral gas to calculate T_g [130]. Of course, particle simulations (DSMC) yield the species temperatures without the need for any assumptions.

In general, boundary conditions are difficult to specify and oftentimes difficult to incorporate into the numerical scheme. Typical boundary conditions used are given in [92, 97, 129, 130, 136]. Boundary conditions for the mass continuity Eq. (22) specify a zero electron density at the wall, or an electron flux equal to the local thermal flux multiplied by an electron reflection coefficient. The ion diffusion flux is set to vanish at the wall, i.e., ion transport is completely by drift at that point. The neutral species boundary condition is set such that the flux is equal to the reaction rate. The latter is often specified through a "sticking" probability. Care should be exercised to account for rarefaction effects at low pressures; see Chantry for a discussion [147]. An "extrapolation" boundary condition (second derivative of velocity equal to zero) is typically specified for the momentum Eq. (23). Boundary conditions for the electron energy equation have included a constant energy at the wall, or adiabatic wall. More often, the energy flux (Eq. 32) is set equal to the electron number flux multiplied by the energy carried by each electron to the wall. Boundary conditions on the Poisson Eq. (34) include a specified potential on metal surfaces (either grounded or RF driven) or a specified current at the metal wall. A current continuity boundary condition is used on insulators [129]. The method of solution of the governing equations varies. Most often finite difference methods are employed [92, 122–124, 130–138]. Finite element methods are preferred for handling complicated geometries [97, 129, 148].

The above set of equations applies as shown to capacitively coupled systems. An identical set applies to the modeling of semiconductor devices [149]. Also, the similarity with the equations governing electrochemical systems (Section 8) is striking. The main difference is that continuity equations for electrons are not necessary, since there are no free electrons in the solution.

The above set of equations is similar to that used in traditional computational fluid dynamics (CFD) (except for the electron energy balance and the EM equations), and advances made in that field can be used to benefit the plasma reactor simulation problem. As an example, the sheath near the wall can be thought of as similar to a boundary layer in fluid flow (chemically reacting or not) [130]. Separating the flow into bulk (inviscid) and boundary layer (viscous) and then "patching" the two solutions (asymptotic analysis) has long been practiced in fluid mechanics and may also be applied to the plasma problem [102, 103, 151].

Application to capacitively-coupled reactors Figure 24a shows the electron temperature distribution in an argon discharge sustained in a one-dimensional parallel plate reactor of the kind shown in Fig. 7. The temperature peaks near the plasma-sheath interface, where the product of the current and electric field (Eq. 31) is highest, and steep gradients develop in that region. Electrons which diffuse towards the electrode during the sheath potential minimum (around $\tau = 0.25$ at left electrode, see also Fig. 8) are pushed back into the plasma as the potential reverses during the cathodic part of the cycle. One can think of electrons 'riding the wave' as the sheath boundary moves away from the electrode during the cathodic part of the cycle. The electric field in the bulk plasma is much weaker than that in the sheath. Hence the bulk electron temperature is lower. The bulk electron temperature in this case is 3.39 eV (this corresponds to an average energy of 5.09 eV).

The excitation rate profiles are shown in Fig. 24b. Substantial modulation is observed even deep in the bulk plasma. The excitation peak moves away from the left electrode and is washed into the bulk plasma as the left electrode potential goes through the negative zero crossing ($\tau = 0.5$), to the maximum negative potential ($t = 0.75$), to the positive zero crossing ($\tau = 0$) and finally to the maximum positive

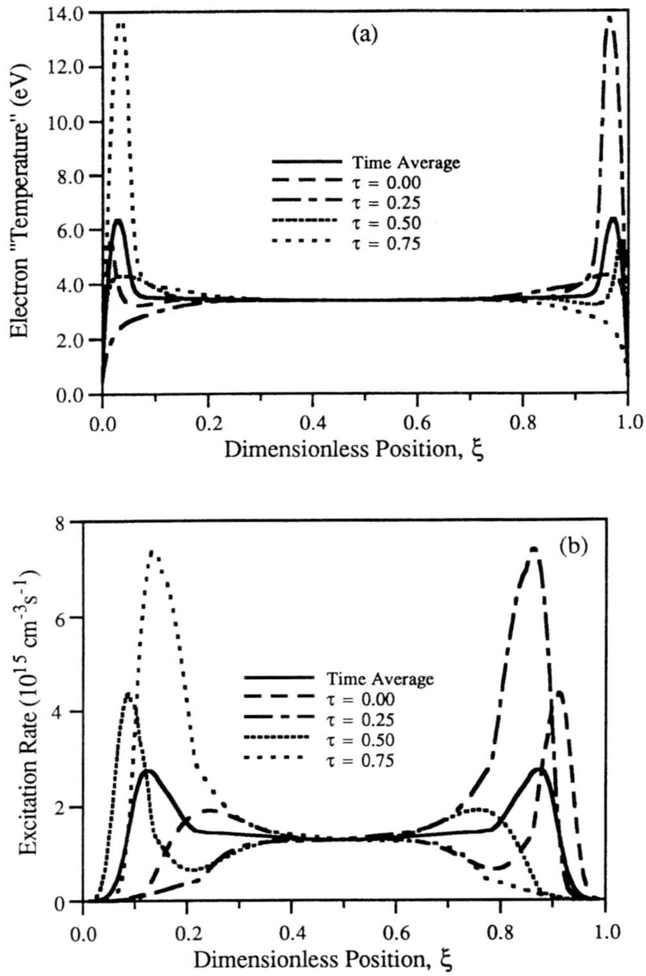

Fig. 24. Spatiotemporal distribution of the electron temperature (top) and excitation rate (bottom) in an argon RF capacitively-coupled diode discharge. Time is normalized with respect to the RF period. After [36]. Pressure 1 torr, electrode spacing 1″, frequency 13.56 MHz, applied RF peak voltage 100 V.

potential ($\tau = 0.25$), Fig. 8. The situation at the right electrode is symmetric but 180° out of phase. The time-average excitation rate (not shown) is negligible right next to the electrodes (in the sheath) and peaks at a distance from the electrodes which decreases with increasing pressure. At high enough pressures (> 100s mtorr), the plasma would appear as two relatively bright disks separated from the electrodes by a dark space, and connected through a less bright bulk plasma. Details of the spatiotemporal distributions of important discharge quantities are found in [36].

Application to inductively-coupled reactors When modeling inductively coupled systems (Fig. 20a), the electric field powering the electrons is azimuthal. In the case of azimuthal symmetry (no variation in the θ-direction), the heating field $\tilde{E}_\theta = -j\omega\tilde{A}_\theta$ is obtained from a one-dimensional Helmoltz equation [129, 150]

$$\frac{1}{r}\frac{\partial}{\partial r}\left(r\frac{\partial \tilde{A}_\theta}{\partial r}\right) + \frac{\partial^2 \tilde{A}_\theta}{\partial z^2} + \left(\frac{\omega^2}{c^2} - r^{-2}\right)\tilde{A}_\theta = -\mu \tilde{J}_\theta \tag{35}$$

where $\tilde{A}_\theta$ is the complex vector potential, and the current density is given by $\tilde{J}_\theta = \tilde{\sigma}\tilde{E}_\theta$ where $\tilde{\sigma}$ is the complex conductivity of the medium. The permeability of the medium μ should not be confused with mobility; c is the speed of light in the medium (plasma, metal wall, etc.). The (time-average) power deposited in the plasma by ohmic heating is then found to be $\overline{W}(r, z) = \frac{1}{2}\{Re(\tilde{\sigma}\tilde{E}_\theta^2)\}$, where Re is the real part of the complex number in parenthesis [129]. Typical results obtained for an oxygen discharge in an inductively coupled reactor with a stovetop coil are shown in Fig. 25 [131]. These results were calculated with a model similar to that described in Section 5.4.1, except that the EVDF was obtained by solving the Boltzmann equation by a Monte Carlo method. This can be thought as a fluid-kinetic (hybrid) model. However, similar results are found by assuming a Maxwellian EEDF [100, 130]. The power is deposited by the electromagnetic fields in the shape of a toroid (Fig. 25, bottom) peaking at about midradius near the coils. The penetration depth of the azimuthal field is a few cm. Since there was no bias voltage applied to the bottom electrode and capacitive coupling from the coils was neglected, the plasma potential is low peaking at 17.5 V (Fig. 25, bottom). Strong potential gradients are seen near the walls where a sheath is formed. The electron production rate (Fig. 25b) is highest near the maximum in power deposition and drops off away from this zone as the electrons cool down. The density profiles of major charged particles (electrons, O^- and O_2^+ ions) all show off axis maxima (Figs. 25 top and middle). An off axis maximum in the positive ion density, for example, can be sustained if there are reactive losses at the center (e.g., by ion-ion recombination of the type of reaction R22 in Table 4) or significant diffusive losses towards the axial walls (see also Fig. 29, below). The negative ions have low energy (near room temperature) and are likely to be pooling around the maxima of the plasma potential.

5.4.2 Judicious Approximations

When dealing with multidimensional geometries and complex chemistries, "full-blown" self-consisted plasma simulations pose a very challenging task. While "brute force" simulations (for example ones that solve the complete set of conservation equations for all particles) are feasible, they are also labor and time consuming. Judicious approximations, based on the physics of the problem, can reduce the simulation times dramatically. Examples are the space-time averaging used in the non-local approach [58, 59], or the approximations used to construct rapid plasma simulation tools [59, 101, 148, 152, 153].

The two-dimensional self-consistent ICP reactor simulation of Wise et al. [101, 148, 152] employs a modular approach similar to that used for capacitively-coupled

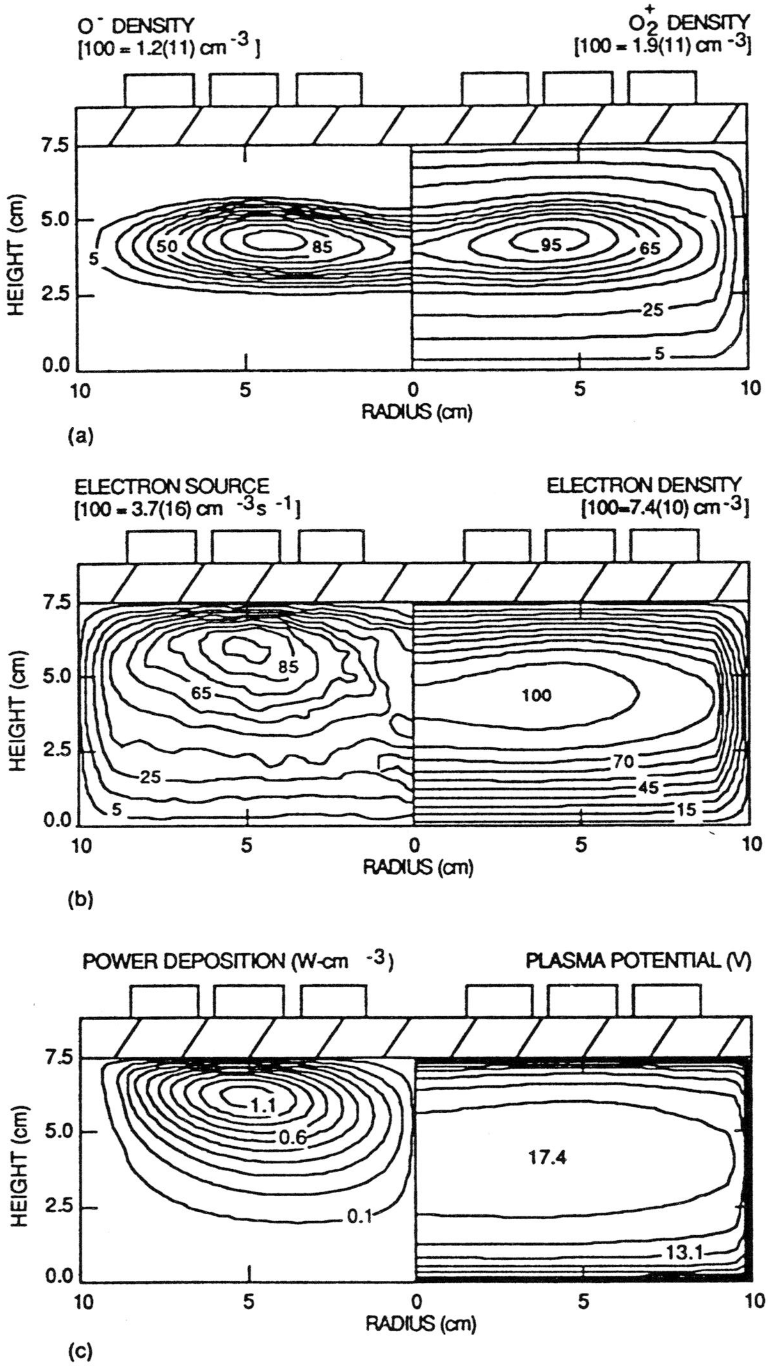

Fig. 25. Calculated profiles of plasma parameters in an ICP in oxygen. O^- and O_2^+ density (top), electron production rate and electron density (middle), and power deposition profiles and plasma potential (bottom). After [131]. Pressure 10 mtorr.

[97] and inductively-coupled [100, 130, 131] systems. The simulation begins by solving Maxwell's equations to determine the electromagnetic fields and power deposition for a specified coil current (Fig. 23, bottom). The power deposition profile is used as input to the electron energy equation to calculate the electron energy or "temperature" and hence the rate coefficients for electron-impact reactions. In turn these are used in the source terms of the neutral and charged species modules. The former is used to calculate the neutral gas composition. The latter is used to calculate the charged particle densities. The calculation is repeated in a cyclic fashion until "convergence". From the converged solution, the self-consistent power deposition, electrostatic potential, electron temperature, charged and neutral species densities and flux, reaction rate and uniformity along the wafer radius can be calculated. This modular approach is essentially an equation splitting approach, in order to overcome the disparate time scales of electron, ion, and neutral transport and chemistry (each of the species is solved in its natural time scale). The Poisson equation can be part of the electromagnetics module, or it can be replaced by Eq. (36) below. In the case of a RF biased substrate, a circuit simulation of the plasma which includes the non-linear impedance of the sheath, can provide the time-dependent plasma potential and information on the energy distribution of ions bombarding the wafer.

The rapid simulation employs a number of simplifying assumptions as described below:

Separation of bulk plasma from sheath For high density plasmas, the sheath thickness is on the order of 10–100s of μm. In contrast, the reactor dimension is typically 10s of cm. This disparity in length scales requires, for any numerical method, a much denser grid near the walls compared to the bulk plasma, or high order approximations. The high degree of spatial stiffness (10^3–10^4) introduces numerical difficulties and results in excessive computation times. For this reason it has become customary to separate the bulk plasma from the sheath [59, 69, 101, 152, 153], or to avoid resolving the sheath [130–132]. It then becomes necessary to device a procedure for "splicing" the sheath to the bulk plasma [102, 103, 151]. Fortunately, the sheath in HDPs is collisionless, since the species mean free path at 10 mtorr is on the order of 1 cm, much larger than the sheath thickness. Thus, if one knows the ion (or electron) distribution function at the sheath edge, and the spatiotemporal distribution of the sheath electric field, one can obtain the ion (or electron) distribution function at the wall.

In order to be consistent with the Bohm criterion for ions, the sheath edge is defined as the point where the ions have been accelerated (presumably by the presheath electric field, Fig. 5) to the Bohm velocity, i.e. the presheath is included as part of the bulk plasma. The Bohm flux also provides a boundary condition (applied at the wall because of the thinness of the sheath) for the positive ion continuity equation. The negative ion density is assumed zero at the walls.

Separation of the bulk solution from the double layer is tacitly assumed in electrochemical reactor analyses. When describing electrochemical systems, one is not interested in resolving the double layer because it is extremely thin. Phenomena within the double layer and on the electrode surface are incorporated into the boundary conditions.

Electroneutrality in the bulk plasma If one is not interested in resolving length scales of the order of the Debye length, the electroneutrality constraint in the bulk plasma is applicable.

$$n_e = \sum_i z_i n_i \tag{36}$$

This is an excellent assumption for the plasmas of interest since the Debye length is exceedingly small (10s of μm) compared to the reactor dimensions. Of course, the electroneutrality constraint can't be applied in the sheath, where the Poisson equation (34) has to be solved. Solving Poisson's equation in the whole reactor domain introduces numerical difficulties. When using the electroneutrality constraint, the electron continuity equation (Eq. 22) is not necessary. The electron density is obtained directly from Eq. (36), having solved for all ion densities. The electroneutrality constraint is routinely used in electrochemical systems outside of the double layer.

Boltzmann electrons If the Poisson equation is used instead of electroneutrality in the bulk, the electrostatic field can be recovered without recourse to any additional assumptions. Since the Poisson equation is not solved in the bulk, the fields are obtained by assuming Boltzmann electrons [6],

$$\boldsymbol{E} = -\frac{\nabla(n_e k T_e)}{e n_e} \tag{37}$$

This expression results from the electron momentum balance (Eq. 23) by assuming inertialess electrons (neglecting the left hand side of the momentum equation) to yield Eq. (25). Now the electron drift and diffusion fluxes nearly balance each other, yielding Eq. (37). This equation simply states that the electric field force balances the electron pressure force. Eq. (37) breaks down when the density of negative ions is large enough such that the negative ion current is an appreciable fraction of the current carried by negatively charged species. Under typical ICP conditions, however, the negative ion density is comparable to the electron density and Eq. (37) is applicable.

Specified electron energy distribution function The EEDF is specified, normally assumed Maxwellian (Eq. 9). The electron energy balance (Eq. 31) is solved assuming an adiabatic condition for electron temperature at the wall. The Maxwellian assumption is very common in the literature [100, 125, 126, 130, 133, 135–137]. Measured EEDFs in ICPs, however, have a Maxwellian bulk (due to electron-electron collisions), and a depleted tail due to inelastic losses and escape of fast electrons to the walls. Thus a bi-Maxwellian distribution may be more appropriate [154]. A Maxwellian distribution is not expected to have a great effect on ion densities since the ionization rate is self-adjusted to balance the loss rate of ions to the walls; and the latter depends only very weakly on the EEDF. The good agreement with experimental data [101, 130, 148, 152] is an indirect evidence that the Maxwellian EEDF is reasonable for obtaining species densities and their distributions. Other forms of the EEDF may also be used, e.g., Druyvesteyn [48]. If accurate determination of the EEDF is necessary one must solve the Boltzmann Eq. (5). Approximations such as the non-local [58, 59] approach (when applicable), offer tremendous advantages, especially for complex multidimensional geometries.

Drift-diffusion for ions The drift-diffusion approximation (Eq. 25) is made for ions, e.g., ion inertia is neglected. This appears to be a good approximation for Kn < 0.2. Eq. (37) provides the field that drives ions in the plasma. Also, a constant mobility is assumed for ions. Actually, when collisions are infrequent (high Kn number), it is better to use a variable mobility model, for which the ion mobility is inversely proportional to the magnitude of the directed velocity of the ion [6]. This model results by assuming a constant mean-free-path of ion-neutral collisions. The constant mobility model results by assuming a constant mean-free-time for ion-neutral collisions.

Diffusive flow for neutrals The importance of convective vs. diffusive flow of neutrals is determined by the Peclet number Pe = uL/D, where L is a characteristic dimension of the system. Away from inlet and exit ports, the characteristic length will be on the order of the reactor dimension. The system will be primarily diffusive when $Pe \ll 1$. For Cl_2 gas in a reactor with $L \sim 0.1$ m and a neutral species diffusivity of $D \sim 5\,m^2\,s^{-1}$ at 20 mtorr, the Peclet number will be $Pe \sim 1$ when $u = 50\,m\,s^{-1}$. Convective gas velocities are not likely to be that high, except for a small region near the gas inlet ports. It follows that gas flow can be approximated as diffusive; this obviates the need for solving the full Navier-Stokes equations which adds to the computational burden. It should be noted that both the diffusivity and the convective velocity scale inversely with gas pressure, so the Pe number is independent of pressure. However, as the pressure is lowered to the point of free molecular flow, the gas diffusion coefficient has no meaning any more. Direct Simulation Monte Carlo (DSMC) [41, 143] can then be applied to solve for the fluid velocity profiles.

In summary the following equations were used for the rapid ICP reactor model: (a) density continuity (Eq. 22) for each and every ion and neutral in the system, (b) the drift-diffusion equation (Eq. 25) for the ion fluxes and diffusion alone for the neutral fluxes, (c) electroneutrality to calculate the electron density, Eq. (36), (d) Boltzmann relation, Eq. (37), to calculate the electrostatic fields, (e) the electron energy Eq. (31) to calculate T_e, and (f) the Helmoltz Eq. (35). The cold plasma approximation was used to calculate the power deposition profiles [150]. This system of equations was supplemented with a table of values for the reaction rate coefficients of electron impact reactions as a function of T_e. Analytical approximations can also be used for this purpose, but table interpolation is generally faster. These equations were incorporated into a simulation code named MPRES (Modular Plasma Reactor Simulator [152]) which can be used as a Technology Computer-Aided Design (TCAD) tool for inductively coupled plasma systems.

The MPRES simulator has been validated by comparing predictions to experimental data taken in a Gaseous Electronics Conference (GEC) reference cell [155]. Predicted [101] (lines) and measured [156] (points) radial profiles of electron density, electron "temperature," and plasma potential for a chlorine plasma are shown in

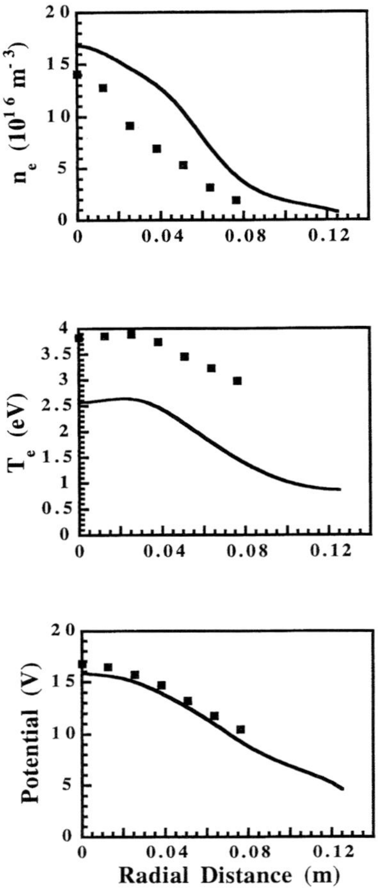

Fig. 26. Comparison of MPRES simulator predictions [101, 152] (solid lines) to experimental data [156] (points). Spatial profiles of electron density (top), electron temperature (middle), and plasma potential (bottom). Pressure 20 mtorr, power 180 W.

Fig. 26. The radial profiles were measured by using a Langmuir probe. The agreement between simulation and experiment is very good, considering that there were no adjustable parameters in the simulation. The width of the electron density profile (Fig. 26, top) is rather small, suggesting a well confined plasma. The predicted electron temperatures (Fig. 26, middle) are ~50% lower than the measured values, possibly due to the assumption of Maxwellian electron energy distribution function used in the simulation. Also, it is well known that the errors in the electron temperature measurement using Langmuir probes can be substantial [157]. The electron temperature gradients are rather small due to the high thermal conductivity of the electron gas at the low operating pressure and high charge density (Eq. 33). Finally, the predicted plasma potentials match the data quite well (Fig. 26, bottom). Overall, the quantitative agreement between simulation and experimental data is remarkable considering that no adjustable parameters were used. Good agreement was also obtained with data in argon discharges [152], and in mixed gas plasmas [158].

The experimentally verified simulation tool was then used to predict plasma

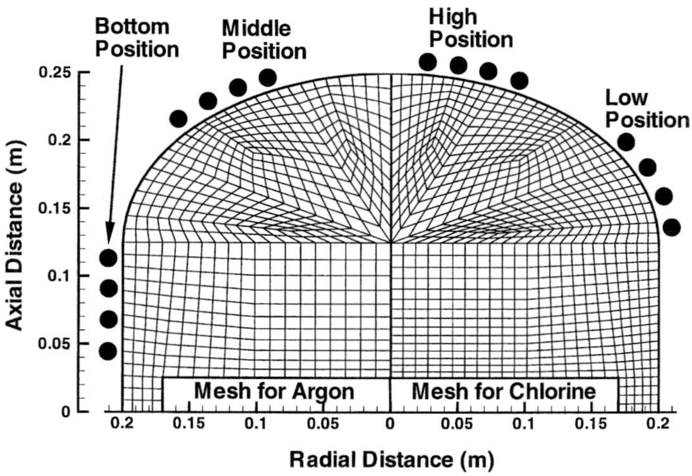

Fig. 27. Dome-shaped (axisymmetric) reactor showing four different placements of the solenoidal coil wound around the reactor (only half of the coil is shown). The finite element meshes used for argon and chlorine discharge MPRES simulations are also shown. After [148].

transport in a dome-shaped reactor [148] that can accommodate 30 cm-diameter wafers, i.e., an industrial-size reactor (Fig. 27). This is an inductively coupled reactor powered by a four-turn solenoidal coil wound around the quartz dome. The coil can be positioned in one of four different locations designated in Fig. 27 as high, middle, low, and bottom positions, in descending order from the top of the dome (for each position only half of the coil is shown in Fig. 27). The chosen base case conditions were a pressure of 10 mtorr, a plasma power of 1000 Watts, no substrate bias, the middle coil position, an excitation frequency of 13.56 MHz, and an inlet gas feedrate of 30 sccm. The finite element grids used for argon and chlorine plasma simulations are also shown in the figure.

The effect of coil location on ion flux uniformity is shown in Figure 28 for the argon discharge. The peak values of the ion flux were 4.20×10^{20}, 4.23×10^{20}, and 4.56×10^{20} and $4.64 \times 10^{20}\ \mathrm{m^{-2}\ s^{-1}}$, respectively, for high, middle, low, and bottom coil positions. The ion flux uniformity improves steadily and dramatically in going from the high to the bottom coil positions. This is a result of altering the location of the plasma generation zone and the ion diffusion length as the coil position is varied. Under the conditions of Fig. 28 the ion density always peaks on axis. However, having the plasma production zone near the wafer edge (when the coil is at the bottom position), results in more uniform ion density profiles. In general, the flux uniformity depends on the Damkohler number (see Section 5.4.5) and the reactor aspect ratio.

The aspect ratio effect is seen in Fig. 29. Given a localized source of species (e.g. radicals), the spatial distribution of these species at steady state will depend on the reactor aspect ratio. Low aspect ratio reactors (tall, small radius) yield a distribution that peaks on axis, while large aspect ratio systems (short, large radius) yield a dis-

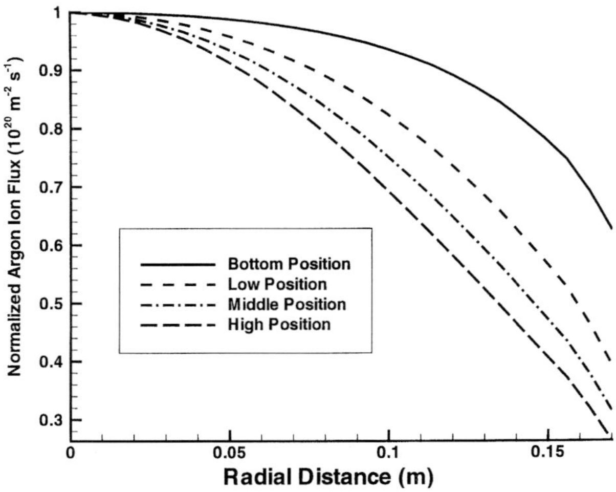

Fig. 28. Ion flux uniformity to the bottom surface of the dome-shaped reactor of Fig. 27 for the four coil locations. The uniformity improves as the coil is moved from high to bottom position.

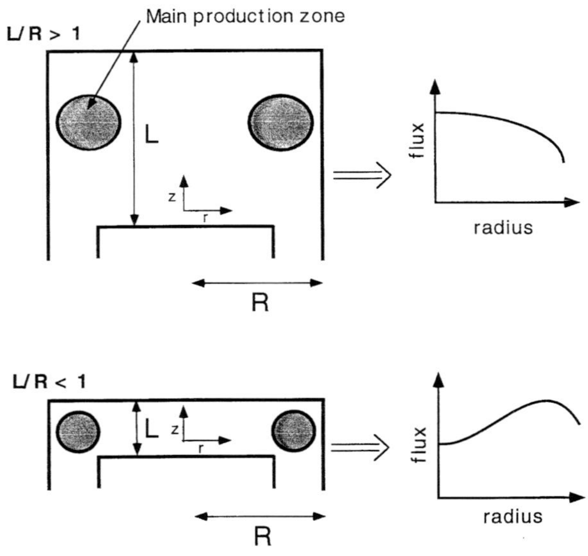

Fig. 29. The effect of reactor aspect ratio on species flux uniformity along the wafer radius.

tribution that peaks off axis. This is because, as the species diffuse radially away from the localized source to try to fill in the central regions of the reactor, they also diffuse towards and lost onto the axial end walls. The final profile is dictated by the relative magnitude of the diffusion lengths, i.e., the aspect ratio. Actually it is the square of the aspect ratio that governs behavior since the diffusion loss "rate coefficient" is inversely proportional to the square of the diffusion length.

5.4.3 Ambipolar Diffusion

Another commonly used approximation is that of ambipolar diffusion. Let us confine our discussion to an electropositive plasma (no negative ions) in which the electron density equals the positive ion density throughout the bulk plasma, $n_+ = n_e = n$. We will also assume that at every point in the plasma $\mathbf{J}_+ = \mathbf{J}_e = \mathbf{J}$ (for singly charged ions). This corresponds to no net current flow in the system. Using the drift-diffusion approximation (Eq. 25) for both particles and solving for the field $\mathbf{E}$ results [6, 15, 18, 42]

$$\boldsymbol{E} = \frac{D_+ - D_e}{\mu_+ + \mu_e}\frac{\nabla n}{n} \tag{38}$$

Substituting this expression for $\mathbf{E}$ into any of the two flux equations (Eq. 25) yields

$$\boldsymbol{J} = -\frac{\mu_+ D_e + \mu_e D_+}{\mu_+ + \mu_e}\nabla n \tag{39}$$

The flux can be written as

$$\boldsymbol{J} = -D_a \nabla n \tag{40}$$

where the ambipolar diffusion coefficient is

$$D_a = \frac{\mu_+ D_e + \mu_e D_+}{\mu_+ + \mu_e} \tag{41}$$

Taking into account that $\mu_e \gg \mu_+$ and using Einstein's relation Eq. (28) results

$$D_a = D_+\left(1 + \frac{T_e}{T_+}\right) \tag{42}$$

It is quite usual to have $T_e \sim 100T_+$ in plasmas of interest, which shows that the ambipolar diffusivity is ~ 100 times the free ion diffusivity. Ambipolar diffusion prevails at high enough plasma densities. At very low plasma densities, the space charge fields are too weak to have any effect on the charged species motion and diffusion is free. The transition from free to ambipolar diffusion as the plasma density increases has been the subject of many investigations and is reviewed in [159]

Using Eq. (40), the mass balance for either electrons or ions can be written as a usual diffusion equation with a variable diffusivity.

$$\frac{\partial n}{\partial t} = \nabla \cdot (D_a \nabla n) + \sum_j R_j \tag{43}$$

Eq. (43) is solved as an eigenvalue problem assuming a zero charge density at the boundary and a constant diffusivity [6, 42, 43]. The eigenvalue corresponding to the dominant eigenmode (diffusion mode) is actually an equation for the electron temperature in terms of the so-called electron *diffusion length* Λ. The latter depends on the geometry of the system and is known for well-defined geometries [6, 42, 43]. The electron temperature can't be set independently. It adjusts itself so that a *self-sustained* discharge results, i.e., the electron production rate equals the electron loss rate (self-organizing system). For the classical positive column, the electron temperature is a function of $N\Lambda$ only [6, 42, 43]. In electrochemical systems, the ion mobilities and diffusivities are comparable and hence the ambipolar diffusivity is about twice the ion diffusivity.

5.4.4 Wafer Heat Transport Model

Wafer temperature affects the rate of spontaneous etching and also polymerization reactions. Wafer temperature control is important in semiconductor manufacturing. The temperature is typically controlled by relatively high pressure (several torr) He cooling from the backside of the wafer. The model below refers to silicon etching in a chlorine plasma as a typical example [98].

An energy balance can be made in the wafer to determine the temperature distribution along the wafer radius; radial uniformity of temperature is important for temperature sensitive processes. Since the thickness of the wafer is much smaller than its radius, and assuming azimuthal symmetry, a one-dimensional analysis suffices. The wafer energy balance reads,

$$\rho_{Si} C_{pSi} \frac{\partial T_s}{\partial t} = K_{Si}\left(\frac{1}{r}\frac{\partial}{\partial r}\left(r\frac{\partial T_s}{\partial r}\right)\right) + \frac{Q_N}{\Delta} \tag{44}$$

where

$$Q_N = \tfrac{1}{2} k_{sr} n_{Cl,s} \Delta H_{rec} + \tfrac{1}{4} ER_{tot} \Delta H_{etc} + J_+ \varepsilon_+ - h_T (T_s - T_g) \\ - h_B (T_s - T_c) - 2\sigma_{SB} \varepsilon_{Si} (T_s^4 - T_c^4) \tag{45}$$

In Eq. (44) the net heat flux to the wafer surface, Q_N, is divided by the surface to volume ratio (wafer thickness Δ) to accommodate the surface term into the one-dimensional model. In Eq. (45) the first two terms are heat generation due to surface recombination (reaction R24 in Table 4, heat of recombination ΔH_{rec}) and etching ($Si + 4\,Cl \rightarrow SiCl_4$, heat of reaction ΔH_{rec}), respectively, which are normally both exothermic reactions. The third term is the rate of energy transfer to the wafer via energetic ion bombardment, assuming that the ions deposit all their energy on the wafer. The fourth and fifth terms are the rates of energy loss from the wafer top and backside respectively. The last term is the energy loss via radiation, where ε_{Si} is the emissivity of silicon. Energy gain and loss mechanisms and model geometry are depicted in Fig. 30.

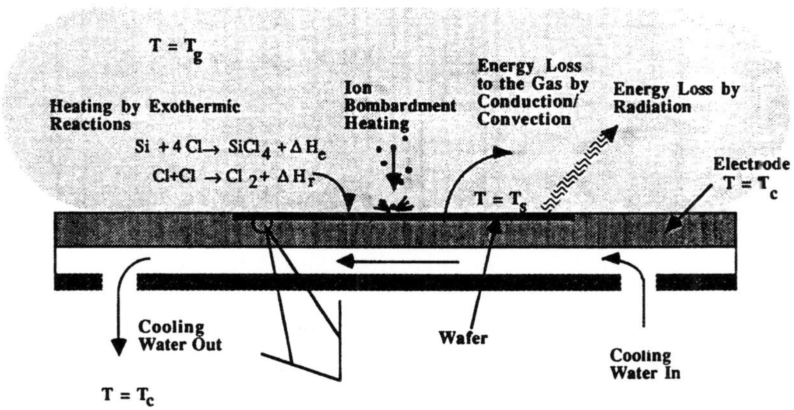

Fig. 30. Mechanisms of heat gain and loss of a silicon wafer during etching in a chlorine plasma. After [98].

Boundary conditions are

$$\frac{\partial T_s}{\partial r} = 0 \quad \text{at } r = 0 \tag{46}$$

$$-K_{Si}\frac{\partial T_s}{\partial r} = h_T(T_s - T_g) \quad \text{at } r = r_w \tag{47}$$

$$T_s = T_c \quad \text{at } t = 0 \tag{48}$$

The initial wafer temperature may be taken equal to the electrode temperature (Eq. 48). The latter can be assumed to be kept constant by cooling water flow.

The wafer energy balance has to be solved simultaneously with the plasma reactor model since the etchant concentration, ion flux, and energy are needed as input to the wafer energy balance [98]. In turn, if the etch rate is temperature sensitive, the flux of product into the plasma (and in turn the plasma composition and output of the plasma reactor model) will depend on wafer temperature. Even for a temperature insensitive etch reaction, the wafer temperature will affect the gas density for an etch with controlled pressure (typical case).

The most important energy loss mechanism is typically backside cooling. The gap between the wafer and the electrode is expected to be some tens of microns. At low pressures, at which the mean free path of gas species is much larger than the wafer-electrode gap, the heat transfer coefficient for backside cooling is independent of the gap width and is given by

$$h_B = \alpha_{acc}\Lambda_{Cl_2}P \tag{49}$$

where

$$\Lambda_{Cl_2} = \frac{1}{8}\frac{(\gamma+1)}{(\gamma-1)}\frac{1}{T_c}\sqrt{\frac{8kT_c}{\pi m_{Cl_2}}} \tag{50}$$

is the free-molecular heat conductivity. It has been assumed that the gas dissociation is very low [50, 160] so that molecular chlorine is by far the dominant gas phase species. In HDP systems a substantial fraction of the gas may be dissociated [130] and adjustments to the model will have to be made. In practice, wafers are clamped to the electrode and the backside is filled with He to a pressure of several torr. In such cases the properties of He must be used in Eq. (50).

When molecules strike a hot surface, complete interchange of energy does not occur at the first collision. Indeed, several collisions may be required before the molecule and the surface equilibrate. To take this effect into account, Knudsen introduced the accommodation coefficient, α_{acc}, which "can be defined as standing for the fractional extent to which those molecules that fall on the surface and are reflected or re-emitted from it have their mean energy adjusted or 'accommodated' toward what it would be if the returning molecules were issuing as a stream out of a mass of gas at the temperature of the wall." Accommodation coefficients have been measured or estimated for some systems [161]. For very rough surfaces, the accommodation coefficient approaches unity. Some attempts have been made to calculate the accommodation coefficient based on molecular collision and scattering concepts. A simple expression is [162]

$$\alpha_{acc} = \frac{2m_1 m_2}{(m_1 + m_2)^2} \tag{51}$$

where m_1 and m_2 are the mass of gas specie and surface atoms, respectively.

Fig. 31 shows the heat transfer coefficient for various wafer-electrode gap widths and He backside pressures, assuming an accommodation coefficient of $\alpha_{acc} = 0.4$ [163]. As mentioned above, the heat transfer coefficient is independent of gap width and increases in proportion to pressure in the Knudsen regime. At high pressures, for which gas conduction predominates, the heat transfer coefficient is independent of pressure and varies inversely proportionately with the gap width. The dots on the curves in Fig. 31 signify the point at which the gap width is equal to the molecule mean free path at the corresponding pressure.

Heat transfer from the topside of the wafer can be described by

$$-K_g \left(\frac{dT_g}{dz} \right)_{z=-L} = h_T (T_s - T_g) \tag{52}$$

For stagnation point flow between a pair of parallel disk electrodes with the gas uniformly distributed through the area of one of the disks (see Fig. 3a), the hydrodynamics is well defined to the point that an analytic solution is possible for low Re [160]. The corresponding heat transfer coefficient can be shown to be

$$h_T = \frac{K_g}{L} Pe_h \left(\frac{3}{16} + \frac{821}{26880} Re \right) \frac{1}{\left(1 - exp\left(-Pe_h \left(\frac{3}{16} + \frac{821}{26880} Re \right) \right) \right)} \tag{53}$$

where $Pe_h = u_w \rho_g C_{pg} L / K_g$, is the Peclet number for heat transfer and $Re = u_w L / 2\nu_g$, is the Reynolds number based on the gas velocity at the showerhead electrode u_w.

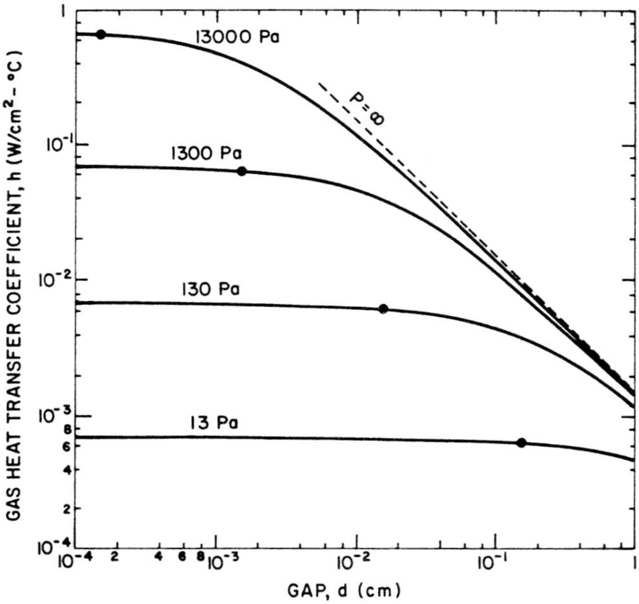

Fig. 31. Heat transfer coefficient between the wafer backside and the electrode for different He gas pressures as a function of the gap width. The point on each isobaric curve gives the gap width which is equal to the corresponding gas mean free path. After [163].

When $Pe_h \ll 1$ and $Re \sim 1$, the heat transfer coefficient can be approximated by K_g/L, that is, for low gas flow rates, heat transfer from the top of the wafer is dominated by conduction.

Figure 32, top shows the space and time resolved etch rate of polysilicon monitored in real time using spatially resolved laser interferometry [164]. The atomic chlorine concentration (measured by OES as the ratio of the 808.7 nm Cl emission to the 811.5 nm Ar emission, Section 7.1.2) as a function of time is also shown. The etch rate increases monotonically with time despite the fact that the atomic chlorine concentration initially decreases with time during etching (from start of etching to point A). The wafer edge etches faster than the wafer center. The edge starts clearing 220 seconds after the plasma is turned on (point A). At this time the atomic chlorine concentration in the reactor starts rising. Points B, C and D indicate the times at which the film clears to radial positions of 28 mm, 12 mm and 0 mm (complete clearing), respectively. As the film edge starts clearing and the atomic chlorine concentration rises, the etch rate increases even more rapidly. This increase in etch rate is partly due to an increase in Cl concentration in the reactor due to loading (Section 6.4): the Cl reactivity on oxide is much lower than that on polysilicon. As the polysilicon film clears and the underlying oxide is exposed the atomic chlorine concentration rises. However, the increase in etch rate before wafer edge clearing (from start of etching to point A) can't be attributed to loading. Since the Cl concentration is actually decreasing during that time, the increase in etch rate is due to an increase in wafer temperature.

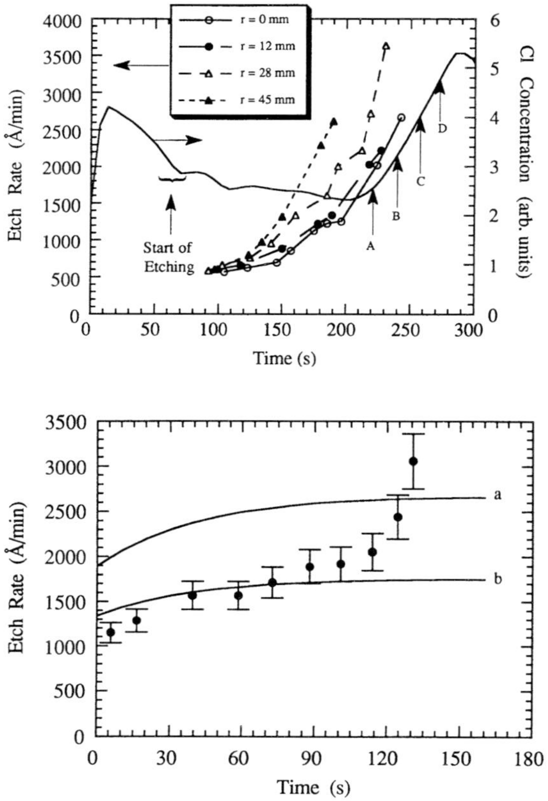

Fig. 32. (top) Spatiotemporal profiles of etch rate of silicon in a chlorine plasma measured by laser interferometry. The measured Cl atom concentration is also shown (bottom) Etch rate of silicon (at the wafer center) in a chlorine plasma as a function of time predicted by a mathematical model (solid lines) and measured under the same conditions (points). Curves a and b correspond to accommodation coefficient values of 0.50 and 0.75, respectively. After [98].

Figure 32, bottom shows the comparison between the experimental (points) and theoretical (lines) etch rate at the wafer center as a function of time. Model predictions are shown for two different thermal accommodation coefficients, α_{acc}, which was the only adjustable parameter in the model. In order to compare theory to experiment, the initiation period required to remove the oxide (as determined by laser interferometry) was subtracted from the real time. The etch rate is higher for the lower thermal accommodation coefficient because the wafer temperature is then higher. The model captures the initial transients as well as the magnitude of etch rate satisfactorily considering the uncertainty in the etch rate expression used. The increase in etch rate after 120 s is due to loading as the polysilicon film is clearing and progressively more of the underlying oxide is exposed to the plasma. Film clearing was not incorporated in the model but can be easily handled [79].

5.4.5 Dimensionless Numbers

It is instructive to look at the set of dimensionless numbers that characterize the (neutral) gas flow in plasma systems. Using typical values for low-pressure high

Table 5. Important dimensionless numbers in neutral transport and reaction in plasma reactors.

Symbol	Name	Expression	Typical value
Da	Damkohler number	$\frac{k_i n_e L^2}{D}$	1–100
Pe	Peclet number	$\frac{uL}{D}$	1
Re	Reynolds number	$\frac{uL}{\nu}$	1
Φ	Thiele Modulus	$\frac{k_s L}{D}$	0.1–10

density systems, and the physical and transport properties of chlorine, one obtains the results shown in Table 5. These results do not apply at very low pressures for which the usual transport properties (gas diffusivity) loose meaning. Also, the flow velocity is that prevailing at the bulk of the reactor; velocities near entrance ports are normally much higher due to small size of the inlet ports. Finally, reactions are assumed to be first order in reactant concentration.

The Reynolds number, which characterizes the importance of inertial forces compared to viscous forces is around unity. This implies that the non-linear terms in the Navier-Stokes equations are weak, easing the task of solving these equations. For some systems, the assumption of Stokes flow may be reasonable, i.e., the inertial terms are set equal to zero; this affords a significant simplification of the fluid flow problem [160]. The Reynolds number is independent of pressure, when everything else is held constant.

The Peclet number shows the relative importance of mass (or heat) transport by convection as compared to molecular (gas-phase) diffusion. Again, typical values of Pe are around unity, implying that the flow is quite diffusive. The Peclet number is independent of pressure, when everything else is held constant.

The Damkholer number shows the relative importance of species production (or loss) by gas-phase mechanisms relative to their ability to diffuse away from the source (or sink). When Da > 1, large concentration gradients may be expected. The Da number in these systems varies depending on the power level (which affects electron density and degree of dissociation) and pressure (which affects neutral density, diffusivity, and the rate coefficient for electron impact reactions).

The Thiele Modulus shows the relative importance of surface reaction compared to gas-phase diffusion. The value of Φ depends critically on the wall reaction rate coefficient, for example the wall recombination probability for Cl radicals (reaction R24 in Table 4). Large values of the Thiele Modulus imply a diffusion-controlled situation and strong density gradients. The Thiele Modulus decreases as pressure is decreased. However, even at pressures as low as 10 mtorr, the Thiele Modulus can be high enough for substantial concentration gradients to develop [41].

Ion flow is highly convective especially near the wall, and $Pe_+ \gg 1$. In low pressure plasmas, ions accelerate from the point of their birth until they strike the wall. Away from the central plane of the discharge (plane of symmetry), the ion drift velocity is much larger than their thermal velocity and both are larger than the gas velocity.

5.5 Sheath Simulations

Formation of the sheath was detailed in Section 2.1. Typical events that can occur in the sheath are shown in Fig. 33, top [163]. Fast electrons from the plasma (event a) can overcome the (decelerating) sheath potential and reach the wall. In fact, such electron current is necessary to neutralize the positive ion current to the wall. Slow electrons (event b) are reflected back into the plasma. Negative ions (event c) can't penetrate as far as electrons (because of the much lower energy of the negative ions) and are also reflected. Positive ions (event d) are accelerated by the sheath field. In their transit through the sheath, positive ions can suffer elastic scattering collisions (event f), or charge exchange collisions (event e) or engage in chemical reactions with neutrals (not shown in Fig. 33). Collisions lower the energy of ions bombarding the electrode to below the sheath potential. Secondary electrons can be emitted as a result of ion bombardment (event g), and are accelerated back into the plasma, performing ionization of neutrals on the way (event h).

One of the simplest sheath models is that of a DC high voltage ($V_{sh} \gg T_e$) sheath that contains no electrons and in which ion flow is collisionless (space-charge limited current). The resulting sheath equation is called the Child-Langmuir law [165].

$$J_+ = \frac{4}{9}\varepsilon_0 \left(\frac{2e}{m_+}\right)^{1/2} \frac{V_{sh}^{3/2}}{s^2} \tag{54}$$

Equation (54) was first used for diodes to calculate the current that can be drawn between two plates of given separation as a function of the potential between the plates. When applied to the plasma sheath, it provides important scaling between the sheath thickness, voltage, and ion current. It turns out that for a constant ion current, the sheath thickness scales as $s \propto \lambda_{De}(V_{sh}/T_e)^{3/4}$ where the Debye length is calculated by using the electron density and temperature at the sheath edge [6]. For a collisional sheath, assuming a constant ion mean free path [6],

$$J_+ = \frac{2}{3}\left(\frac{5}{3}\right)^{3/2} \varepsilon_0 \left(\frac{2e\lambda_+}{\pi m_+}\right)^{1/2} \frac{V_{sh}^{3/2}}{s^{5/2}} \tag{55}$$

For fixed sheath voltage and ion current, the sheath thickness scales as $s \propto \lambda_+^{1/5}$, i.e., a weak dependence on pressure.

A variety of models of the DC sheath have been published that make different assumptions regarding ion flow (collisional vs. collisionless), presence or absence of electrons (including secondary electrons), and the form of the boundary conditions. Both fluid [78, 166, 167] and kinetic models [168, 169] have been developed. Some of these are used to describe the high frequency RF sheath, when ions respond to the time-average (a DC) voltage [78]. Figure 33, bottom shows typical results of a fluid

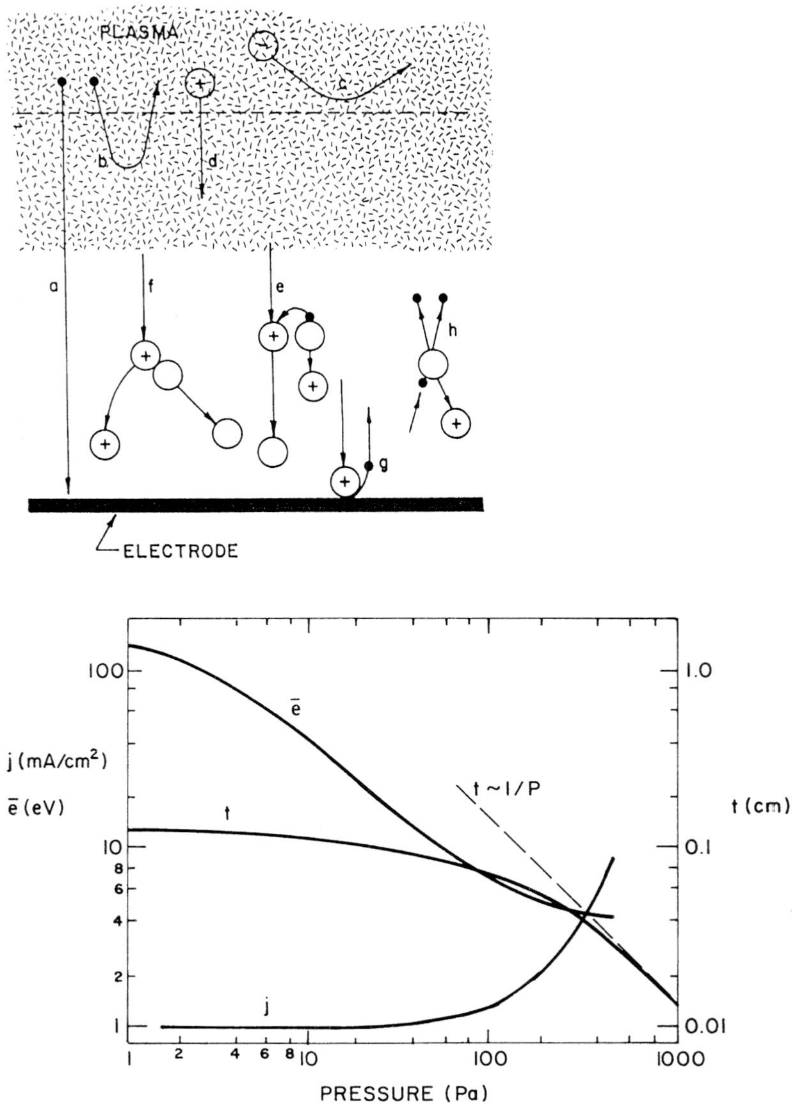

Fig. 33. (top) Representation of phenomena occurring in the sheath: (a) fast electrons can overcome the potential barrier of the sheath thus escaping to the electrode, (b) slow electrons are reflected by the potential barrier, (c) negative ions are also reflected by the potential barrier, (d) positive ions are continuously injected into the sheath, (e) positive ions can suffer charge exchange with neutrals, or (f) elastic scattering by neutrals; (g) ion bombardment of the electrode can release secondary electrons, which (h) can induce ionization in the sheath. After [163]. (bottom) Calculated DC sheath behavior versus pressure for the conditions given in the text. Here t is the sheath thickness, j is the ion flux striking the wall and $\bar{e}$ is the average energy of ions bombarding the wall. After [166].

model of a DC sheath [166]. The sheath thickness, ion current density, and mean ion bombardment energy, are shown as a function of pressure. These results were calculated for a sheath voltage of 200 V, an ion current injected at the sheath boundary (assumed constant) of 1 mA/cm^2, and a secondary electron emission coefficient of 0.2 (also assumed constant). This means that for every 10 ions striking the electrode there are 2 electrons ejected. The ion bombardment energy approaches asymptotically the sheath potential at low pressures for which there are few, if any, collisions in the sheath. At low pressures the sheath thickness decreases slowly with pressure ($s \propto \lambda_+^{1/5}$) as predicted by Eq. (55) assuming that the ion mean free path is independent of the ion velocity. As pressure increases, the secondary electrons cause ionization in the sheath resulting in a rapid increase of the ion flux to the electrode. The larger ion current and associated space charge cause the sheath to contract as predicted again by the collisional version of the Child law Eq. (55). At high pressures, the sheath thickness is inversely proportional to pressure. Since the ion mean free path is also inversely proportional to pressure, the mean number of collisions ions experience in the sheath remains constant. Hence the mean ion bombardment energy levels out at high pressures.

As explained in Section 3.2, the ion energy distribution at the wafer is an important factor in determining the rate of surface reactions in plasma processing. Ions falling through a DC sheath without collisions will acquire the full sheath potential. For a sheath potential much larger than the electron temperature (the latter expressed in volts) the ions bombarding the wall will be almost monoenergetic. Ion-neutral collisions will lower the ion impact energy and result in an ion energy distribution (IED) function.

Plasma processing reactors normally operate with the wafer biased at radio frequencies, typically in the range 0.1 to 13.56 MHz. Even if the ions injected at the sheath edge were monoenergetic, an IED would result in an RF (time-dependent) sheath, even in the absence of collisions. The literature on RF sheaths is voluminous. Both fluid [170–175] and kinetic (e.g., Monte Carlo) [176–180] simulations have been reported. One of the most important results of such simulations is the IED. The ion angular distribution (IAD) [74, 75] and sheath impedance (for use in equivalent circuit models) [32] are also of importance.

The critical parameter that controls ion modulation in collisionless RF sheaths is $\omega\tau_+$, where ω is the frequency of the applied field, and τ_+ is the ion transit time through the sheath. The ion transit time can be estimated by assuming a collisionless Child-Langmuir sheath (Eq. 54), and neglecting the velocity of ions entering at the sheath edge (Bohm velocity). The result is [181]

$$\tau_+ = 3s\sqrt{m_+/(2eV_{sh})} \tag{56}$$

When $\omega\tau_+ \ll 1$ ions traverse the sheath in a short time compared to the field oscillations. Under this condition, an ion traversing the sheath experiences the sheath voltage prevailing at the time the ion entered the sheath. In the absence of collisions, the IED function will reflect precisely the variation of the sheath voltage with time. This quasi steady-state condition of $\omega\tau_+ \ll 1$ is satisfied for low RF frequencies or short ion transit times, i.e., thin sheaths (low sheath voltage or small Debye length),

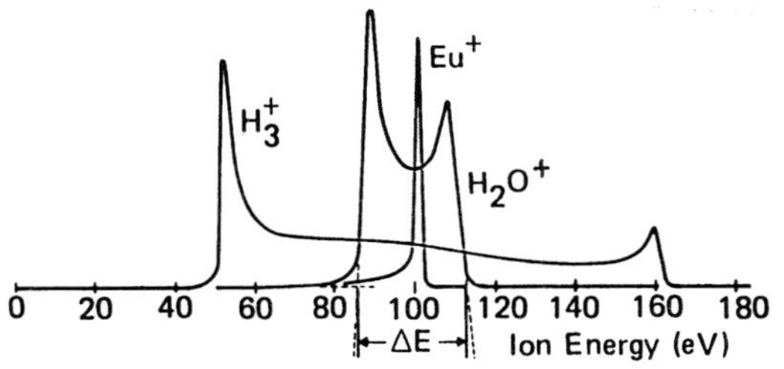

Fig. 34. Measured ion energy distributions (IED) at the wall for ions of different masses crossing a RF sheath over that wall. The energy spread is reduced as the ion mass increases. After [28].

or ions of small mass. At the other extreme of $\omega\tau_+ \gg 1$, ions experience many field oscillations while in transit through the sheath. Ions will then respond to the time-average sheath potential, and the IED function will have a smaller spread over energy.

Fig. 34 shows the energy distribution of several ions impinging on the grounded electrode of a 13.56 MHz RF discharge [28]. In this case the sheath potential is identical to the plasma potential. Heavy ions (Eu^+) have a long transit time corresponding to $\omega\tau_+ > 1$. Their IED is narrow and corresponds to the time-average plasma potential (100 V in this case). As the ion mass and transit time decrease ($\omega\tau_+ < 1$), the IED becomes wider and reflects more and more the time-dependence of the plasma potential. Theory predicts that the energy spread ΔE should scale as $m_+^{-0.5}$ [182]. The measurements are in reasonable agreement with this expectation.

The two extreme conditions discussed above are more amenable to analysis since, in both cases, the sheath can be described as a DC sheath; actually a series of DC sheaths at the different moments in time during the RF cycle when $\omega\tau_+ \ll 1$, and a DC sheath at the time-average voltage when $\omega\tau_+ \gg 1$. The most difficult situation to analyze is when $\omega\tau_+ \sim 1$. Monte Carlo simulations have been performed in this intermediate regime [179, 180] but the boundary conditions for ion injection are not well understood, and comparison with experimental data is lacking. A fluid model developed by Miller and Riley [32] attempts the bridge the gap between the low and high frequency sheaths.

In general the sheath current can be decomposed into conduction current (due to charged particle flow through the sheath) and displacement current (due to the oscillations of the electron cloud near the wall, Fig. 7). A generalized sheath diagram is shown in Fig. 35 [183]. Different regimes of operation are shown. In general, under the condition $\omega\tau_+ \ll 1$, the sheath is resistive, meaning that the ion conduction current dominates over the displacement current. In this case, the capacitor can be dropped but the resistor (or current source) has to be retained in the equivalent circuit of the sheath (Fig. 10). At the other extreme $\omega\tau_+ \gg 1$ the sheath is capacitive, meaning that the displacement current dominates over the ion conduction current. In this case, the resistor (or current source) can be dropped, but the capacitor has to be retained in the equivalent circuit of the sheath.

In capacitively coupled systems (Fig. 7) the ion bombardment energy increases with decreasing frequency (Fig. 36). This is because the sheath voltage increases at lower frequency [184]. Hence a larger fraction of the applied power is dissipated for accelerating ions in the sheath rather than for producing radicals in the bulk plasma.

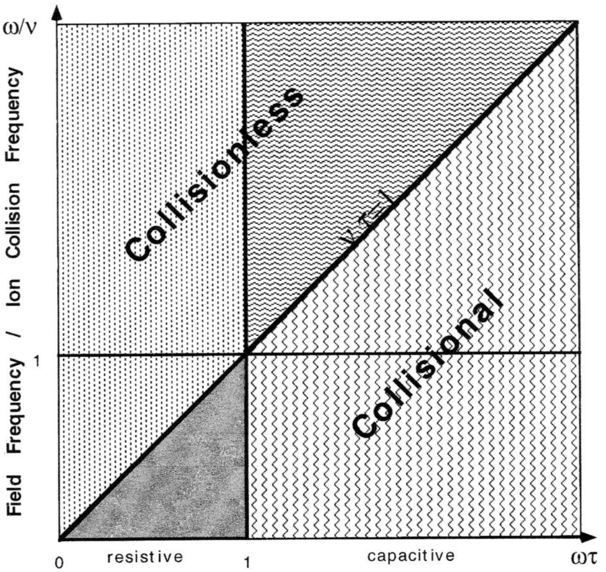

Fig. 35. Classification of sheath behavior based on the applied frequency (ω), ion transit time through the sheath (τ) and ion-neutral collision frequency (ν). After [183].

Higher ion energy generally implies better anisotropy, but selectivity and radiation damage worsen.

5.6 Microfeature Simulations

5.6.1 Profile Evolution

The ultimate goal of plasma etching is to obtain controlled wall profiles of microscopic features etched into a film. The minimum feature width is rapidly falling below 0.25 μm in modern microelectronic devices, and the typical aspect ratio of features will soon become greater than of 10:1. Etching narrow, high aspect ratio features in a reproducible manner in a manufacturing environment is a very challenging task. The problem is complicated by the fact that, frequently, features of different aspect ratios have to be etched on the same wafer in the same plasma. If the etch rate depends on aspect ratio (Fig. 40, below), overetching of some regions of the wafer will be required which can lead to profile distortion and device damage.

The shape of an etched feature depends on a plethora of geometric, material, and plasma parameters. The shape evolution problem is coupled to the plasma reactor through the sheath (Fig. 23, top). Feedback from the feature to the reactor occurs through the flux of product species coming out of the feature and the varying surface area available for reaction as the feature shape evolves. The former affects the plasma gas composition and in turn the flux and energy distribution of species incident on

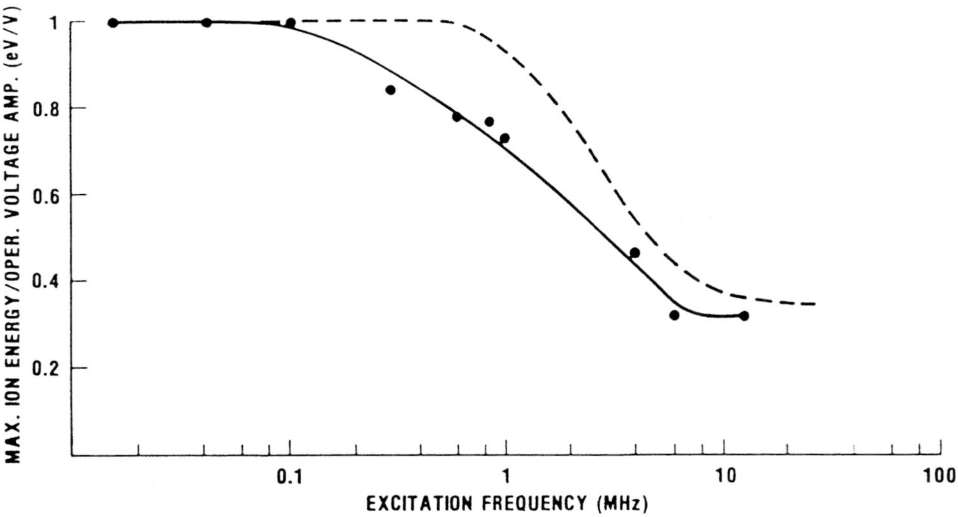

Fig. 36. Dependence of the maximum ion bombardment energy on the excitation frequency in a capacitively coupled chlorine discharge at a pressure of 0.3 torr. At low frequencies ions traverse the sheath in a short time compared to the period of the field. The maximum ion energy is then equal to the peak of the sheath voltage. At high frequencies, the ion transit time is many RF cycles; ions respond only to the time-average sheath voltage. The transition happens at the ion plasma frequency which for the conditions of this Figure is about 1 MHz. After [184].

the feature. The latter can be important, for example, when *loading* is present (see Section 6.4).

Another issue which is becoming more important as the feature aspect ratio increases is that of charging damage resulting from ion and electron bombardment of the wafer. One form of damage is the so-called *notching* [185–187] observed in etching of polysilicon lines over oxide (Fig. 37). Because ions coming out of a plasma are much more directional than electrons, the sidewalls of a microfeature charge negatively while its bottom charges positively. Further ions coming into the feature are deflected by the charge deposited on the surface, and their trajectories are bent to the point that these ions induce etching of the polysilicon sidewall. This distortion of the sidewall profile (notching) is highly undesirable. Charging and ion deflection are also suspected to be responsible for the so-called *RIE-lag* or ARDE in which narrow, high-aspect-ratio features etch slower than wide, low-aspect-ratio ones (see Section 5.6.2). Another form of serious charging damage is *gate oxide breakdown* because of large currents flowing through the oxide in a non-uniform plasma. In order to simulate charging damage and profile evolution of microfeatures, the multidimensional electric field profiles near the trench must be calculated.

It is instructive to first consider the forward problem from the reactor to the feature, ignoring the coupling back into the reactor. Because of the disparity of the length scales involved, it helps to break down the problem into several pieces. An approach proposed by Economou and Alkire [89] is shown in Fig. 38. The near wafer space is separated into two regions, Region I and Region II. Region I contains the

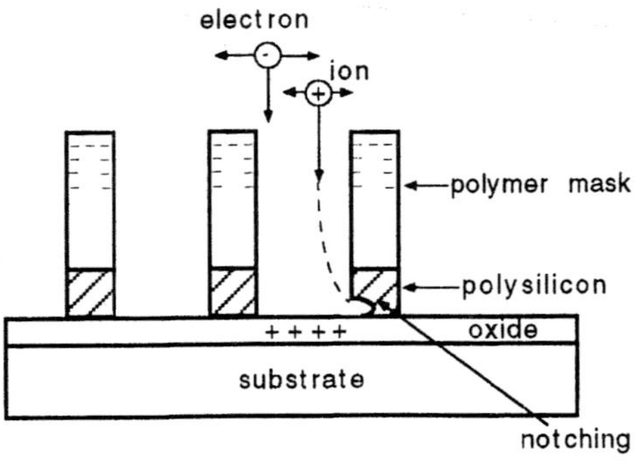

Fig. 37. The notching effect. When the polysilicon film clears, the oxide charges up to the point of deflecting oncoming ions which then etch the sidewall. Electrons can't reach the bottom of the trench to neutralize the positive charge.

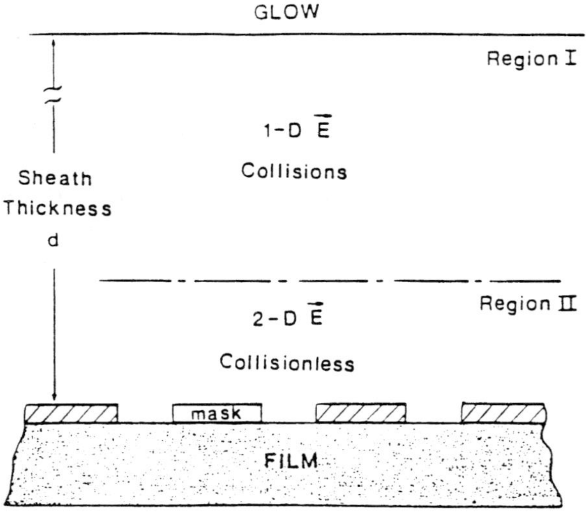

Fig. 38. The two region model for doing feature charging calculations. After [89].

sheath except for the immediate vicinity to the features which is designated as Region II. Since the sheath is much thicker than the feature size and, at the same time, much thinner than the wafer size, a one-dimensional sheath model suffices for Region I. Region I contains a net space charge, hence the Poisson equation has to be solved for that Region. The "flow" of the calculation is shown in Fig. 39 [187]. The boundary conditions at the sheath/plasma interface are the electron and ion densities, electron temperature (or the velocity distribution functions of these species), and the plasma potential relative to the macroscopic surface potential. A one-dimensional sheath simulation yields the density, energy, and angular distribution of the charged species leaving Region I and entering Region II. Region II, in the immediate vicinity of the feature, has to be described by at least a two-dimensional electric field model. The species trajectories, as influenced by this field, are followed in Region II until species strike the wall and deposit their charge. Owing to the very low charge density near the feature, the Poisson equation can be replaced by Laplace's equation in that region. As charges accumulate on the wall, the wall potential is modified and Laplace's

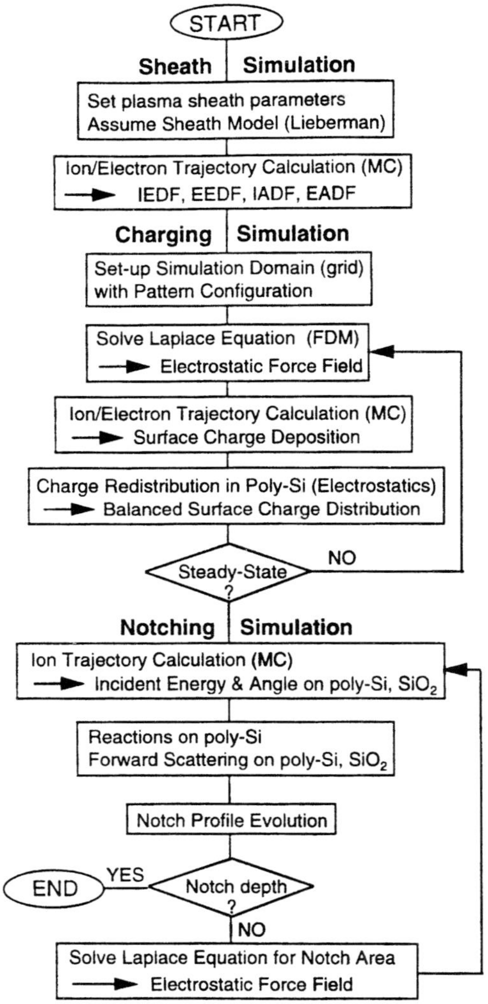

Fig. 39. Flow chart for topography-dependent charging simulation. After [187].

equation has to be solved iteratively in Region II until a steady state is reached. The time constant to approach the steady state is a few ms, much faster that the time scale of shape evolution [188]. Knowing the flux and energy of ions striking the walls, the local etch rate can be calculated (assuming a reaction yield), the wall profile advanced, and the charging calculation repeated until the feature is etched to the desired depth [188–190].

In general, in order to perform a shape evolution calculation one needs to have information on the following:

(a) the flux, energy distribution, and angular distribution of ions and neutrals along the wall as a function of the instantaneous shape of the feature. As explained above, charged species trajectories can be deflected by local charging of the walls.

Neutral species flux variations can also exist along the walls. For example, if the neutrals react with high probability, the neutral flux will be determined by the line of sight flux reaching the surface (neutral shadowing); or Knudsen diffusion will lower the concentration of neutrals deep inside high aspect ratio trenches.

(b) etch rate of the materials involved (film to be etched, mask) for given neutral-to-ion flux ratio and angle of incidence of the ions.

(c) the energy and angular distribution of energetic ions (and possibly energetic neutrals) reflected off of the walls. These species may have substantial energy (especially when they strike the wall at glancing angles) and they can do further chemistry when striking the wall again.

(d) the reaction or sticking probability of products evolving from the feature walls.

(e) the surface diffusivity of adsorbed species and surface conduction of charge.

The shape evolution of microfeatures is followed using a number of techniques. The string model was popular at first [191], but profile advancement based on the method of characteristics [192, 193] is more robust. In another method [194, 195] the solid is divided into a large number of elements (digitized). The volume "digits" are removed (etching) based on the local etch rate and a local mass balance. If deposition takes place a volume element is added locally. Unfortunately, items (b)–(e) above are largely unknown for almost all material systems. Surface chemistry in particular is one of the major limitations of current models of profile evolution. One is then forced to use a number of adjustable parameters that limit the predictive value of the model.

5.6.2 Aspect Ratio Dependent Etching (ARDE)

ARDE refers to a situation commonly observed in etching of high aspect ratio (depth:width) trenches [196]. It has been found that the etch rate decreases as the trench aspect ratio increases (also known as RIE lag). Fig. 40 shows a typical situation [197]. ARDE can occur due to a combination of effects. As the trench becomes deeper, for example, ions may impact the sidewall instead of the trench bottom. This can be due to the angular distribution of the incoming ion flux, or the decollimation of ions due to electric fields developed by non-uniform charging of the mask or the trench sidewalls. As the flux of ions impinging on the trench bottom decreases, the etch rate follows suit. Also, transport limitations may reduce the flux of neutrals reaching the bottom of the trench for high aspect ratio trenches. Finally, in a less common situation, the etch rate increases as the trench deepens (inverse RIE lag).

5.7 Atomistic Simulations

Atomistic simulations provide information at the molecular level. Such information is useful for understanding the progress of an event occurring in the gas phase or on the surface, for delineating reaction mechanisms and pathways, for calculating rate coefficients of difficult to measure reactions, and for enhancing one's understanding of the process by visualizing the atomic scale events. Molecular modeling techniques

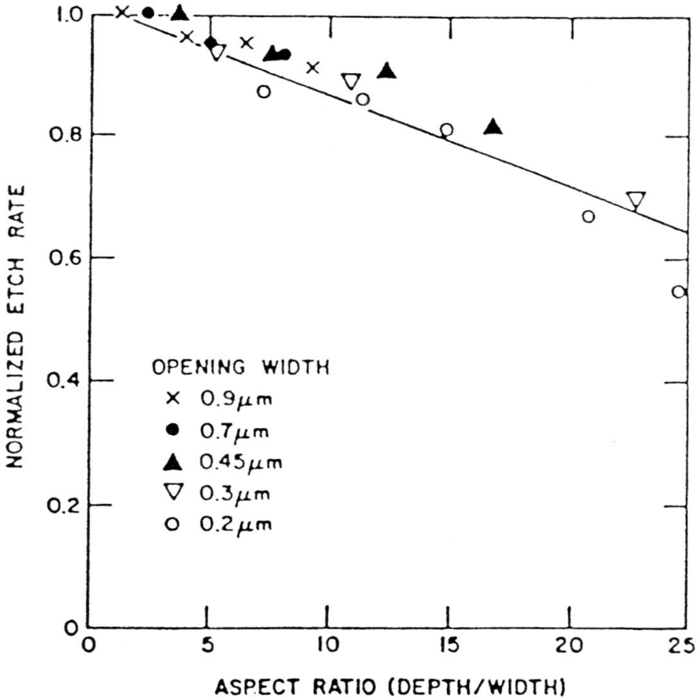

Fig. 40. Aspect ratio dependent etching of silicon in a CCl_2F_2/O_2 gas mixture. The etch rate is a function of the aspect ratio of the feature (depth/width). After [197].

include Ab initio methods (e.g., density functional theory, DFT) of electronic structure and total-energy calculations [198], and semi-empirical and empirical atomic-scale simulations (e.g., Molecular Dynamics, MD [199], and Monte Carlo, MC [200]).

MD simulations are suitable for studying the interaction of energetic ions with solids. MD follows the trajectory of each atom in the simulation cell as a function of time for several ps of the ion-solid interaction by solving Newton's equation of motion for each atom. Physical quantities of interest may be calculated by time or ensemble averaging. The most critical input is the interatomic potential that is used to calculate the force on each atom. Reliable empirical potentials exist for common systems (e.g., silicon), but such potentials are, in general, not readily available. The MD simulation cell can contain millions of atoms, i.e., length scales >100 Å can be addressed. The current practical limit of time scales is ~ 100 ps. MD simulations have been used to study physical sputtering [201], physical vapor deposition [202], and the reaction of energetic neutrals with a surface in the absence of ion bombardment [203]. MD simulations have also been used to study the interaction of ions with Si under simultaneous exposure to chlorine [204], and fluorine [205], as would be the case in plasma etching. Finally, MD was employed to understand atomic scale events and surface reaction mechanisms during atomic layer etching of silicon [206, 207].

Kinetic Monte Carlo (MC) methods are also useful [208], but one needs to have a fairly complete idea of the physics and chemistry of the problem to apply them. For example, the types of possible events and the probability of occurrence of each event must be known, before the time evolution of the process can be simulated. MC can address much longer time scales compared to MD (e.g., diffusional or adsorption time scales). Combinations of MD and MC may be useful in simulating the range of time scales from atomic vibrations to surface chemical reactions.

6 Plasma Etching and Deposition Systems

Table 2 provides a list of materials, and gases used for etching these materials. Silicon, oxide and metal are among the most important materials in microelectronics. Salient features of etching these materials are discussed below. Etching of other materials such as compound semiconductors [209], polymers [210] and other metals [211] is also of importance.

6.1 Etching of Silicon

Etching of any material in a reactive gas plasma results in the formation of a modified near surface layer where reactions occur. A visualization [212] of such layer formed during crystalline silicon (c-Si) etching in a chlorine plasma is shown in Fig. 41, top. This layer can be detected by spectroscopic ellipsometry, XPS, or laser-desorption/laser-induced-fluorescence (LD-LIF) measurements [213]. The surface layer is modeled by a top layer containing amorphous silicon (a-Si) and $SiCl_x$ species, and an interfacial layer containing a-Si, c-Si and $SiCl_x$. The c-Si substrate that was not affected by ion bombardment is shown underneath. Under a set of typical operating conditions (100 sccm chlorine gas flow, 2 mtorr pressure, $\sim$200 eV ion bombardment energy) a steady-state surface layer forms within a few seconds after plasma exposure. The thickness of the surface layer, as determined by spectroscopic ellipsometry, is a linearly increasing function of ion (Cl^+ and Cl_2^+) bombardment energy as shown in Fig. 41, bottom. The ion energy is approximately equal to the sum of the plasma potential ($\sim$20–30 V) and the absolute value of the dc bias. As ion energy increases, ions are able to penetrate deeper inside the silicon lattice, thereby creating a thicker near surface (damaged) layer. The surface layer is in the range of 10–35 Å under these conditions. The "equivalent' thickness of the top and interfacial layers as a function of ion bombardment energy are also shown in Fig. 41, bottom. The results of XPS measurements agree fairly well with the ellipsometric data. Surface films are all too common in electrochemical engineering, especially in corrosion.

The silicon etching yield (silicon atoms removed per incident ion) was found to depend on the ratio of neutral to ion fluxes. A set of beam experiments under high vacuum conditions (no plasma) is shown [70] in Fig. 42. An undoped polysilicon sample was bombarded simultaneously by separate beams of Cl radicals and Cl^+ ions

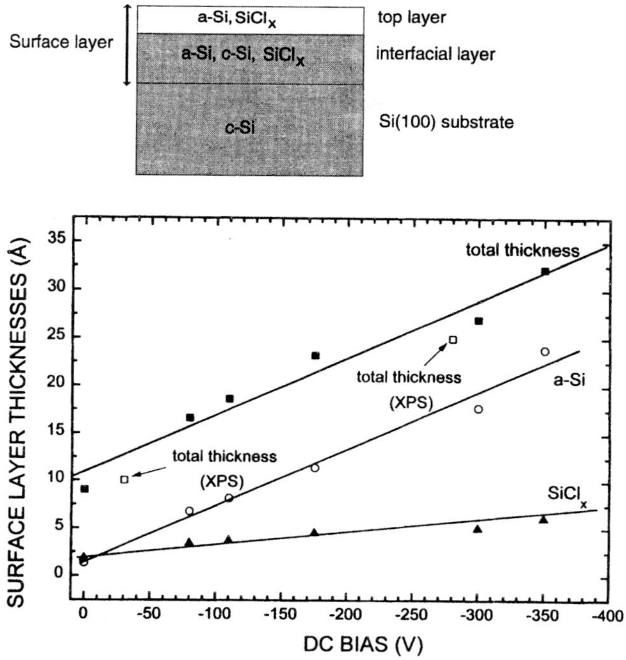

Fig. 41. (top) Model of surface layer formed during plasma etching of silicon with chlorine. (bottom) Surface layer thickness as a function of the DC bias voltage (ion energy). After [212].

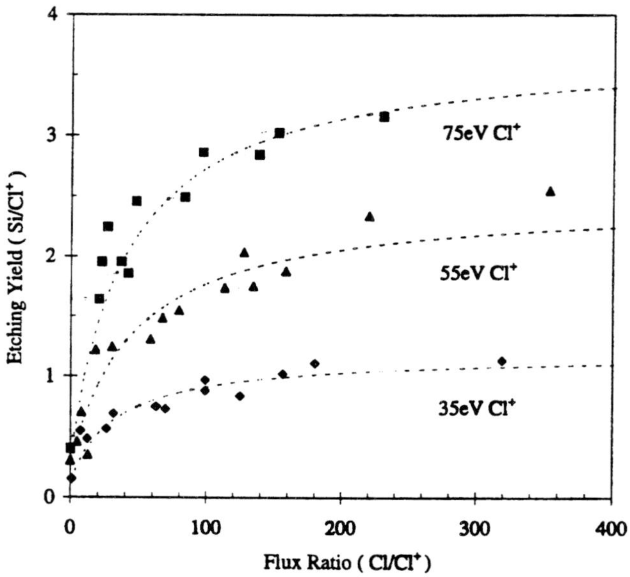

Fig. 42. Etching yield of silicon bombarded by a Cl atom beam simultaneously with a Cl^+ ion beam of controlled energy. At low values of the flux ratio, etching is limited by Cl atom supply. At the other extreme, etching is limited by the ion flux. Dashed lines are the result of a Langmuir-Hinshelwood model of the surface reaction. After [70].

of controlled flux and energy, similar to the experiments shown in Fig. 17. Exposure of the sample to the radical beam alone did not produce any appreciable etching under these low temperature conditions. The etching yield increases with ion energy; it turns out that the ion energy dependence is given by Eq. 17. For a given ion energy the yield is linear with flux ratio (FR) for low values of FR, and saturates for high FR. For low FR, etching is limited by the neutral flux; there is simply not enough reactant (chlorine) to form product, and the Cl surface coverage is expected to be low. At high FR, the surface is expected to be fully covered with chlorine, and etching is limited by the ion flux. Actually, since ion bombardment forms a damaged layer on top of silicon (Fig. 41), the chlorine coverage is more than a monolayer. The beam experiment shown in Fig. 42 is most relevant to polysilicon etching in high density plasma sources where the molecular chlorine dissociation is high. Hence, most of the reactant is Cl and the corresponding ion in the plasma is Cl^+. However, similar effects are observed [70, 214, 215] when silicon is bombarded by a molecular chlorine beam (instead of Cl) and an argon or Cl_2^+ beam (instead of Cl^+), except that in these cases the etch yield is lower.

For given ion energy and neutral to ion flux ratio, the polysilicon etching yield decreases monotonically with angle of incidence of the Cl^+ ion beam measured from the surface normal [70]. This behavior is to be contrasted with the angular dependence of the sputtering yield [15], when the sample is bombarded by an inert gas ion beam (e.g., Ar^+) alone. The angular dependence of the yield is needed to calculate the evolution of the profile of microfeatures during etching.

Silicon can be etched by chlorine atoms without any ion bombardment. Oxide is not etched by Cl and, therefore, a very high selectivity can be obtained. The chemical etch rate of n-type single crystal Si or polysilicon is given by [216]

$$R_{Si} = 4.04 \times 10^{-18} n_{Cl} N_D^{\gamma} \sqrt{T}\, exp(-E_A/kT) \tag{57}$$

where N_D is the dopant density. The activation energy E_A and the exponent γ depend on the type of the substrate (polysilicon or single crystal silicon, see Ref. 216). Wall passivation is necessary to achieve anisotropy in this system (Fig. 16d). Eq. (57) predicts a *doping effect*. In fact, heavily n-type doped silicon etches faster than undoped silicon which in turn etches faster than heavily p-type doped silicon. The effect is much stronger in Cl etching (orders of magnitude) compared to F etching of silicon (several times). The doping effect has been attributed to the variation in charge density at the surface as doping is varied. It is believed that Cl^- or F^- are ultimately involved in etching, hence the dependence on charge density [217].

Silicon can also be etched by F atoms via a chemical mechanism (without ion bombardment). The etch rate of single crystal Si (100) has been measured to be [218]

$$R_{Si} = 2.86 \times 10^{-12} n_F \sqrt{T}\, exp(-1250/T) \tag{58}$$

Finally silicon can be etched in fluorocarbon plasmas (e.g., CF_4). The addition of oxidation (oxygen) or reducing (hydrogen) agents to these plasmas has been used to control polymer formation and thereby affect the etch rate and selectivity. Fig. 43

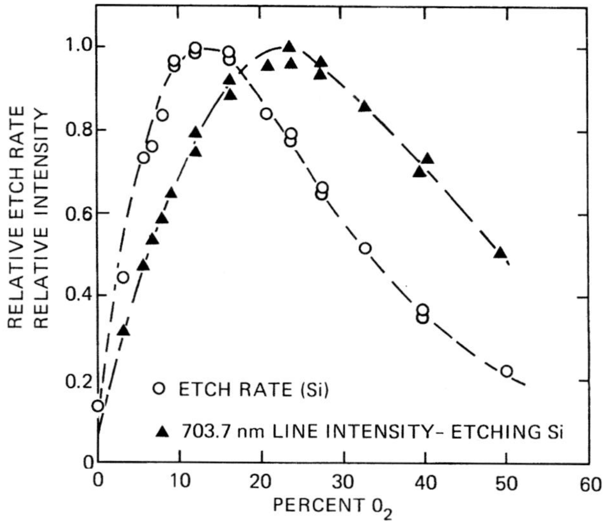

Fig. 43. Silicon relative etch rate and relative intensity of the 703.7 nm F-atom emission as a function of oxygen additions to a CF_4 discharge. After [219].

shows the effect of oxygen addition [219]. At low % oxygen, the formation of extra F is favored via reactions of the form

$$O/O_2 + CF_x \rightarrow COF_2, CO, CO_2 + F/F_2 \tag{59}$$

This enhances the etch rate of both silicon and oxide. However, at high % oxygen, two factors play a role: (a) the etchant is diluted, and (b) the Si surface is oxidized, and oxide etches much slower than Si. Hence the Si etch rate drops at high oxygen additions. The optimum oxygen addition is $\sim$10% for silicon. Oxide already has some oxygen, so the optimum is shifted to higher values $\sim$20 %. Similar results have been obtained by oxygen addition in other F-containing gases such as SF_6 and NF_3. Hydrogen addition to CF_4 [220] can have a more dramatic effect (Fig. 44). Hydrogen ties up F atoms in reactions such as

$$H, H_2 + F, CF_x \rightarrow HF, H, CF_{x-1} \tag{60}$$

thereby increasing the concentration of unsaturates. Hence the Si etch rate drops as hydrogen is added. At higher hydrogen additions, polymer formation blocks etching of silicon. However, oxygen from silicon dioxide can compensate the effect of hydrogen and prevent polymer formation, allowing etching of the oxide to proceed, provided there is energetic ion bombardment of the surface. Thus very high selectivity of etching oxide over silicon can be achieved. Coburn and Winters used the concept of F/C ratio to describe etching in fluorocarbon plasmas [221].

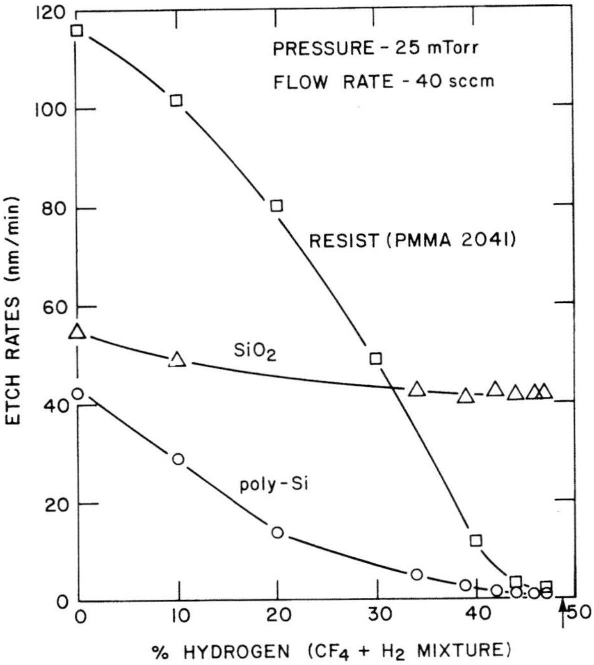

Fig. 44. Etch rate of polysilicon, oxide, and photoresist as a function of hydrogen addition to a CF4 discharge. Etch rate stops on non-oxide materials due to polymer build up at high % H_2. Oxide continues to etch due to polymer removal by the available oxygen. This yields high selectivity of etching oxide over silicon. After [220].

6.2 Etching of Silicon Dioxide (Oxide)

Oxide etching is performed in fluorocarbon plasmas, e.g., CF_4, CHF_3, C_2F_6, C_4F_8, or their mixtures. Selectivity with respect to silicon and silicon nitride (called simply nitride) is important. Selectivity is obtained by choosing the chemistry and operating conditions that favor etching of oxide but deposition of a fluorocarbon film that blocks etching when the silicon or nitride interface is reached. The chemistry of oxide etching is very complicated. Feedstock gases are dissociated in the plasma to produce fluorocarbon radicals (CF_x) and ions (CF_x^+). Radicals tend to deposit a fluorocarbon polymer layer on non-oxide surfaces (e.g., silicon), or on oxide surfaces not exposed to energetic ion bombardment (e.g., the sidewalls of a trench). The polymer does not form at the bottom of the trench on the oxide because ion bombardment induces reaction of the polymer-forming radicals volatilizing carbon as CO/CO_2 and also forming silicon fluoride product, thus etching oxide. In contrast, there is no efficient mechanism for carbon removal from silicon or nitride since they do not contain oxygen.

Oxide etching is ion driven. The etch rate of a flat (no features) bare oxide (no photoresist) surface as a function of ion energy in a CHF_3 inductively coupled plasma (Fig. 20a) is shown in Fig. 45 [222]. The ion energy was varied independently by applying a bias to the substrate electrode. All other conditions (including coil power generating the plasma) were kept fixed. At low ion bombardment energies

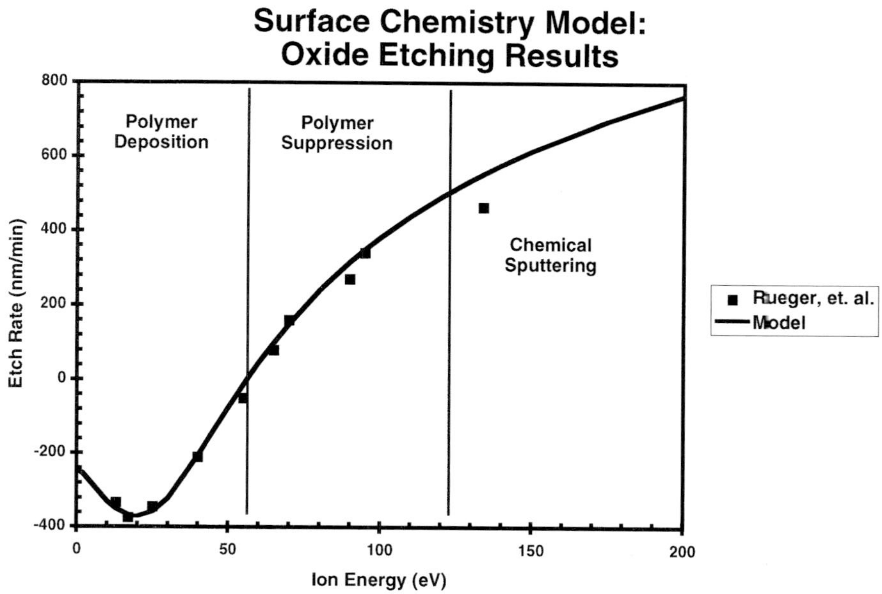

Fig. 45. Etch rate of oxide in fluorocarbon plasma vs. ion bombardment energy. Points are experimental data from [222].

(polymer deposition regime) there is actually polymer depositing on the oxide. Interestingly, initially there is an increase in the polymer deposition rate as the ion energy is increased. This corresponds to ion-enhanced deposition [223]. Apparently low energy ions create more dangling bonds (surface sites) on the polymer surface that enhance the deposition rate. As the ion energy is increased further (polymer suppression regime), polymer deposition competes with polymer (chemically-enhanced) sputtering; eventually sputtering wins and beyond an ion energy of $\sim$60 eV there is net oxide etching. In the etching regime, the oxide etch rate increases with ion energy following approximately the square root law (Eq. 17). A visualization of the surface as the ion energy is increased is shown in Fig. 46 [224]. At low ion energies the whole surface is covered by a polymer layer. Active sites on the polymer induced by ion bombardment are shown as crosses. As ion energy is increased, the polymer layer breaks apart into patches that eventually disappear. The surface is now covered by F and CF_x type adsorbates. The surface coverage is reduced as the ion energy is increased since more of the adsorbates are used up to form reaction products (the etch rate increases with ion energy, see Fig. 45). In reality, reaction does not happen strictly on the surface but in an ion-induced mixing layer that can be 10s of Å thick (compare to Fig. 41). This layer contains C_yF_x species and oxide etching byproducts mixed into the layer. Ion bombardment promotes the formation of species that are sputtered more easily and/or desorb spontaneously into the gas phase resulting in net etching of oxide.

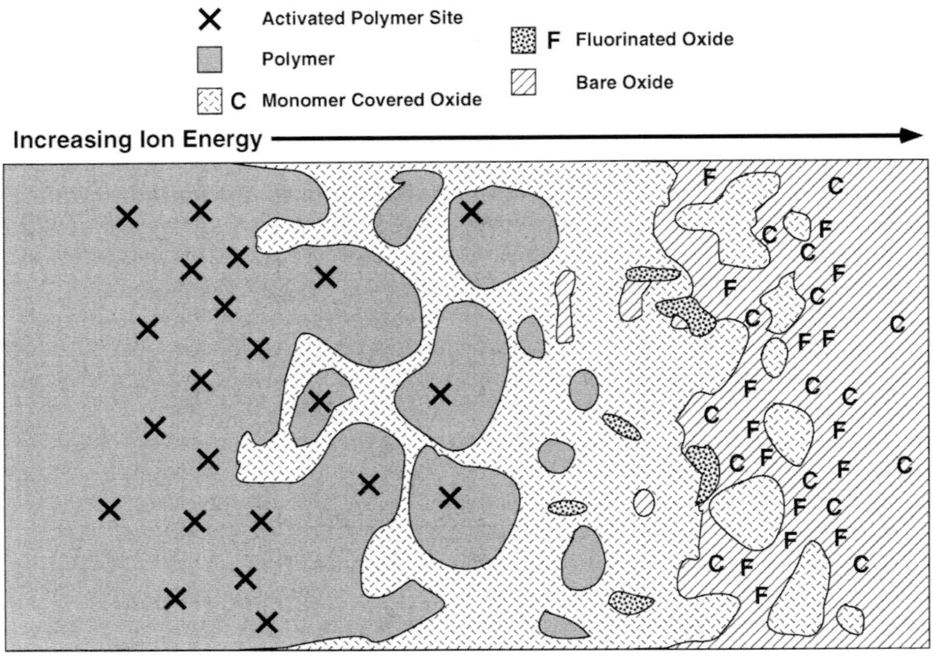

Fig. 46. Visualization of the surface of oxide during etching in fluorocarbon plasma as a function of energy of the impinging ions. After [224].

Fluorine atoms F also etch SiO_2 in the absence of ion bombardment with a rate [218]

$$R_{SiO_2} = 0.614 \times 10^{-12} n_F \sqrt{T}\, exp(-1890/T) \tag{61}$$

6.3 Etching of Aluminum

Aluminum is used as a metallization material in current microelectronic devices. It can be etched by chlorine containing plasmas, provided that the native aluminum oxide is removed. This is accomplished with oxide scavengers such as BCl_3 added to the plasma. Aluminum etching is isotropic; in fact a clean aluminum surface reacts with atomic and molecular chlorine even in the absence of a plasma [225]. Anisotropy is achieved by the wall passivation mechanism (Fig. 16d).

Aluminum is often used as an alloy with a few atomic % Si and/or Cu to reduce electromigration (Fig. 1). This alloy is difficult to etch because copper does not form readily volatile compounds [73]. Heating of the wafer to some 200 °C and intense ion bombardment to sputter off involatile products are then necessary to effect etching. Residual chlorine remaining on the wafer after aluminum etching can be deleterious because it can promote corrosion, especially in Al-Cu alloys. Hence wafer rinsing with DI water is common. Dry passivation by exposure to a fluorocarbon plasma has also been reported to protect aluminum by converting aluminum chloride to fluoride, which can then be removed by a nitric acid wet clean. Corrosion engineering is of critical importance for this material system.

Lately, copper metallization has become a viable alternative to aluminum [2, 226]. Copper is preferred over aluminum because of the higher conductivity and greater electromigration resistance of copper. Copper films can be grown by electro-deposition or by electroless deposition. The advantage of electrochemical methods over dry processes is the conformal coverage and complete filling of high aspect ratio sub-quarter-micron features. These are very difficult to fill with physical vapor deposition (PVD) [227] although ionized PVD has shown better results [228]. One drawback of copper is that a reliable etch process is not yet available. However, this is not a serious problem as chemical mechanical polishing used in the so-called *dual damascene* process can produce sub-quarter-micron patterns without the need for etching.

6.4 Loading

Loading refers to a situation whereby the etch rate varies inversely with the area of etchable material exposed to the plasma, Fig. 47 [229, 230]. The reason is that, for given plasma conditions, the etchant generation rate is substantially independent of the area. As the etchable area increases, the consumption rate of the etchant also increases, thereby depressing the etchant concentration and in turn the etch rate. One can distinguish between global loading (averaged over the whole reactor) and local

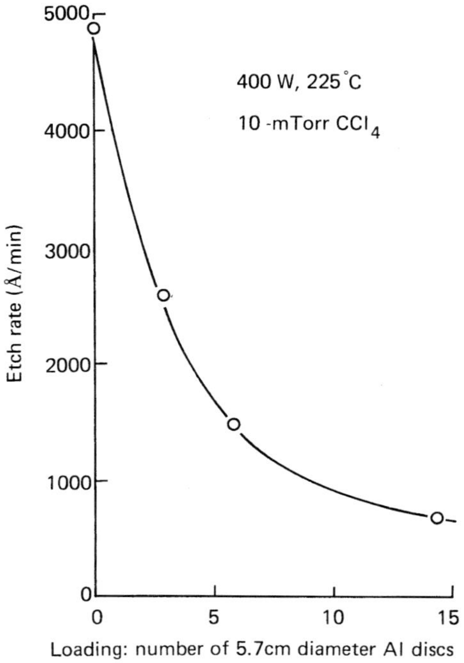

Fig. 47. The loading effect. The etch rate of aluminum decreases as the area available for etching increases. After [229].

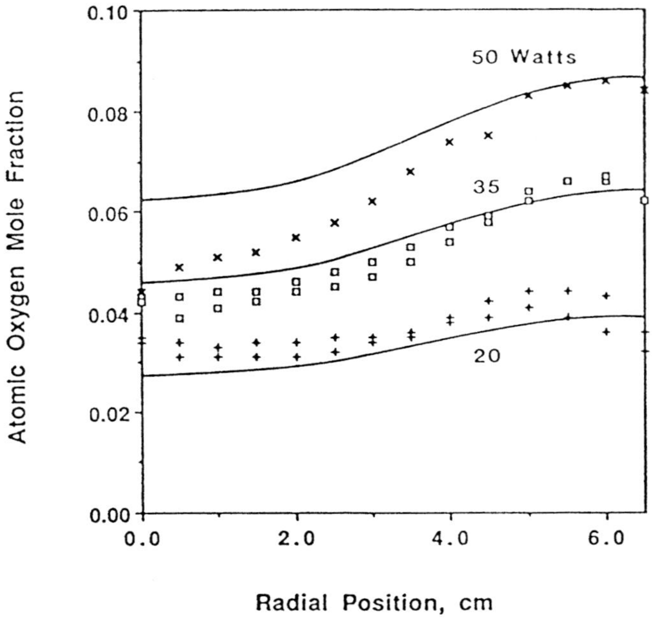

Fig. 48. Atomic oxygen concentration, measured by optical emission actinometry, as a function of radius in a 13.56 MHz oxygen discharge sustained in a diode reactor. The electrode is covered with a reactive film up to a radius of 3.75 cm. This film acts as a sink for atomic oxygen (loading) resulting in significant radial concentration gradients. Such gradients are responsible for etch non-uniformity. Solid lines show the result of mathematical model predictions. After [231]. Pressure 2 torr, gas flow 100 sccm.

loading such as can be observed near the interface between etchable and non-etchable surfaces; for example at the wafer edge or at the perimeter of an individual die on the wafer. In the latter case, loading is due to a difference in reactivity between the two surfaces [231, 232]. At endpoint, the etchable film surface area will decrease, which can lead to an increase in etch rate and perhaps rapid mask undercut. In addition, changes in surface area can affect the rate of recombination of radicals, again affecting the plasma gas composition.

Fig. 48 shows the atomic oxygen concentration profiles along the electrode radius in an oxygen plasma [231]. These profiles were measured by optical emission actinometry (see Section 7.1.2). The electrode has a reactive central region of 3.75 cm radius. The rest of the electrode is relatively inert. Oxygen atoms are consumed over the reactive region, but build up over the inert region. A gradient in the radical density is thus established, which is most important around the periphery of the active region. This gradient would lead to a so-called *bullseye* clearing pattern of a film, whereby the etch rate decreases from the periphery to the center of the wafer. This "local" loading can be enhanced at higher power input to the plasma.

Loading can also occur at the microscopic feature scale (*microloading*). It has been observed, for example, that isolated features etch faster than dense patterns. This is due to local reactant depletion over the dense pattern caused by greater consumption of the reactant (greater etchable surface area exposed by the dense pattern).

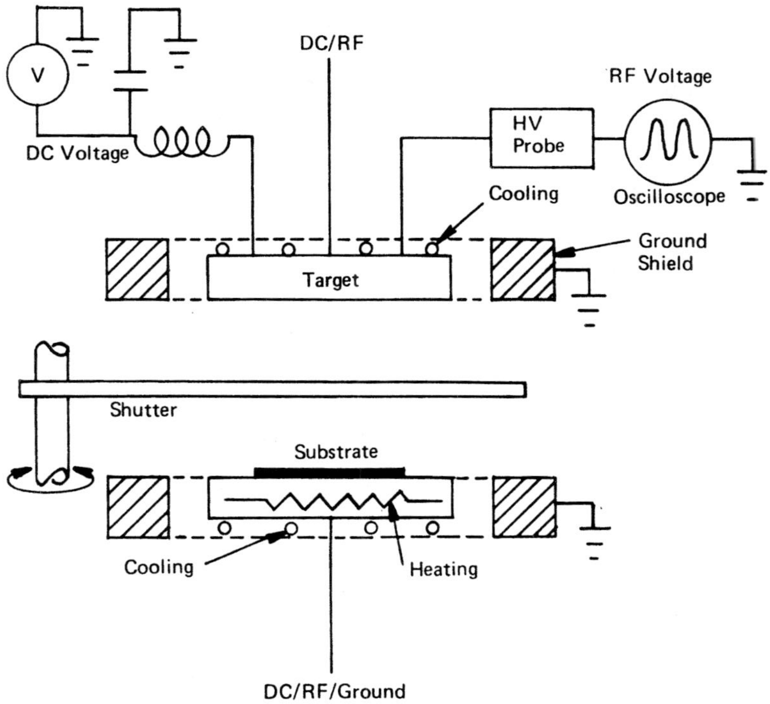

Fig. 49. Schematic of sputter deposition reactor. After [15].

If reactant transport is not adequately fast to alleviate concentration gradients over the length scale of the feature cluster (few mm), then loading can manifest itself. Similar pattern dependent effects are observed in CVD [90, 91].

6.5 Sputter Deposition

Sputter deposition [15] is used for growing elemental films (e.g., aluminum) and also compounds (e.g., TiN, oxides, etc.). In physical sputtering, an inert gas (usually argon) is used to sputter the material of interest from a target. Sputtered material then deposits on all surfaces it can encounter, including the substrate. A typical reactor configuration is shown in Fig. 49 [15]. The gas pressure is selected low enough ($\lesssim 10$ mtorr) to minimize scattering collisions of the sputtered material with the gas. Growth rates are low in the range of 100–1000 Å/min. The morphology of sputtered films is a strong function of gas pressure and substrate temperature relative to the melting point of the film [233]. Since plain RF or DC discharges are not efficient at these low pressures, magnetic fields are used to confine the electrons. Film growth uniformity and target erosion uniformity are important issues, particularly in the presence of a magnetic field [5].

In reactive sputtering, a reactive gas is mixed in with the inert gas. For example,

TiN is grown using a titanium target and a mixture of argon with nitrogen as the plasma gas. Titanium landing on the substrate reacts with N atoms formed in the plasma to yield TiN. Interesting hysteresis effects can occur in these systems as the partial pressure of nitrogen is varied, because atomic nitrogen can also convert the target to TiN which has a smaller sputtering rate compared to pure titanium [234].

6.6 Plasma Enhanced Chemical Vapor Deposition (PECVD)

Plasma enhanced chemical vapor deposition (PECVD) [14] is a process where feed-stock gases are decomposed in a plasma to from radicals. These radicals may undergo further reactions in the gas phase before adsorbing and reacting on the wafer surface to deposit a thin solid film. Since plasma electrons (rather than heat) are used to initiate chemistry, PECVD is performed at much lower temperatures than (thermal) CVD [20]. Typical substrate temperatures range from ambient to 400 °C. In general, the substrate temperature only weakly affects the deposition rate; in contrast it greatly affects the film properties including composition and morphology. Examples of PECVD are hydrogenated amorphous silicon (a-Si:H), silicon dioxide and silicon nitride films. a-Si:H films are used for solar cells and thin film transistors for flat panel displays [235]. Both applications require deposition over large areas. a-Si:H films are deposited from silane containing plasmas, often diluted with hydrogen or an inert gas such as Ar [236]. Silane is dissociated in the plasma to form deposition precursors such as SiH_2 and SiH_3. The former radical has a high sticking coefficient on the growing film and is thought to result in deposits of poor quality. SiH_3 has a lower sticking coefficient allowing the radical to diffuse on the surface thereby filling in voids in the growing film, resulting in better film quality. Hence, the deposition system must be designed carefully to maximize the SiH_3/SiH_2 ratio. Ion-driven surface reactions are also contributing to film growth and properties and hence a uniform ion flux to the substrate is necessary to achieve high quality films. Furthermore, design of the gas flow distribution in the reactor is important to achieve uniform deposition over large areas.

Silicon dioxide (oxide) is used as interlevel dielectric, for device isolation, and as a masking layer for etching and ion implantation. PECVD oxide is deposited from $SiH_4/Ar/N_2O$ plasmas or tetraethoxysilane (TEOS) $Si(OC_2H_5)_4/O_2$ or O_3 plasmas [237]. The silicon containing feedstock gases are decomposed in the plasma to form radicals that adsorb on the film surface. Atomic oxygen, also generated in the plasma by molecular oxygen dissociation, oxidizes the adsorbed species leading to oxide formation. In the case of TEOS, oxygen dilution is very high (e.g., 99%), in part to ensure that carbon and hydrogen in TEOS are converted to volatile carbon oxides and water, and are not incorporated to any significant extend in the growing film. Also, in the case of TEOS, ion bombardment has been shown to significantly enhance the deposition rate [223].

Silicon nitride is deposited by PECVD using typically SiH_4/NH_3 mixtures [238]. As with any PECVD process, the film composition and properties are a strong function of the feedstock gas composition and the plasma parameters (pressure, gas flow, power, excitation frequency, etc.). The hydrogen content depends on substrate temperature with less hydrogen at higher temperatures. Film stress can be varied by varying the ion bombardment energy, e.g., by lowering the plasma excitation frequency.

All of the above films have typically been deposited using RF capacitively-coupled parallel plate (diode) systems. Lately, high density plasmas with gas injection downstream of the main plasma are being introduced. To deposit oxide in such a system, for example, one would feed oxygen through the plasma and SiH_4 downstream (see Fig. 20) some distance above the substrate. Many other kinds of films have been deposited by PECVD. An example is diamond films [239].

6.7 Step Coverage

An issue with any film deposition over topography is the conformality of the deposit along the surface contour. Ideally, the film thickness should be uniform along the surface (Fig. 50a [1]). Since the mean free path of species is much larger than typical microfeature sizes, the species flux striking a surface element will depend on the cosine of the angle subtended by that element. If the reaction probability of the deposition precursor is high, as in the case of SiH_2, non-uniform deposits are expected (Fig. 50b). For low reaction probabilities, the deposition precursor has a better chance to diffuse along the surface or desorb and strike the surface a number of times before

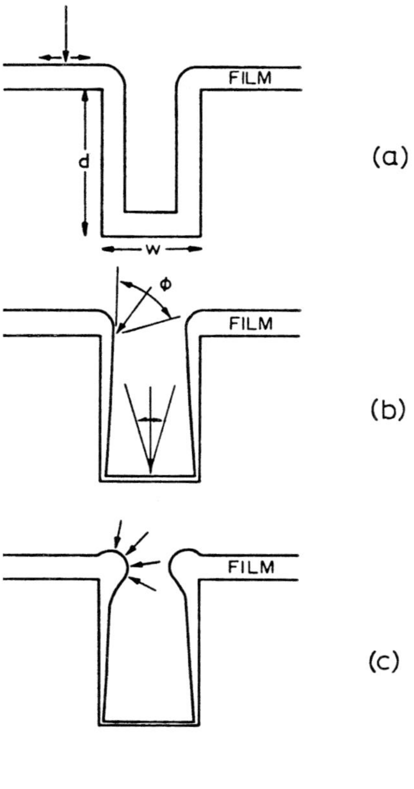

Fig. 50. The step coverage problem. Good step coverage (a), poor step coverage under long mean free path (low pressure) conditions (b), and poor step coverage under short mean free path (high pressure) conditions (c). After [1].

eventually depositing. Such a scenario will lead to better conformality of the deposit. One of the reasons the TEOS-based chemistry is preferred over silane-based chemistry for oxide deposition is that the former yields deposition precursors with much lower reaction probability (e.g., 0.05 vs. 0.5). Fig. 50c shows the resulting film thickness when the species mean free path is smaller than the feature opening (e.g., atmospheric pressure CVD). Non-conformal step coverage is unacceptable since, if the deposit were metal in a via for example, the contact will be prone to electromigration failure [1].

7 Plasma and Surface Diagnostics

Diagnostics are of critical importance in plasma processing. They serve as probes of the plasma or wafer state, helping to understand the process and test theories, mathematical models, and numerical simulations. They also serve as sensors useful for designing, controlling, and optimizing the process. Plasma diagnostics probe the gas-phase, whereas surface diagnostics probe the solid surface. Real time diagnostics are most useful as they provide information while etching or deposition is taking place. Other desirable characteristics are: high spatial and temporal resolution, high accuracy and sensitivity, capability for quantitative analysis, simplicity and low cost. Reviews of plasma and surface diagnostics are available [240].

7.1 Plasma Diagnostics

Important plasma diagnostics include: Langmuir probes, optical emission spectroscopy, laser induced fluorescence, absorption spectroscopy, mass spectrometry, ion flux and energy analysis, and plasma impedance analysis. A plasma reactor equipped with several of these diagnostics is shown in Fig. 51 [35, 160]. A capacitively coupled plasma is sustained between the parallel plates of the upper (etching) chamber. The lower (analysis) chamber is differentially pumped and communicates with the etching chamber through a pinhole on the lower electrode.

7.1.1 Langmuir Probe

This is a thin metallic wire immersed in the plasma and biased with respect to a reference potential (normally ground). The current drawn by the probe I is recorded as a function of bias potential, V. An ideal I-V characteristic of a single Langmuir probe is shown in Fig. 52 [15]. The floating potential V_f (corresponding to no net current to the probe) and the plasma potential V_p are shown. The ion and electron saturation currents are obtained at large negative and positive potentials, respectively. These correspond to the mass transfer limiting current of an electrochemical system. The electron saturation current is much larger (in absolute value) than the ion saturation

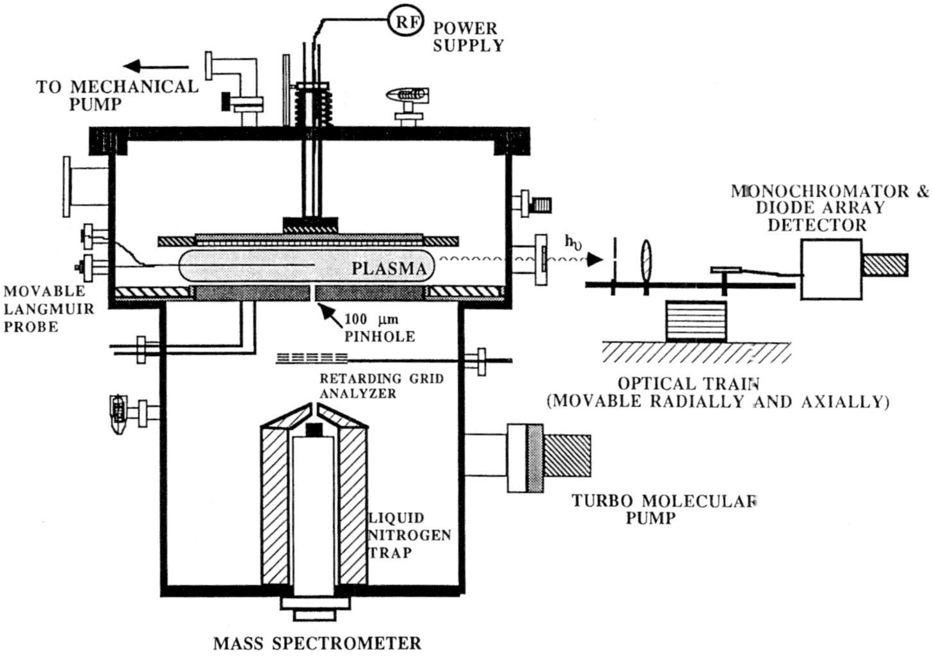

Fig. 51. A RF capacitively-coupled diode with a variety of plasma diagnostics, including Langmuir probe, optical emission spectroscopy, molecular beam mass spectrometry, and ion flux/energy analysis systems. After [35].

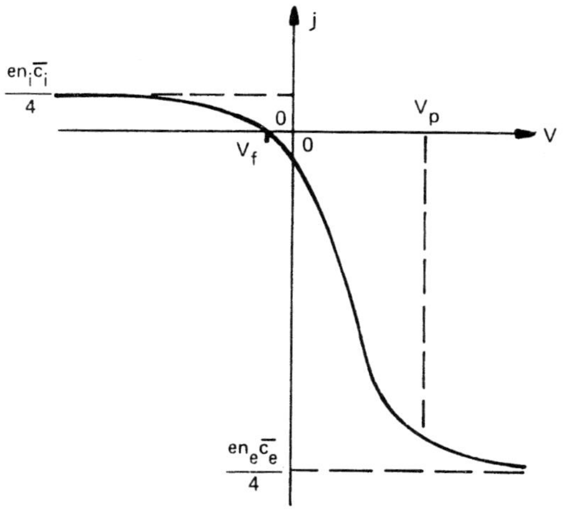

Fig. 52. I–V characteristic of a single Langmuir probe. The limiting currents due to electrons (lower right) and ions (upper left) are shown. After [15].

current because of the much higher electron mobility. From the ion saturation current (the electron saturation current is often difficult to obtain experimentally), one can estimate the positive ion density. The plasma potential is found as the potential at the knee of the I–V curve. The electron energy distribution function can also be obtained from the Langmuir probe [157]. Using the EEDF, the electron mean energy (or electron temperature for a Maxwellian distribution), and absolute electron density can be found. Several reviews of the Langmuir probe method have been published [157, 241–243].

Because of its simplicity, Langmuir probes have found widespread use for diagnosing plasmas. However, care must be exercised when doing the measurements, especially in RF plasmas, as the time-varying plasma potential can corrupt the probe characteristic. Also, a theory is needed to analyze the characteristic and extract the plasma parameters. Such theory is not always available, depending on plasma parameters and probe design. Finally, the probe must introduce as small as possible a disturbance to the plasma, and this is not easy to achieve. Currently, rather sophisticated Langmuir probe setups exist that can record I–V characteristics and extract the plasma parameters with substantial spatial and especially time (μs) resolution.

Instead of being immersed in the plasma, the probe can cover part of the electrode surface (with the appropriate electrical isolation). In this case the probe can be used to measure the ion current striking the surface. A sectioned electrode can also be used to measure the current distribution much like in electrochemical engineering. The ion current as a function of radius [244] provides important information about the plasma uniformity and in turn the etch uniformity.

7.1.2 Optical Emission Spectroscopy (OES)

OES is used extensively in plasma processing because it is a simple and non-intrusive technique [245]. It is based on the light emitted by plasma species after these species have been excited by electron impact (reaction R15 in Table 4). Plasma emission is collected and dispersed into its wavelengths by a monochromator. The light intensity as a function of wavelength is measured with a photomultiplier tube. Photodiode array detectors and charge coupled device (CCD) cameras are also used to provide rapid acquisition of spatially and spectrally resolved emission. The OES spectrum is used to identify plasma species which emit in the wavelength range monitored. Common species monitored by OES and their corresponding wavelength are given in [246]. OES is most often used for qualitative characterization of the plasma. Semi-quantitative results can be obtained by applying a technique known as actinometry [246, 247]. OES is also used widely for end-point detection [248]. This is based on monitoring the signal from a reactant or product the concentration of which changes substantially at end-point. For example, the concentration of F-atoms in a fluorocarbon plasma etching silicon may increase as the silicon film is etched away to expose oxide underneath. In fact the slope of the F-atom OES signal vs. time can be used to infer the etch uniformity as well. The slope of the trace becomes steeper as uniformity improves.

7.1.3 Absorption Spectroscopy

In this technique, the attenuation of light intensity as the probe beam passes through the plasma is related to the density of the absorbing species (Beer's law). Quantitative analysis is possible. Multiple passes are usually required to obtain an acceptable signal-to-noise ratio, and this limits the spatial resolution of the technique. Tunable diode laser absorption [249] has gained popularity as the probe beam wavelength can be controlled with high spectral resolution. This means more selective measurements.

7.1.4 Laser Induced Fluorescence (LIF)

LIF uses a tunable laser to perform selective excitation of plasma species to a higher energy level [250, 251]. Upon de-excitation the species emits flucrescence which is a measure of the species density. LIF is more powerful than OES because it can monitor directly the ground state species, which are the most abundant and often the most important etchant species. In contrast, OES monitors the excited state species density and indirectly (when a set of rather restrictive conditions are satisfied) the ground state density. LIF can offer high selectivity, spatial (<1 mm) and temporal (<10 ns) resolution, and can be made quantitative. However, LIF can be applied only to a limited number of species (having fluorescing excited states that can be reached by an optically allowed transition), and it is more complex and expensive compared to OES. LIF has been used to monitor metastable atoms in noble gas discharges, light atoms (O, Cl, H, etc.) which are thought to be very important in a number of plasma systems, metals (e.g., Al), and a variety of radicals (BCl, SiF_2, CF_2, etc.).

7.1.5 Mass Spectrometry

Generally speaking, mass spectrometry has been used in two configurations [252]: effluent gas analysis and molecular beam. In the former case, the mass spectrometer is attached to the exhaust port of the plasma reactor. In this configuration only stable gaseous species can be detected. Radicals and ions normally have ample chance to recombine on the walls of the apparatus before reaching the mass spectrometer. Effluent gas analysis can provide valuable information about the process.

In molecular beam mass spectrometry, a sample is drawn directly from the plasma into the instrument as a molecular beam. The goal is to minimize or totally eliminate collisions which can alter the sample composition. Most often the spectrometer is housed in a separate chamber which has a line of sight communication with the plasma through a pinhole (Fig. 51). Neutrals effusing from the plasma can be detected before suffering collisions provided a beam chopper and phase sensitive detection are employed to differentiate against the background gas. By turning the ionizer of the mass spectrometer off, ions exiting the plasma (which would otherwise strike a wafer resting on the lower electrode) can also be detected to obtain a qualitative picture of the ion flux on the electrode. Quantitative measurements can be made if careful corrections of the sensitivity factors of the mass spectrometer vs. ion energy are made. In another application of this configuration called *appearance mass spectrometry* (AMS) [253], the energy of the electron beam of the mass spectrometer is varied to achieve selective ionization of neutrals exiting the plasma. This way, one can determine the concentration of radicals that are difficult to detect otherwise. The technique has been applied extensively in chemically complex systems such as fluorocarbon plasmas. Quadrupole mass spectromety (QMS) can also be used for real time process monitoring and is a candidate sensor for process control [254].

7.1.6 Retarding Grid Energy Analysis

A retarding grid energy analyzer/Faraday cup assembly placed under a pinhole on the substrate electrode (Fig. 51) can be used to measure the ion bombardment flux and energy distribution function. A typical analyzer consists of three high-transmission metal grids above a metal plate which serves as a Faraday cup. The top grid is kept at the potential of the substrate electrode to form a field-free drift region between the electrode and the grid. The energy distribution of ions bombarding the electrode is determined by applying a variable positive potential V_{ap} to the middle grid (energy selector), and measuring the current I_+ at the Faraday cup. The bottom grid is kept at a negative potential to suppress secondary electron emission from the Faraday cup plate. The ion energy distribution function is calculated from the Faraday cup current-voltage characteristics using

$$f_{IED}(\varepsilon_+) = -\frac{1}{I_{+0}} \frac{dI_+(\varepsilon_+)}{d\varepsilon_+} \tag{62}$$

where $\varepsilon_+ = \mathrm{e}V_{ap}$. The total ion current I_{+0} through the pinhole is measured using a picoammeter by biasing all grids and the cup at a large negative potential (say –100 volts). The total ion current is the sum of the currents collected by the grids and the Faraday cup.

7.1.7 Impedance Analysis

Impedance analysis is used to estimate important plasma properties such as electron density, time-averaged sheath capacitance, thickness, and voltage as well as the bulk plasma resistance [32–35, 255]. The technique is illustrated here for a case when the bulk plasma can be represented as a resistor and each sheath can be represented by a capacitor (a simplification of Fig. 10). This is a fairly typical case for parallel plate capacitively coupled reactors operated at frequencies above the ion plasma frequency [255]. The *current and voltage waveforms* are measured at the RF feedthrough and corrected for stray capacitance and line inductance [32–35]. The current is measured using a wide band current transformer, and the voltage is sampled through a high voltage probe. Both current and voltage waveforms can be recorded using a digitizing oscilloscope. The actual power dissipated in the plasma is calculated from the product of the corrected current and voltage waveforms, accounting for the phase shift. The plasma resistance and sheath capacitance can be calculated from the measured current and voltage waveforms and the phase shift ϕ between the two using

$$R_p = \frac{V_0}{I_0} \cos\phi \tag{63}$$

$$C_s = -\frac{I_0}{\pi \nu_m V_0 \sin\phi} \tag{64}$$

Here V_0 and I_0 are the amplitudes of the measured voltage and current waveforms, assumed sinusoidal.

7.2 Surface diagnostics

Surface diagnostics are of particular importance because the ultimate goal of plasma processing is to modify the surface of the wafer. Non-real-time diagnostics either in-situ or ex-situ can provide a wealth of information. Such surface diagnostics include many of the standard techniques that have been used in surface science [256]. Real time, in-situ diagnostics are most useful since they provide information while the wafer is being processed. Such diagnostics can therefore be used for real time process control, whereby faults are corrected hopefully before they have an impact on the wafer. Such diagnostics include: (a) ellipsometry [257] to monitor film thickness and surface conditions with nanometer scale resolution, (b) laser interferometry to monitor etch or deposition rate and uniformity in real time [164, 258], (c) attenuated total internal reflection infrared spectroscopy to monitor IR active adsorbates on the surface [259, 260], (d) quartz crystal microbalance to monitor film thickness changes with submonolayer accuracy [261], and (d) scatterometry to measure sub-micron linewidths [262].

8 Plasma vs. Electrochemical Engineering

Determining the current distribution in the reactor is a central problem in electrochemical engineering [263, 264]. The problem is to find the reaction rate distribution along a macroscopic surface (electrode) or along the walls of a microscopic feature [88]. In general, this is a coupled problem of potential field with transport and reaction of charged and neutral species, much like the plasma reaction engineering problem. However, simplifications can be applied in a variety of practical situations by decoupling the equations. Some of these parallel the approximations discussed in Section 5.4.2. Assuming Newtonian incompressible fluid and dilute solutions, the governing equations may be written as (neglecting natural convection)

$$\frac{\partial \boldsymbol{u}}{\partial t} + \boldsymbol{u} \cdot \nabla \boldsymbol{u} = -\frac{1}{\rho_l} \nabla P + \nu_l \nabla^2 \boldsymbol{u} \tag{65}$$

$$\nabla \cdot \boldsymbol{u} = 0 \tag{66}$$

$$\frac{\partial c_i}{\partial t} + \nabla \cdot \boldsymbol{J}_i = \sum_j R_{ij} \tag{67}$$

$$\boldsymbol{J}_i = -D_i \nabla c_i + z_i F \mu_i c_i \boldsymbol{E} + \boldsymbol{u} c_i \tag{68}$$

Eq. (65) is the Navier-Stokes equation for the fluid velocity field. Eq. (66) is the continuity equation for the incompressible fluid, and Eq. (67) is the mass balance for species i. Here $\mathbf{u}$ is the fluid velocity vector, P is pressure, v_l is kinematic viscosity, ρ_1 is solution density, and c_i and z_i are the concentration and charge number of species i, respectively. In Eq. (67) R_{ij} accounts for volume reactions (solution chemistry, e.g., complexation reactions). The total flux of species i ($\mathbf{J}_i$) in Eq. (68) is the sum of component fluxes due to diffusion, migration and convection. In regions of uniform concentration (well-mixed reactor, primary or secondary current distribution), the solution potential satisfies Laplace's equation

$$\nabla^2 V = 0 \tag{69}$$

When excess supporting electrolyte is present, Laplace's equation can still be used to solve for the tertiary current distribution (mass transport of ions plays a role). In such case the potential field is coupled to the concentration field at the electrode surface [264]. In the more general case of concentration gradients, the solution electroneutrality condition is imposed instead of Laplace's equation (assuming spatially invariant dielectric constant)

$$\sum_i z_i c_i = 0 \tag{70}$$

In this case the concentration and potential fields are coupled throughout the solution volume. The boundary conditions can be complex in the case of electrodeposition in the presence of additives or when a salt film precipitates on the surface during electrochemical dissolution.

The above set of equations must be augmented by an energy balance for the solution and/or the solid phase if temperature effects are important. An example is high rate etching or deposition effected by a laser beam [265]. Also, potential depended transport of charge carries (electrons and holes) in the semiconductor must be accounted for in photochemical and photoelectrochemical etching [266, 267].

At this point it is instructive to compare the equations governing the flow, concentration, and potential fields in plasma and electrochemical reactors. The equations are (of course) identical except that an equation equivalent to the electron energy balance (Eq. 31) is not needed for electrochemical systems. The electroneutrality assumption is often made in plasmas (see Section 5.4.2) as is done in electrochemical engineering.

Plasma and electrochemical systems share many common characteristics including:

1. Potential distribution coupled with charged and neutral species transport and reaction, both at the microscopic feature scale ($\sim$0.1–10s of μ) and at the reactor scale ($\sim$10s of cm). Similar sets of equations (compare Eqs. (65)–(70) to Eqs. (22)–(34)) describe both systems, and the same approximations can be often applied (electroneutrality, ambipolar diffusion, etc.).
2. Complex homogeneous and heterogeneous chemistry leading to surface modification of materials.

3. Complex boundary conditions especially when the surface is covered by thin films (sidewall passivation or damaged layers in plasma (Fig. 41), salt films in electrochemical engineering).
4. A space charge region near electrodes and other walls. This region corresponds to the sheath (dark space) in plasmas, and to the double layer in wet electrochemical systems. Due to the large disparity in the Debye length, the size of these regions differs by many orders of magnitude between the two systems (0.01–1 cm in plasmas, 10–100 Å in electrochemical systems).
5. The surface chemistry is not temperature driven. It is driven by ion bombardment in plasmas and by the surface potential in electrochemical systems.

Plasma reactors differ from electrochemical reactors in the following:

1. At low pressures (or more precisely at high values of the Knudsen number $Kn > 0.2$) the continuum approximation can no longer be used, and rarefied gas dynamics has to be invoked for plasma flow. For practical plasma reactors, this becomes necessary at pressures below ~ 10 mtorr.
2. Plasmas are strongly non-equilibrium systems with "hot" light particles (electrons) and "cold" heavy particles (neutrals and ions in the bulk plasma). Charged particles can achieve kinetic energies of 100s of eV in the sheath (Fig. 11). In this respect, electrochemical systems are closer to thermal plasmas [268] for which local thermodynamic equilibrium (LTE) may be assumed (i.e., all particles are at the same "temperature"), and the pressure (~ 1 atm) is well above the limit of applicability of the continuum approximation.

9 Summary and Future Outlook

Plasma processing will continue to be of critical importance for fabricating ULSI devices [269]. More complex patterns with finer dimensions down to a few hundred Å, as well as films with properties tailored to more specific applications are expected in the future. These stringent requirements will necessitate integrated processing and process automation with sophisticated smart sensors for real-time process control. In particular, real-time surface sensors will be indispensable.

Multidimensional simulations of plasma reactors can serve as powerful tools for improving our understanding of reactive plasmas, and for helping in the design of new and improved plasma processes. Despite their young age, such simulations are already quite advanced. Further understanding of plasma physics and chemistry as well as developments in numerical methods and in parallel computing will have a profound impact on multidimensional simulations of glow discharges coupled with neutral transport and reaction, culminating in a *virtual plasma reactor*. This virtual reactor will be based on an *integrated system* approach in which the reactor design, control, and optimization problems are not solved separately but simultaneously. In general, a combination of well-characterized experiments with mathematical modeling and simulation will continue to be the best approach in unraveling the intricacies of plasmas and the plasma-surface interactions.

Understanding at the molecular level will become increasingly important. The push for lower pressure to improve uniformity and ion directionality will necessitate the further use of "particle" simulations. An example is direct simulation by Monte Carlo (DSMC), and Particle in Cell (PIC) methods. Surface processes including ion-assisted (or beam assisted in general) reaction kinetics, deposit nucleation and growth, and adhesion must be better understood to engineer materials and microstructures with tailored properties (e.g., superlattices). Hybrid Molecular Dynamics-Monte Carlo simulations (to cover both the ps time scale of events upon ion impact and the relaxation of the lattice after impact) coupled with well defined beam experiments, will provide much insight into the ion-assisted chemistry happening on surfaces exposed to plasmas. Atomic resolution techniques with capability of real time, in-situ monitoring [270, 271] such as Scanning Tunneling Microscopy (STM) and Atomic Force Microscopy (AFM) will find increasingly more applications. Atomic layer processing [272, 273], with the capability to control film deposition and etching with monolayer accuracy, will be further developed. Understanding the relation between processing-microstructure-properties of deposited films will result in much improved tailored materials [274].

Efficient treatment of the disparate length and time scales will be the key to the further development of predictive computational models for both plasma and electrochemical engineering. This is not limited to the plasma or liquid-phase processes. It includes the solid materials produced or modified by these technologies; from the atomic level, to the microstructure to the bulk material properties. Finally, integration of modeling/simulation with design, sensors, control, optimization, safety and reliability will result in an ultimate integrated system which is based on molecular principles and which extends all the way to the factory scale!

Acknowledgments. The author would like to thank NSF (CTS 9216023 and CTS 9713262), Sandia/SEMATECH, The Welch Foundation and the State of Texas (Texas Advanced Technology Program) for financial support of this work. Many thanks to Prof. R. C. Alkire for his continuing encouragement on the subject of bridging plasma with electrochemical engineering.

10 References

1. S. M. Sze, ed., VLSI Technology, 2nd ed., McGraw Hill, New York, NY (1988).
2. P. Singer, Semiconductor International, p. 57, August 1994.
3. S. K. Ghandhi, VLSI Fabrication Principles: Silicon and Gallium Arsenide, second edition, J. Wiley & Sons, New York, NY (1994).
4. J. L. Vossen and W. Kern, editors, Thin Film Processes II, Academic Press, San Diego, CA (1991).
5. S. M. Rossnagel, J. J. Cuomo, and W. D. Westwood (eds.), Handbook of Plasma Processing Technology, Noyes Publications, Park Ridge, NJ (1990).

6. M. A. Leiberman and A. J. Lichtenberg, "Principles of Plasma Discharges and Materials Processing," John Wiley and Sons, New York, NY(1994).
7. A. Grill, Cold Plasma in Materials Fabrication: From Fundamentals to Applications, IEEE Press, New York, NY (1994).
8. VLSI Electronics: Microstructure Science, Vol. 8, Plasma Processing for VLSI, N. G. Einspruch and D. M. Brown, (eds.) Academic Press, San Diego, CA (1984).
9. D. J. Economou in: Electronic Materials Processing, H. B. Pogge (ed.), Marcel Dekker, New York, NY (1996).
10. J. W. Coburn, Plasma Etching and Reactive Ion Etching, AVS Monograph Series, N. Rey Whetten, Series editor, American Institute of Physics, New York, NY (1982).
11. J. A. Mucha, D. W. Hess, and E. S. Aydil in: Introduction to Microlithography, 2nd ed., L. F. Thompson, C. G. Willson, and M. J. Bowden (eds.), ACS Professional Reference book, American Chemical Society, Washington, DC (1994).
12. D. L. Flamm in Plasma Etching: An Introduction, D. M. Manos and D. L. Flamm (eds.), Academic Press, San Diego, CA (1989).
13. S. J. Fonash, J. Electrochem. Soc. 137, 3885 (1990).
14. M. Konuma, Film Deposition by Plasma Techniques, Springer, New York, NY (1992).
15. B. Chapman, Glow Discharge Processes, J. Wiley, New York, NY (1980).
16. H. Yasuda, Plasma Polymerization, Academic Press, San Diego, CA (1985).
17. F. F. Chen, Introduction to Plasma Physics and Controlled Fusion, Vol. 1, Plasma Physics, 2nd ed., Plenum, New York, NY (1984).
18. Y. P. Raizer, Gas Discharge Physics, Springer-Verlag, Berlin, 1991.
19. R. J. Goldston and P. H. Rutherford, Introduction to Plasma Physics, Institute of Physics Publishing, Bristol (1995).
20. A. Sherman, Chemical Vapor Deposition for Microelectronics, Noyes Publications, Park Ridge, NJ (1987).
21. S. A. Campbell, The Science and Engineering of Microelectronic Fabrication, Oxford University Press, New York, NY (1996).
22. D. Bohm in: The Characteristics of Electrical Discharges in Magnetic Fields, A. Guthrie and R. K. Wakerling (eds.), McGraw Hill, New York, NY (1949).
23. R. N. Franklin, Plasma Phenomena in Gas Discharges, Clarendon, Oxford (1976).
24. V. A. Godyak, Soviet Radio Frequency Discharge Research, Delphic Associates, Inc., Falls Church, VA (1986).
25. C. G. Goedde, A. J. Lichtenberg and M. A. Lieberman, J. Appl. Phys. 64, 4375 (1988).
26. D. P. Lymberopoulos, Modeling and Multidimensional Simulation of Plasma Processing Reactors, Ph.D. Thesis, University of Houston (1995).
27. H. R. Koenig and L. I. Maisel, IBM J. Res. Develop. 14, 168 (1970).
28. J. W. Coburn and E. Kay, J. Appl. Phys. 43, 4965 (1972).
29. K. Kohler, J. W. Coburn, D. E. Horne, E. Kay, and J. H. Keller, J. Appl. Phys. 57, 59 (1985).
30. P. Bletzinger and M. J. Flemming, J. Appl. Phys. 62, 4688 (1987).
31. P. Andries, G. Ravel, and L. Peccoud, J. Vac. Sci. Technol. A 7, 2774 (1989).
32. P. A. Miller and M. E. Riley, J. Appl. Phys. 82, 3689 (1997).
33. M. Sobolewski, IEEE Trans. Plasma Sci. 23, 1 (1995); N. St. J. Braithwaite, Plasma Sources Sci. Technol. 6, 133 (1997).
34. J. W. Butterbaugh, L. D. Baston, and H. H. Sawin, J. Vac. Sci. Technol. A 8, 916 (1990).
35. E. S. Aydil and D. J. Economou, J. Electrochem. Soc. 139, 1406 (1992).
36. D. P. Lymberopoulos and D. J. Economou, J. Appl. Phys. 73, 3668 (1993).
37. N. L. Bassett and D. J. Economou, J. Appl. Phys. 75, 1931 (1994).
38. R. A. Gottscho, G. R. Scheller, T. Intrator, and D. B. Graves, J. Vac. Sci. Technol. A 6, 1393 (1988).
39. S. A. Cohen, in Plasma Etching: An Introduction, Dennis M. Manos and D. L. Flamm (eds.) Academic Press, San Diego, CA (1989).

40. T. J. Sommerer and M. J. Kushner, J. Appl. Phys., 70, 1240 (1991).
41. D. J. Economou, T. J. Bartel, R. S. Wise, and D. P. Lymberopoulos, IEEE Trans. Plasma Sci. 23, 581 (1995).
42. A. T. Bell, Fundamentals of Plasma Chemistry, in Techniques and Applications of Plasma Chemistry, J. R. Hollahan and A. T. Bell, editors, J. Wiley, New York, NY (1974).
43. B. E. Cherrington, Gaseous Electronics and Gas Lasers, Pergamon Press, Oxford (1979).
44. E. Meeks and J. Shon, JEEE Trans. Plasma Sci. 23, 539 (1995).
45. L. G. Christophorou, ed., Electron-Molecule Interactions and their Applications, Vols. 1 and 2, Academic Press, New York, NY (1984).
46. L. C. Pitchford, B. V. McKoy, A. Chutjian, and S. Trajmar (eds.), Swarm Studies and Inelastic Electron-Molecule Collisions, Springer, New York, NY (1987).
47. E. W. McDaniel, Atomic Collisions: Electron and Photon Projectiles, J. Wiley & Sons, New York, NY (1989).
48. V. E. Golant, A. P. Zhilinsky, and I. E. Sakharov, Fundamentals of Plasma Physics, J. Wiley, New York, NY (1980).
49. I. P. Shkarofsky, T. W. Johnston, and M. P. Bachynski, The Particle Kinetics of Plasmas, Addison-Wesley, Reading, MA (1966).
50. E. S. Aydil and D. J. Economou, J. Electrochem. Soc. 139, 1396 (1992).
51. P. A. Sa, J. Loureiro, and C. M. Ferreira, J. Phys. D: Appl. Phys. 25, 960 (1992).
52. Y. Weng and M. J. Kushner, Phys. Rev. A 42, 6192 (1990).
53. P. Jiang and D. J. Economou, J. Appl. Phys. 73, 8151 (1993).
54. M. Capitelli, C. Gorse, R. Winkler, and J. Wilhelm, Plasma Chem. Plasma Process. 8, 399 (1988).
55. E. A. Karoulina and Y. A. Lebedev, J. Phys. D: Appl. Phys. 21, 411 (1988); L. D. Tsendin, Sov. Phys. Tech. Phys. 22, 925 (1977).
56. I. B. Bernstein and T. Holstein, Phys. Rev. 94, 1475 (1954).
57. L. D. Tsendin, Sov. Phys. JETP 39, 805 (1974).
58. V. I. Kolobov and V. A. Godyak, IEEE Trans. Plasma Sci. 23, 503 (1995).
59. U. Kortshagen, I. Pukropski, and L. D. Tsendin, Phys. Rev. E 51, 6063 (1995).
60. E. W. McDaniel, Collision Phenomena in Ionized Gases, J. Wiley & Sons, New York, NY (1964).
61. S. C. Brown, Basic Data of Plasma Physics, Technology Press and J. Wiley, New York, NY (1959).
62. E. W. McDaniel, J. B. A. Mitchell, and M. E. Rudd, Atomic Collisions: Heavy Particle Projectiles, J. Wiley & Sons, New York, NY (1993).
63. Ch. Steibruchel, Appl. Phys. Lett. 55, 1960 (1989).
64. D. C. Gray, I. Tepermeister, and H. H. Sawin, J. Vac. Sci. Technol. B, 11, 1243 (1993).
65. C. G. Hill, An Introduction to Chemical Engineering Kinetics and Reactor Design, J. Wiley, New York, NY (1977).
66. J. W. Coburn and H. F. Winters, J. Appl. Phys. 50, 3189 (1979).
67. Q. Gao, C. C. Cheng, P. J. Chen, W. J. Choyke, and T. J. Yates, Jr., J. Chem. Phys. 98, 8303 (1993).
68. P. Singmund in Inelastic Ion-Surface Collisions, Tolk et al. (eds.), Academic Press, New York, NY (1977).
69. D. J. Economou and R. C. Alkire, J. Electrochem. Soc. 135, 2786 (1988).
70. J. P. Chang, J. C. Arnold, G. C. H. Zau, H.-S. Shin, and H. H. Sawin, J. Vac. Sci. Technol. A, 15, 1853 (1997).
71. U. Gerlach-Meyer, Surf. Sci. 103, 525 (1981).
72. R. H. Bruce in Plasma Processing, R. G. Frieser and C. J. Mogab (eds.), p. 243, The Electrochem. Soc., Pennington, NJ (1981).
73. H. F. Winters, J. Vac. Sci. Technol. A 3, 786 (1985).

74. J. R. Woodworth, M. E. Riley, D. C. Meister, B. P. Aragon, M. S. Le, and H. H. Sawin, J. Appl. Phys. 80, 1304 (1996).
75. M. J. Kushner, J. Appl. Phys. 58, 4024 (1985).
76. A. C. Adams and C. D. Capio, J. Electrochem. Soc. 128, 366 (1981).
77. G. S. Oehrlein, J. F. Rembetzki, and E. H. Payne, J. Vac. Sci. Technol. B 8, 1199 (1990).
78. D. J. Economou, D. Evans, and R. C. Alkire, J. Electrochem. Soc. 135, 756 (1988).
79. D. J. Economou and R. C. Alkire, J. Electrochem. Soc. 132, 648 (1985).
80. J. F. Battey, J. Electrochem. Soc. 124, 437 (1977).
81. V. Vukanovic, G. A. Takacs, E. A. Matuszak, F. D. Egitto, F. Emmi, and R. S. Horwath, J. Vac. Sci. Technol. B 6, 66 (1988).
82. N. Hayasaka, H. Okono, and Y. Horiike, Solid State Technology, 31, 127 (1988).
83. M. A. Lieberman and R. A. Gottscho, Design of High Density Plasma Sources for Materials Processing, in Physics of Thin Films, M. Francombe and J. Vossen (eds.), Academic Press, San Diego, CA (1993).
84. J. Hopwood, Plasma Sources Sci. Technol. 1, 109 (1992).
85. J. H. Keller, J. C. Foster, and M. S. Barnes, J. Vac. Sci. Technol. A 11, 2487 (1993).
86. R. J. Hoekstra and M. J. Kushner, J. Appl. Phys. 79, 2275 (1996).
87. T. Kessler and R. C. Alkire, Plating Surf. Finish. 63, 22 (1978).
88. B. DeBecker and A. C. West, J. Electrochem. Soc. 143, 486 (1996).
89. D. J. Economou and R. C. Alkire, J. Electrochem. Soc. 135, 941 (1988).
90. M. K. Gobbert, T. P. Merchant, L. J. Borucki, and T. S. Cale, J. Electrochem. Soc. 144, 3945 (1997).
91. S. T. Rodgers and K. F. Jensen, J. Appl. Phys., 83, 524 (1998).
92. M. Meyyappan, ed., Computational Modeling in Semiconductor Processing, Artech House, Boston, MA (1994).
93. L. Kline and M. J. Kushner, Crit. Rev. in Solid State and Materials Science 16, 1, (1989).
94. A. M. Popov, A. T. Rakhimov, and T. V. Rakhimova, Plasma. Phys. Rep. 19, 651 (1993).
95. See papers in special issue of IEEE Trans. Plasma Science, vol. 23 (1995).
96. See papers in special issue of IEEE Trans. Plasma Science, vol. 19 (1991).
97. D. P. Lymberopoulos and D. J. Economou, J. Vac. Sci. Technol. A 12, 1229 (1994).
98. E. S. Aydil and D. J. Economou, J. Electrochem. Soc. 140, 1471 (1993).
99. D. P. Lymberopoulos, V. Kolobov, and D. J. Economou, J. Vac. Sci. Technol. A 16, 564 (1998).
100. D. P. Lymberopoulos and D. J. Economou, IEEE Trans. Plasma Sci. 23, 573 (1995).
101. R. S. Wise, D. P. Lymberopoulos, and D. J. Economou, Appl. Phys. Lett. 68, 2499 (1996).
102. T. E. Nitschke and D. B. Graves, IEEE Trans. Plasmas Sci. 23, 717 (1995).
103. M. J. Grapperhaus and M. J. Kushner, J. Appl. Phys. 81, 569 (1997).
104. C. Lee, D. B. Graves, M. A. Lieberman, and D. W. Hess, J. Electrochem. Soc., 141, 1546 (1994).
105. E. Meeks and J. W. Shon, IEEE Trans. Plasmas Sci. 23, 539 (1995).
106. S. C. Deshmukh and D. J. Economou, J. Appl. Phys. 72, 4597 (1992).
107. L. E. Kline, W. D. Partlow, and W. E. Bies, J. Appl. Phys. 65, 70 (1989).
108. I. C. Plump and K. R. Ryan, Plasma Chem. Plasma Proc. 6, 205 (1986).
109. M. J. Kushner, W. Z. Collison, M. J. Grapperhaus, J. P. Holland, and M. S. Barnes, J. Appl. Phys. 80, 1337 (1996).
110. M. J. Kushner, J. Appl. Phys. 82, 5312 (1997).
111. M. Dalvie and K. F. Jensen, J. Vac. Sci. Technol. A 8, 1648 (1990).
112. S.-K. Park and D. J. Economou, J. Electrochem. Soc. 137, 2624 (1990).
113. D. B. Graves and K. F. Jensen, IEEE Trans. Plasma Sci. 14, 78 (1986).
114. S.-K. Park and D. J. Economou, J. Appl. Phys. 68, 3904 (1990).

115. A. D. Richards, B. E. Thompson, and H. H. Sawin, Appl. Phys. Lett. 50, 492 (1987).
116. M. Meyyappan and T. R. Govindan, J. Vac. Sci. Technol. A 10, 1344 (1992).
117. A. P. Paranjpe, J. P. McVittie, S. A. Self, J. Vac. Sci. Technol. A 8, 1654 (1990).
118. G. R. Misium, A. M. Lichtenberg and M. A. Lieberman, J. Vac. Sci. Technol. A 7, 1007 (1989).
119. W. N. G. Hitchon, T. J. Sommerer and J. E. Lawler, IEEE Trans. Plasma Sci. 19, 113 (1991).
120. V. A. Feoktistov, A. M. Popov, O. B. Popovicheva, A. T. Rakhimov, T. V. Rakhimova, and E. A. Volkova, IEEE Trans. Plasma Sci. 19, 163 (1991).
121. N. Sato and H. Tagashira, IEEE Trans. Plasma Sci. 19, 102 (1991).
122. E. Gogolides and H. H. Sawin, J. Appl. Phys. 72, 3971 (1992).
123. T. J. Sommerer and M. J. Kushner, J. Appl. Phys. 71, 1654 (1991).
124. M. S. Barnes, T. J. Cotler and M. E. Elta, J. Comp. Phys. 77, 53 (1988).
125. M. H. Wilcoxson and V. I. Manousiouthakis, J. Comp. Phys. 115, 376 (1994).
126. Y. Oh, N. Choi, and D. Choi, J. Appl. Phys. 67, 3264 (1990).
127. D. P. Lymberopoulos and D. J. Economou, Appl. Phys. Lett. 63, 2478 (1993).
128. P. L. G. Ventzek, T. J. Sommerer, R. J. Hoekstra, and M. J. Kushner, Appl. Phys. Lett. 62, 3247 (1993).
129. D. P. Lymberopoulos and D. J. Economou, J. Res. Natl. Inst. Stand. Technol. 100, 473 (1995).
130. J. D. Bukowski and D. B. Graves, J. Appl. Phys., 80, 2614 (1996).
131. P. L. G. Ventzek, R. J. Hoekstra and M. J. Kushner, J. Vac. Sci. Technol. B 12, 461 (1994).
132. P. L. G. Ventzek, M. Grapperhaus and M. J. Kushner, J. Vac. Sci. Technol. B 12, 3118, (1994).
133. M. Dalvie, M. Surendra, G. S. Selwyn, Appl. Phys. Lett. 62, 3207 (1993).
134. F. F. Young and C.-H. (John) Wu, IEEE Trans. Plasma Sci. 21, 312 (1993).
135. J. P. Beouf and L. C. Pitchford, Phys. Rev. E, accepted.
136. M. H. Wilcoxson and V. I. Manousiouthakis, in Proceedings of the First International Symposium on Process Control, Diagnostics, and Modeling in Semiconductor Manufacturing, M. Meyyappan, D. J. Economou, and S. W. Butler, (eds.), vol. 95-4, The Electrochemical Society Inc., 1995; J. Vac. Sci. Technol. A, 16, 2162 (1998).
137. J. D. P. Passchier and W. J. Goedheer, J. Appl. Phys. 74, 3744 (1993).
138. Y. Yang and H. Wu, J. Appl. Phys. 80, 3699 (1996).
139. C. K. Birdsall, IEEE Trans. Plasma Sci. 19, 65 (1991).
140. V. Vahedi, G. DiPeso, C. K. Birdsall, M. A. Lieberman, and T. D. Ronglien, Plasma Sources Sci. Technol. 2, 261 (1993).
141. M. Surendra and M. Dalvie, Physical Rev. E 48, 3914 (1993).
142. M. D. Kilgore, H. M. Wu, and D. B. Graves, J. Vac. Sci. Technol. B 12, 494 (1994).
143. V. Serikov and K. Nanbu, J. Vac Sci. Technol. A 14, 3108 (1996).
144. M. Surendra, Plasma Sources Sci. & Technol. 4, 56 (1995).
145. M. Surendra and D. Vender, Appl. Phys. Lett. 65, 153 (1994).
146. M. M. Turner, Plasma Sources Sci. & Technol. 5, 159 (1996).
147. P. J. Chantry, J. Appl. Phys. 62, 1141 (1987).
148. D. J. Economou, J. Feldsien, and R. S. Wise in Electron Kinetics and Applications, U. Kortshagen and L. D. Tsendin (eds.), Plenum, New York, NY (1998).
149. S. J. Polak, C. Den Heijer, W. H. A. Schilders, and P. Markowich, Int. J. Num. Meth. Eng. 24, 763 (1987).
150. E. F. Jeager and L. Berry, Phys. Plasmas 2, 2597 (1995).
151. K.-U. Riemann, Phys. Plasmas 4, 4158 (1997).
152. D. J. Economou, R. S. Wise and D. P. Lymberopoulos in Plasma Processing XI, G. Mathad and M. Meyyappan (eds.) vol. 96-12, The Electrochemical Society Inc. (1996).

153. R. P. Brinkman, R. F. Furst, C. Werner, and M. Hierlemann, in Proceedings of the First International Symp. on Control, Diagnostics, and Modeling in Semiconductor Manufacturing, M. Meyyappan, D. J. Economou, and S. W. Butler (eds.), vol. 95-2, The Electrochemical Society (1995).
154. V. I. Kolobov, D. P. Lymberopoulos, and D. J. Economou, Physical Rev. E 55, 3408 (1997); V. A. Godyak, V. P. Meytlis, and H. R. Strauss, IEEE Trans. Plasma Sci. 23, 728 (1995).
155. P. J. Hargis et al., Rev. Sci. Instr. 65, 140 (1994).
156. P. A. Miller, G. A. Hebner, K. E. Greenberg, P. D. Pochan, and B. P. Aragon, J. Res. Natl. Inst. Stand. Technol. 100, 427 (1995).
157. V. A. Godyak, R. B. Piejak, and B. M. Alexandrovich, Plasma Sources Sci. & Technol. 1, 36 (1992).
158. S. J. Choi and R. Veerasingam, J. Vac. Sci. Technol. A, 16, 1873 (1998).
159. A. V. Phelps, J. Res. Natl. Inst. Stand. Technol. 95, 407 (1990).
160. Eray Aydil, Theoretical and Experimental Investigation of Chlorine Glow Discharges and Polysilicon Etching, Ph.D. Thesis, University of Houston (1991).
161. M. N. R. Ashford and C. Rettner, Dynamics of Gas Surface Collisions, Royal Society, London (1991).
162. S. Dushman in Scientific Foundations of Vacuum Technique, 2nd edition, J. M. Lafferty, editor, J. Wiley, New York, NY (1962).
163. D. L. Smith in VLSI Electronics: Microstructure Science, Vol. 8, Plasma Processing for VLSI, N. G. Einspruch and D. M. Brown, (eds.) Academic Press, San Diego, CA (1984).
164. D. J. Economou, E. Aydil, and G. Barna, Solid State Technology, 34, 107 (1991).
165. C. D. Child, Phys. Rev. 32, 492 (1911).
166. W. B. Pennebaker, IBM J. Res. Dev. 23, 16 (1979).
167. R. T. Farouki, M. Dalvie, L. F. Pavarino, J. Appl. Phys. 68, 6106 (1990).
168. K.-U. Riemann, Phys. Fluids 24, 2163 (1981).
169. C. A. Ordonez, Phys. Fluids B 4, 778 (1992).
170. V. A. Godyak and N. Sternberg, Phys. Rev. A 42, 2299 (1990).
171. V. A. Godyak and N. Sternberg, IEEE Trans. Plasma Sci 18, 159 (1990).
172. A. Metze, D. W. Ernie, and H. J. Oskam, J. Appl. Phys., 60, 3081 (1986).
173. M. A. Lieberman, IEEE Trans. Plasma Sci 17, 338 (1989).
174. F. R. Myers and T. S. Cale, J. Electrochem. Soc. 139, 3587 (1992).
175. P. M. Vallinga, P. M. Meijer, and F. J. De Hoog, J. Phys. D: Appl. Phys. 22, 1650 (1989).
176. W. J. Goedheer and P. M. Meijer, IEEE Trans. Plasma Sci 19, 245 (1991).
177. R. T. Farouki, S. Hamaguchi, and M. Dalvie, Phys. Rev. A 45, 5913 (1992).
178. R. T. C. Tsui, Phys. Rev. 168, 107 (1968).
179. M. J. Kushner, J. Appl. Phys., 58, 4024 (1985).
180. M. S. Barnes, J. C. Foster, and J. H. Keller, IEEE Trans. Plasma Sci 19, 240 (1991).
181. M. Lieberman, private communication.
182. P. Benoit-Cattin and L. C. Bernard, J. Appl. Phys. 39, 5273 (1968).
183. V. Kolobov and D. J. Economou, unpublished.
184. V. Donnelly, D. Flamm, and R. Bruce, J. Appl. Phys. 58, 2135 (1985).
185. N. Fujiwara, T. Maruyama, and M. Yoneda, Jpn. J. Appl. Phys. 35, 2450 (1996).
186. T. Nozawa, T. Kinoshita, T. Nishizawa, A. Narai, T. Inoue, and A. Nakaue, Jpn. J. Appl. Phys. 34, 2107 (1995).
187. G. S. Hwang and K. P. Giapis, J. Vac. Sci. Technol. B 15, 70 (1997).
188. T. Kinoshita, M. Hane, and J. P. McVittie, J. Vac. Sci. Technol. B 14, 560 (1996).
189. J. C. Arnold and H. H. Sawin, J. Appl. Phys. 70, 5314 (1991).
190. H. Ootera, T. Oomori, M. Tuda, and K. Namba, Jpn. J. Appl. Phys. 33, 4276 (1994).

191. E. W. Scheckler and A. R. Neureuther, IEEE Trans. Computer-Aided Design 13, 219 (1994).
192. S. Hamaguchi, M. Dalvie, R. T. Farouki, and S. Sethuraman, J. Appl. Phys. 74, 5172 (1993).
193. B. Abraham-Shrauner and C. D. Wang, J. Electrochem. Soc. 143, 672 (1996).
194. D. G. Coronell and K. F. Jensen, J. Computer-Aided Materials Design 1, 3 (1993).
195. R. J. Hoekstra, M. J. Grapperhaus, and M. J. Kushner, J. Vac. Sci. Technol. A 15, 1913 (1997).
196. R. A. Gottscho, C. W. Jurgensen, and D. J. Vitkavage, J. Vac. Sci. Technol. B 10, 2133 (1992); E. S. Shaqfeh and C. W. Jurgensen J. Appl. Phys. 66, 4664 (1989).
197. D. Chin, S. H. Dhong, and G. J. Long, J. Electrochem. Soc. 132, 1705 (1985).
198. H. F. Schaefer III, editor, Modern Theoretical Chemistry, Vols. 3 and 4, Plenum, New York NY (1977); M. C. Payne, M. P. Teter, D. C. Allan, T. A. Arias, and J. D. Joannopoulos, Reviews of Modern Physics 64, 1045 (1992).
199. J. M. Haile, Molecular Dynamics Simulation: Elementary Methods, John Wiley, New York, NY (1992).
200. M. P. Allen and D. J. Tildesley, Computer Simulation of Liquids, Oxford University Press (1990); M. H. Kalos and P. A. Witlock, "Monte Carlo Methods. Part I: Theory", J Wiley & Sons, New York, NY (1986).
201. H. H. Andersen, Nucl. Instr. Meth. Phys. Res. B 18, 321 (1987).
202. K. H. Muller, Phys. Rev. B 35, 7906 (1987).
203. T. A. Schoolcraft and B. J. Garrison, J. Am. Chem. Soc. 113, 8221 (1991).
204. H. Feil, J. Dieleman, and B. J. Garrison, J. Appl. Phys. 74, 1303 (1993).
205. M. E. Barone and D. B. Graves, J. Appl. Phys. 77, 1263 (1995).
206. S. Athavale and D. J. Economou, J. Vac. Sci. Technol. A 13, 966 (1995).
207. N. A. Kubota, D. J. Economou and S. Plimpton, J. Appl. Phys. 83, 4055 (1998).
208. S. Das Sarma, J. Vac. Sci. Technol. A 8, 2714 (1990); C. C. Battaile, D. J. Srolovitz, and J. E. Butler, J. Appl. Phys. 82, 6293 (1997).
209. S. J. Pearton, F. Ren, T. R. Fullowan, A. Katz, W. S. Hobson, U. K. Chakrabarti, and C. R. Abernathy, Mater. Chem. Phys. 32, 215 (1992).
210. M. A. Hartney, D. W. Hess, and D. S. Soane, J. Vac. Sci. Technol. B 7, 1 (1989).
211. T. J. Cotler and M. E. Elta, IEEE Circuits and Devices 6, 38 (1990).
212. N. Layadi, V. M. Donnelly, J. T. C. Lee, and F. P. Klemens, J. Vac. Sci. Technol. A 15, 604 (1997).
213. C. C. Cheng, K. V. Guinn, V. M. Donnelly, and I. P. Herman, J. Vac. Sci. Technol. A 12, 2630 (1994); J. Y. Choe, I. P. Herman, and V. M. Donnelly, J. Vac. Sci. Technol. A 15, 3024 (1997).
214. J. P. Chang and H. H. Sawin, J. Vac. Sci. Technol. A 15, 610 (1997); J. P. Chang, A. P. Mahorowala, and H. H. Sawin, J. Vac. Sci. Technol. A 16, 217 (1998).
215. J. A. Levinson, E. S. G. Shaqfeh, M. Balooch and A. V. Hamza, J. Vac. Sci. Technol. A 15, 1902 (1997).
216. E. A. Ogryzlo, D. E. Ibbotson, D. L. Flamm, and J. A. Mucha, J. Appl. Phys. 67, 3115 (1990).
217. H. F. Winters and J. W. Coburn, Surf. Sci. Rep. 14, 161 (1992).
218. D. L. Flamm, V. M. Donnelly and J. A. Mucha, J. Appl. Phys. 52, 3633 (1981).
219. C. J. Mogab, A. C. Adams, and D. L. Flamm, J. Appl. Phys. 49, 3796 (1978).
220. L. M. Ephrath and E. J. Petrillo, to appear in J. Electrochem. Soc.
221. J. W. Coburn and H. F. Winters, J. Vac Sci. Technol. 16, 391 (1979).
222. N. R. Rueger, J. J. Beulens, M. Schaepkens, M. F. Doemling, J. M. Mirza, T. E. F. M. Standaert, and G. S. Oehrlein, J. Vac Sci. Technol. A 15, 1881 (1997).
223. S. M. Han and E. S. Aydil, J. Vac. Sci. Technol. A 14, 2062 (1996).
224. J. Feldsien and D. J. Economou, unpublished.

225. D. A. Danner and D. W. Hess, J. Appl. Phys. 59, 940 (1986); D. A. Danner, M. Dalvie and D. W. Hess, J. Electrochem. Soc. 134, 669 (1987).
226. J. Li, T. E. Seidel, and J. W. Mayer, MRS Bull. 19, 15 (1994); S. P. Murarka and S. W. Hymes, Crit. Rev. Solid State Mater. Sci. 20, 87 (1995).
227. K. Wasa and S. Hayakawa, Handbook of Sputter Deposition Technology, Noyes Publications, Park Ridge, NJ (1992).
228. M. J. Grapperhaus, Z. Krivokapic, M. J. Kushner, J. Appl. Phys. 83, 35 (1998).
229. P. M. Schaible, W. C. Metzger, and J. P. Anderson, J. Vac. Sci. Technol. 15, 334 (1978).
230. C. J. Mogab, J. Electrochem. Soc. 124, 1262 (1977).
231. D. J. Economou, S.-K. Park, and G. Williams, J. Electrochem. Soc. 136, 188 (1989).
232. A. G. Nagy, J. Electrochem. Soc. 131, 1871 (1984).
233. J. A. Thornton, J. Vac. Sci. Technol. A 4, 3059 (1986).
234. S. Berg, H. O. Blom, M. Moradi, C. Nender, and T. Larsson, J. Vac. Sci. Technol. A 7, 1225 (1989).
235. J. L. Crowley, Solid State Technology, 35, 94 (1992).
236. M. Hack, E. A. Schiff, S. Wagner, A. Matsuda, and R. Schropp (eds.), Amorphous Silicon Technology, Materials Research Society, Pittsburgh, PA (1996).
237. D. L. Smith and A. S. Alimonda, J. Electrochem. Soc. 140, 1496 (1993).
238. J. Kanicki and N. Yoke, Mat. Res. Soc. Symp. Proc., 68, 167 (1986).
239. K. E. Spear and J. P. Dismukes, Synthetic Diamond: Emerging CVD Science and Technology, J. Wiley, New York, NY (1994).
240. O. Auciello and D. L. Flamm (eds.), Plasma Diagnostics, Vols. 1 and 2, Academic Press, San Diego, CA (1989).
241. J. D. Swift and M. J. R. Schwar, Electrical Probes for Plasma Diagnostics, Elsevier, New York, NY (1969).
242. N. Hershkowitz in Plasma Diagnostics, Vol. 1, O. Auciello and D. L. Flamm (eds.), Academic Press, San Diego, CA (1989).
243. F. F. Chen in Plasma Diagnostics, R. H. Huddlestone and S. L. Leonard, Academic Press, New York, NY (1965); B. E. Cherrington, Plasma Chem. Plasma Proc. 2, 113 (1982).
244. G. Mümken and U. Kortshagen, J. Appl. Phys. 80, 6639 (1996).
245. V. M. Donnelly in Plasma Diagnostics, Vol. 1, O. Auciello and D. L. Flamm (eds.), Academic Press, San Diego, CA (1989).
246. F. D'Agostino and F. Illusi in Deposition, Treatment, and Etching of Polymers, F. D'Agostino ed., p. 95, Academic Press, New York, NY (1990).
247. J. W. Coburn and M. Chen, J. Appl. Phys. 51, 3134 (1980).
248. S. F. Bergeron, A. S. Bergendahl, B. F. Duncan, and D. L. Harmon, Microelectronic Manufacturing and Testing, August 1983.
249. K. Miyata, M. Hori, and T. Goto, J. Vac. Sci. Technol. A 15, 568 (1997).
250. R. A. Gottcho and T. A. Miller, Pure and Appl. Chem. 56, 189 (1984).
251. R. W. Dreyfus, J. M. Jasinski, R. E. Walkup, and G. S. Selwyn, Pure and Appl. Chem. 57, 1265 (1985); W. Demtroder, Laser Spectroscopy, Springer-Verlag, Berlin, 1988.
252. M. J. Vasile and H. F. Dylla in Plasma Diagnostics, Vol. 1, O. Auciello and D. L. Flamm (eds.), Academic Press, San Diego, CA (1989).
253. H. Sugai and H. Toyoda, J. Vac. Sci. Technol. A 10, 1193 (1992).
254. A. I. Chowdhury, W. W. Read, G. W. Rubloff, L. L. Tedder, and G. N. Parsons, J. Vac. Sci. Technol. B 15, 127 (1997).
255. C. B. Zarowin, J. Electrochem Soc. 130, 1144 (1983).
256. J. C. Riviere and S. Myhra, Handbook of Surface and Interface Analysis, Marcel Dekker, New York, NY (1998).
257. D. E. Aspnes and R. P. H. Chang, in Plasma Diagnostics, Vol. 2, O. Auciello and D. L. Flamm (eds.), Academic Press, San Diego, CA (1989).

258. J. P. Roland, P. J. Marcoux, G. W. Ray, and G. H. Rankin, J. Vac. Sci. Technol. A 3, 631 (1986).
259. T. Iwasita and F. C. Nart in Advances in Electrochemical Science and Engineering, H. Gerischer and C. W. Tobias (eds.), Vol. 4, p. 123, VCH (1995).
260. Z.-H. Zhou, E. S. Aydil, R. A. Gottscho, Y. J. Chabal, and R. Reif, J. Electrochem. Soc. 140, 3316 (1993).
261. J. W. Coburn, in Plasma Diagnostics, Vol. 2, O. Auciello and D. L. Flamm (eds.), Academic Press, San Diego, CA (1989).
262. S. A. Coulombe, B. K. Minhas, C. J. Raymond, S. Sohail, H. Naqvi, and J. R. McNeil, J. Vac. Sci. Technol. B 16, 80 (1998).
263. N. Ibl, in Comprehensive Treatise of Electrochemistry, Vol. 6, p. 329, Plenum Press, New York, NY (1983).
264. J. S. Newman, Electrochemical Systems, Prentice Hall, NJ (1973).
265. R. J. von Cutfeld, J. Opt. Soc. Am. B, 4, 272 (1987).
266. D. V. Podlesnik, H. H. Gilgen, A. E. Willner, and R. M. Osgood, Jr., J. Opt. Soc. Am. B 3, 775 (1986).
267. M. M. Carrabba, N. M. Nguyen, and R. D. Rauh, J. Electrochem. Soc. 134, 1855 (1987).
268. P. Fauchais, J. F. Coudert, and M. Vardelle in Plasma Diagnostics, Vol. 1, O. Auciello and D. L. Flamm (eds.), Academic Press, San Diego, CA (1989).
269. Plasma Processing of Materials: Scientific Opportunities and Technological Challenges, NRC Report, National Academy Press, Washington, DC, 1991.
270. P. Allongue in Advances in Electrochemical Science and Engineering, H. Gerischer and C. W. Tobias (eds.), Vol. 4, p. 1, VCH (1995).
271. E. D. Eliadis, R. G. Nuzzo, A. A. Gewirth, and R. C. Alkire, J. Electrochem. Soc. 144, 96 (1997).
272. T. F. Kuech, P. D. Dapkus, and Y. Aoyagi (eds.) Atomic Layer Growth and Processing, Materials Research Society Symp. Proceedings, Vol. 222, MRS, Pittsburgh, PA (1991).
273. S. Athavale and D. J. Economou, J. Vac. Sci. Technol. B 14, 3702 (1996).
274. J. M. E. Harper and K. P. Rodbell, J. Vac. Sci. Technol. B 15, 763 (1997).

Index